Philosophical Studies Series

Volume 124

Editor-in-Chief
Luciano Floridi, University of Oxford, Oxford Internet Institute, United Kingdom
Mariarosaria Taddeo, University of Oxford, Oxford Internet Institute, United Kingdom

Executive Editorial Board
Patrick Allo, Vrije Universiteit Brussel, Belgium
Massimo Durante, Università degli Studi di Torino, Italy
Phyllis Illari, University College London, United Kingdom
Shannon Vallor, Santa Clara University

Board of Consulting Editors
Lynne Rudder Baker, University of Massachusetts at Amherst
Stewart Cohen, Arizona State University, Tempe
Radu Bogdan, Tulane University
Marian David, University of Notre Dame
John M. Fischer, University of California at Riverside
Keith Lehrer, University of Arizona, Tucson
Denise Meyerson, Macquarie University
François Recanati, Institut Jean-Nicod, EHESS, Paris
Mark Sainsbury, University of Texas at Austin
Barry Smith, State University of New York at Buffalo
Nicholas D. Smith, Lewis & Clark College
Linda Zagzebski, University of Oklahoma

More information about this series at http://www.springer.com/series/6459

Mariarosaria Taddeo • Ludovica Glorioso
Editors

Ethics and Policies for Cyber Operations

A NATO Cooperative Cyber Defence Centre of Excellence Initiative

Springer

Editors
Mariarosaria Taddeo
Oxford Internet Institute
University of Oxford
Oxford, UK

Ludovica Glorioso
Security Force Assistance Centre in Cesano
Rome, Italy

Philosophical Studies Series
ISBN 978-3-319-45299-9   ISBN 978-3-319-45300-2   (eBook)
DOI 10.1007/978-3-319-45300-2

Library of Congress Control Number: 2016957346

© Springer International Publishing Switzerland 2017
This work is subject to copyright. All rights are reserved by the Publisher, whether the whole or part of the material is concerned, specifically the rights of translation, reprinting, reuse of illustrations, recitation, broadcasting, reproduction on microfilms or in any other physical way, and transmission or information storage and retrieval, electronic adaptation, computer software, or by similar or dissimilar methodology now known or hereafter developed.
The use of general descriptive names, registered names, trademarks, service marks, etc. in this publication does not imply, even in the absence of a specific statement, that such names are exempt from the relevant protective laws and regulations and therefore free for general use.
The publisher, the authors and the editors are safe to assume that the advice and information in this book are believed to be true and accurate at the date of publication. Neither the publisher nor the authors or the editors give a warranty, express or implied, with respect to the material contained herein or for any errors or omissions that may have been made.

Printed on acid-free paper

This Springer imprint is published by Springer Nature
The registered company is Springer International Publishing AG Switzerland

Contents

1. Deterrence and the Ethics of Cyber Conflict 1
   Paul Cornish

2. Blind Justice? The Role of Distinction in Electronic Attacks 17
   Jack McDonald

3. Challenges of Civilian Distinction in Cyberwarfare 33
   Neil C. Rowe

4. Towards a Richer Account of Cyberharm: The Value of Self-Determination in the Context of Cyberwarfare 49
   Patrick Taylor Smith

5. Just Information Warfare 67
   Mariarosaria Taddeo

6. Regulating Cyber Operations Through International Law: In, Out or Against the Box? 87
   Matthew Hoisington

7. Military Objectives in Cyber Warfare 99
   Marco Roscini

8. Defining Cybersecurity Due Diligence Under International Law: Lessons from the Private Sector 115
   Scott J. Shackelford, Scott Russell, and Andreas Kuehn

9. Cyber Warfare and Organised Crime. A Regulatory Model and Meta-Model for Open Source Intelligence (OSINT) 139
   Pompeu Casanovas

10. A Model to Facilitate Discussions About Cyber Attacks 169
    Jassim Happa and Graham Fairclough

11	**Strategies of Cyber Crisis Management: Lessons from the Approaches of Estonia and the United Kingdom**...............................	187
	Jamie Collier	
12	**Lessons from Stuxnet and the Realm of Cyber and Nuclear Security: Implications for Ethics in Cyber Warfare**...........................	213
	Caroline Baylon	
13	**NATO CCD COE Workshop on 'Ethics and Policies for Cyber Warfare' – A Report**...	231
	Corinne N.J. Cath, Ludovica Glorioso, and Mariarosaria Taddeo	

Index 243

About the Editors

Mariarosaria Taddeo works at the Oxford Internet Institute, University of Oxford and Faculty Fellow at the Alan Turing Institute. Her recent work focuses mainly on the ethical analysis of cyber security practices and information conflicts. Her area of expertise is Information and Computer Ethics, although she has worked on issues concerning Philosophy of Information, Epistemology, and Philosophy of AI. She published several papers focusing on online trust, cyber security and cyber warfare and guest-edited a number of special issues of peer-reviewed international journals: Ethics and Information Technology, Knowledge, Technology and Policy, Philosophy & Technology. She also edited (with L. Floridi) a volume on 'The Ethics of Information Warfare' (Springer, 2014) and is currently writing a book on 'The Ethics of Cyber Conflicts' under contract for Routledge. Dr. Taddeo is the 2010 recipient of the Simon Award for Outstanding Research in Computing and Philosophy and of the 2013 World Technology Award for Ethics. She serves editor-in-chief of Minds & Machines, in the executive editorial board of Philosophy & Technology. Since 2016, Dr Taddeo is Global Future Council Fellow for the Council on the Future of Cybersecurity of the World Economic Forum.

Ludovica Glorioso is Captain (Cpt ITA A) of the Italian Army and a legal advisor to the Italian Armed Forces, currently working at the Defence General Staff. Prior to her current position, she served as Legal Researcher at the NATO Cooperative Cyber Defence Centre of Excellence (NATO CCD CCOE, 2012–2016), as legal advisor in NATO Peacekeeping Operations in the Balkans and Afghanistan. She holds an MA in Law from University of Palermo (Italy), an LL.M in European and Transitional Law from The University of Trento (Italy) and she is admitted to the Italian bar. Her research and activities focus on International Humanitarian Law, Law of Armed Conflict, and International Law applied to cyber warfare.

Introduction

Abstract A relation of mutual influence exists between the way conflicts are waged and the societies waging them. As Clausewitz remarked, more than an art or a science, conflicts are a social activity. And much like other social activities, conflicts mirror the values of societies while relying on their technological and scientific developments. In turn, the principles endorsed to regulate conflicts play a crucial role in shaping societies.A relation of mutual influence exists between the way conflicts are waged and the societies waging them. As Clausewitz remarked, more than an art or a science, conflicts are a social activity. And much like other social activities, conflicts mirror the values of societies while relying on their technological and scientific developments. In turn, the principles endorsed to regulate conflicts play a crucial role in shaping societies.

Think about the design, deployment, and regulation of weapons of mass destruction (WMDs). During World War II, WMDs were made possible by scientific breakthroughs in nuclear physics, which was a central area of research in the years leading to the War. Yet, their deployment proved to be destructive and violent beyond what the post-war world was willing to accept. The Cold War that followed and the nuclear treaties that ended it defined the modes in which nuclear technologies and WMDs can be used, drawing a line between conflicts and atrocities. In doing so, treaties and regulations for the use of WMDs contributed to shape contemporary societies as societies rejecting the belligerent rhetoric of the early twentieth century and to striving for peace and stability.

The same mutual relation exists between information societies and cyber conflicts, making the regulation of the latter a crucial aspect, which will contribute to define current and future societies. In the short term, regulations are needed to avoid a digital wild west, as remarked by Harold Hongju Koh, the former Legal Advisor U.S. Department of State. For this reason, over the past few years, efforts have been devoted to analysing and interpreting the existing corpus of laws to guide states in engaging in international cyber conflicts.

Interpretations often highlight that existing norms raise substantial barriers to the use of cyber weapons and to the use of force to defend cyberspace. It is claimed that international law contains coercive means of permitting lawful responses to cyber provocations and threats of any kind. The legal framework that is referred to mainly encompasses the four Geneva Conventions and their first two Additional Protocols, the international customary law and general principle of law, the Convention restricting or prohibiting the use of certain conventional weapons, and judicial decisions (Glorioso 2015). Arms control treaties, such as the Nuclear Non-Proliferation Treaty and the Chemical Weapons Convention, are often mentioned as providing guidance for action in the case of kinetic cyber attacks (Schmitt 2013). At the same time, coercive measures addressing economic violations are generally considered legitimate in the case of cyber attacks that do not cause physical damage (Lin 2012; O'Connell 2012).

However, the problem at stake is not whether cyber conflicts can be interpreted in such a way as to fit the parameters of kinetic conflicts, economic transgressions, and conventional warfare, and hence whether they fall within the domain of international humanitarian law, as we know it. The problem rests at a deeper level and questions the very normative and conceptual framework of international humanitarian law and its ability to *satisfactorily* and *fairly* accommodate in the medium- and long-term the changes prompted by cyber conflicts (Floridi and Taddeo 2014; Taddeo and Floridi 2014).

In the medium- and long-term, regulations need to be defined so to ensure security and stability of societies, and avoid risks of escalation. To achieve this end, efforts to regulate cyber conflicts will have to rely on an in-depth understanding of this new phenomenon; identify the changes brought about by cyber conflicts and the information revolution (Floridi 2014; Taddeo and Buchanan 2015); and define a set of shared values that will guide the stakeholders operating in the international arena.

Efforts to regulate cyber conflicts cannot afford to be future-blind and disregard questions concerning the impact of these new forms of conflicts on future information societies, on their values, the rights, and security of their citizens, and on national and international politics. Conceptual and ethical questions need to be addressed now, while efforts to regulate this phenomenon are still nascent, to ensure fair and effective regulations, which will contribute to shaping open, pluralistic, peaceful information societies.

Regulation of cyber conflicts need to be developed consistently to (a) Just War Theory, (b) human rights, and (c) international humanitarian laws. However, applying (a)–(c) to the case of cyber conflicts proves to be problematic given the changes in military affairs that they prompted (Dipert 2010; Taddeo 2012a; Floridi and Taddeo 2014). When compared to kinetic warfare, cyber conflicts show fundamental differences: their domain ranges from the virtual to the physical; the nature of their actors and targets involves artificial and virtual entities alongside human beings and physical objects; and their level of violence may range from non-violent to potentially highly violent phenomena. These differences are redefining our understanding of key concepts such as harm, violence, target, combatants, weapons, and attack, and pose serious challenges to any attempt to regulate conflicts in cyberspace (Dipert 2010; Taddeo 2012b, 2014a, b; Floridi and Taddeo 2014).

The Just War Theory principle of proportionality, as specified in the ethical tradition, offers a good example of the case in point. The principle prescribes a balance identifying the necessary and sufficient means to achieve a legitimate goal. Enforcing and respecting this principle is thus crucial while planning and waging cyber conflicts. However, proportionality rests on an assessment of the gains and damages received and caused by a cyber operation, and this assessment is highly problematic. The conventional conceptual framework for calculating pain and gains accounts for casualties, physical and economic damages, and territorial advantages, but proves to be challenging when endorsed to assess damage to virtual objects in cyberspace (Dipert 2013). This highlights an ontological hiatus between the entities involved in cyber conflicts and those taken in consideration in conventional wars (Taddeo 2016). This hiatus demands immediate attention as it encroaches our understanding of cyber conflicts, and any attempts to regulate them.

Such attempts are further complicated when we consider state and non-state actors operating in cyberspace. The use of a state's coercive power is coupled with the concept of state's sovereignty over a given territory. However, the absence of clear national boundaries, the distributed and interconnected nature of cyberspace, as well as the global sharing of information that it enables, make it difficult to define state sovereignty in this domain (Brenner and Susan 2009; Chadwick and Howard 2009; Cornish 2015).

This has serious implications for the definition of state authority and military power and hence on our understanding of the state and non-state actors involved in conflicts, as well as for the definition of lawful and unlawful conducts and the body of law that should be applied. Understanding such conceptual changes, and identifying their medium- and long-term impact on international relations and military strategies is a preliminary and necessary step to any effort for regulating cyber conflicts.

Things are not less problematic when considering ethical issues. Cyber conflicts bring about three kinds of problems, concerning risks, rights, and responsibilities (3R problems) (Taddeo 2012a, b). The more contemporary societies are dependent on ICTs, the more the 3R problems become pressing and undermine ethically blind attempts to regulate cyber conflicts. Consider, for example, the risks of escalation. Estimates indicate that the cyber security market will grow from US$106 billion in 2015 to US$170 billion by 2020, posing the risk of a progressive weaponization and militarisation of cyberspace. At the same time, the reliance on malware for state-run cyber operations (like Titan Rain, Red October, and Stuxnet) risks sparking a cyber arms race and competition for digital supremacy, hence increasing the possibility of escalation and conflicts (MarketsandMarkets 2015). Regulations of cyber conflicts need to address and reduce this risk by encompassing principles to foster cyber stability, trust, and transparency among states (Arquilla and Borer 2007; Steinhoff 2007; European Union 2015).

At the same time, cyber threats are pervasive. They can target, but can also be launched through, civilian infrastructures, e.g. civilian computers and websites.

This may (and in some cases already has) initiate policies of higher levels of control, enforced by governments in order to detect and deter possible threats. In these circumstances, individual rights, such as privacy and anonymity may come under sharp, devaluating pressure (Arquilla 1999; Denning 2007). Ascribing responsibilities also prove to be problematic when considering cyber conflicts. Cyberspace affords a certain level of anonymity, often exploited by states or state-sponsored groups and non-state actors. Difficulties in attributing attacks allow perpetrators to deny responsibility, and pose an escalatory risk in cases of erroneous attribution. These risks have been faced, for example, by the international community in 2014, when malware initially assessed as capable of destroying the content of the entire stock exchange was discovered on Nasdaq's central servers and allegations were made of a Russian origin for the software[1]; and later in 2015, when cyber attacks against TV5 Monde were initially attributed to ISIL/Da'esh.[2]

The volume is multidisciplinary as it collects contributions by leading experts in international law, war studies, philosophy and ethics of war, philosophy of law, information and computer ethics, as well as policy-makers focusing on of the problems prompted by cyber conflicts. The 11 chapters of this volume are either invited contributions or papers presented during the workshop "Ethics and policies for cyber warfare" organised by the NATO Cooperative Cyber Defence Centre of Excellence (CCD COE) in collaboration with the University of Oxford and held at the Magdalen College Oxford, in November 2014. This was the second workshop organised by the Centre (and chaired by the editors of this volume) with the goal of identifying and defining the most pressing issues concerning the regulation of cyber conflicts.[3]

Each chapter provides a detailed analysis of a key problem concerning the ethical or legal implications of cyber conflicts. In more details, the book offers an analysis of the following topics: the conceptual novelty of cyber conflicts and the ethical problems that this engenders; the applicability of existing conceptual and regulatory frameworks to cyber conflicts; the analysis of models to foster cooperation in managing cyber crises; and how to regulate cyber operations through international law.

Just War Theory is a central point of analysis in Cornish's chapter, which opens the volume. The chapter first delves on the understanding of key concepts, such as those of violence, attack, and cyberspace. The attention is devoted to ethical issues and to the analysis of cyberspace as an artificial environment, which is pre-political, pre-strategic, and therefore pre-ethical. The chapter argues that as such cyberspace is not yet susceptible to the forms of political organisation with which we are familiar, it is resistant to the normative constraints with which we expect to manage and moderate traditional conflicts and organised violence. The chapter concludes by

---

[1] http://arstechnica.com/security/2014/07/how-elite-hackers-almost-stole-the-nasdaq/

[2] http://www.alphr.com/security/1000604/isis-hacks-french-broadcaster-tv5-monde

[3] The first workshop 'Ethics of Cyber Conflict' was held in Rome at the 'Centre for High Defence Studies' in Rome, https://ccdcoe.org/multimedia/workshop-ethics-cyber-conflict-proceedings.html ; a special issue of Philosophy & Technology has been published collecting the papers presented during that meeting (Glorioso 2015).

stressing that the regulation of cyber conflicts requires serious, balanced public policy discourse and a new consideration of cyberspace as an arena in which diplomacy, negotiation, bargaining, compromise, concession and, therefore, moral judgement, are all considered possible and proper.

McDonald's analyses three main challenges to maintaining the norm of distinction in cases of cyber attacks: (i) the significant chance that attempts of computer network exploitation may unwittingly target non-military systems; (ii) the use of civilian infrastructure or third party networks that may not be willing participants but would be targeted regardless; (iii) the design and deployment of autonomous software programmes (e.g., 'viruses', 'malware') unable to distinguish targets in all circumstances. The analysis of (i)–(iii) offers the basis for questioning the direct applicability of the Just War Theory principle of distinction to cyber attacks involving computer network exploitation.

The applicability of Just War Theory to the case of cyber conflicts and of the principle of discrimination is also focal aspect in Rowe's contribution. This chapter analyses different ways in which a military cyber attack could hit a civilian target. It focuses on both dual-uses targets and on the military advantage that may result from intentionally target civilian infrastructures and objects. This analysis highlights a vicious dynamics, i.e. cyber attacks targeting civilians objects and infrastructures encourage counter-attacks on similar targets, as such they are close to *perfidy*, which is outlawed by the laws of armed conflict. The chapter concludes with proposed principles for ethical conduct of cyber conflicts to minimize unnecessary harm to civilians, and suggests that focusing on cyber coercion and deterrence rather than cyber warfare will reduce harm to civilians.

Taylor Smith's contribution focuses on the concept of cyber harm. The chapter first offers a definition of cyber harm, whereby it occurs when the normal or intended functioning of a computer network is disrupted in ways that undermine or violate significant human interests or entitlements. The chapter argues that this view is more sophisticated than the one claiming that cyber attacks only occur in presence of *physical* harm or damage. Yet, it is not as complex as the view maintaining that the disruption of a computer system is *per se* a form cyber harm. In the second part, the provided definition of cyber harm is used to identify those occurrences of cyber attacks that count as *casus belli*, justifying unilateral military action in self-defence.

Taddeo's contribution is a reprint of an article (Taddeo 2016) delving on the applicability of Just War Theory to cyber conflicts. It proposes an ethical analysis of cyber warfare with the twofold goal of filling the theoretical vacuum surrounding this phenomenon and providing the conceptual grounding for the definition of new ethical regulations for this phenomenon. The chapter first argues that Just War Theory is a necessary but not sufficient instrument for considering the ethical implications of cyber warfare and that a suitable ethical analysis of this kind of warfare is developed when Just War Theory is merged with Information Ethics. In the initial part, the chapter describes cyber warfare and its main features and highlights the problems that arise when Just War Theory is endorsed as a means of addressing ethical problems engendered by this kind of warfare. In the final part, the main aspects of Information Ethics are provided along to three principles for a just cyber warfare resulting from the integration of Just War Theory and Information Ethics.

With Hoisington's chapter the focus shifts to the analysis of existing laws for the regulations of cyber conflicts. This contribution distinguishes three approaches, namely *'in, out or against'* existing regulatory framework, for the regulation of cyber conflicts. It then highlights the problems that each of these approaches may rise when considering the different stakeholders acting in cyberspace. Attempts to regulate cyber conflicts relying on principles existing within *jus ad bellum* and *jus in bello* are shaken when confronted with the difficulties of applying the general principles such as proportionality and military necessity in cyberspace. The second approach, i.e. finding principles for the regulation of cyber conflicts outside the set of existing laws, describes a new set of international regulations and legal structures, including the interaction with the private sector. The third approach rests on a rebuttal of the existing framework and requires a revision of the existing rules. This analyses stresses the difficulty to amend the UN Charter, while at the same time suggesting the development of new concepts and vocabulary describing cyber operations and the need to involve new actors in the elaboration of the rules. The chapter concludes remarking that international lawyers need to be vigilant to the new developments of the cyber capabilities in order to develop the understanding and the applicability of international law in the cyber context.

Roscini's chapter explores the application of the Law of Armed Conflict's principle of distinction between military objectives and civilian objects in cyberspace. It starts by looking at what type of cyber operations are subject to the law of targeting, i.e. those qualifying as 'attacks' under Article 49(1) of Protocol I Additional to the 1949 Geneva Conventions on the Protection of Victims of War. It then applies the components of the definition of 'military objective', contained in the same Protocol, to cyberspace: 'effective contribution to military action' and 'definite military advantage'. The chapter concludes that, despite current challenges posed by cyber conflicts, existing rules are flexible enough to be applied in a new domain like cyberspace.

Shackelford's, Russell's, Kuehn's chapter analyses nations' due diligence obligations to their respective private sectors and to each another in the international arena. The chapter starts off considering what steps nations and companies under their jurisdiction have to take under international law to secure their networks, and what the rights and responsibilities of transit states are. This chapter reviews the arguments surrounding the creation of a cyber security due diligence norm and argues for a proactive regime that takes into account the common but differentiated responsibilities of public and private sector in cyberspace. The analogy is drawn to cyber security due diligence in the private sector and the experience of the 2014 National Institute of Standards and Technology Framework to help guide and broaden the discussion.

Casanovas' contribution to this volume builds on the findings of two European projects-*(O)SI for (Open) Social Intelligence, and PbD for Privacy by Design* (OSINT) and *Collaborative information, Acquisition, Processing, Exploitation and Reporting* (CAPER)–devoted to embed the legal and ethical issues raised by the

General Data Reform Package in Europe into security and surveillance platforms. The contribution to this volume describes a procedure to flesh out ethical principles through semantic web regulatory models that can be applied to the case of cyber warfare. New ways of designing political institutions and the possibility to build up a meta-rule of law are also discussed.

Happa's and Fairclough's chapter addresses the need for establishing a methodologies supporting different stakeholders in discussing the regulation of cyber conflicts. In particular, the chapter focuses on the difficulties law- and policy-makers, as well as military experts and ethicists and of cyber security analysts to share information, coordinate, and collaborate in defining effective regulation for cyber attacks. This chapter proposes a model enabling a collective, multidisciplinary, and collaborative approach to understand and discuss cyber attacks.

Collier's chapter compares the cyber crisis management strategies of Estonia and the UK. It argues that the two countries' strategies differ significantly. This divergence reflects broader political, historical, and cultural differences between Estonia and the UK, all of which influence the respective national cyber crisis management strategies of the two countries. The variables that affect national strategies include the countries' history, size, political views, digital dependency, and the nature of the threats and adversaries they each face in cyberspace. The chapter concludes that, given the importance of these relative factors in determining their national cyber crisis management strategies, it is difficult to draw from these cases generalizable recommendations that apply to other states. Instead, the importance of creating a cyber crisis strategy appropriate to the specific political, historical, and cultural climate should be recognised. Although cyber attacks may be highly technical in nature, this chapter argues that a successful organisational response to the threat has significant political components.

Baylon contributed a commentary to this volume describing some of the key findings of a Chatham House 18-month project on Cyber and Nuclear Security. The project examined the challenges that ICTs pose for the nuclear industry, which include ethical problems. Using Stuxnet as a case study, the project analysed whether the deployment of this computer worm could be considered an attack on a sovereign state-thus violating Iran's right to develop a peaceful civilian nuclear energy programme-and whether Iran has a legitimate grievance against the US and Israel and would be fully entitled to retaliate.

The volume ends with a report by Cath, Glorioso, and Taddeo of the NATO CCD COE workshop "Ethics and Policies for Cyber Warfare". The report describes the discussion among ethicists, policy-makers, international lawyers, and military experts on the existing regulatory gap concerning cyber warfare and ethical problems underpinning it. The report is divided in three parts. The first one describes the discussion on the extent to which current international legal structures are able to develop cyber security norms. The second part focuses on the applicability of current legal mechanisms of warfare to cyberspace, looking specifically at the issues of deterrence, proportionality, perfidy, and *casus belli*. The final parts describes the debate occurred among the workshop delegates concerning the different mechanisms for developing ethical norms and universal principles that could be applied to cyber conflicts.

Before leaving the reader to the contributions in this volume, we would like to express our gratitude to the NATO CCD COE for supporting the organisation of the workshop 'Ethics and Policies for Cyber Conflicts'. This project allowed us to gather a number of international experts discussing cutting edge conceptual and applied problems and to identify and define the most pressing needs concerning the regulation of cyber conflicts.

Oxford, UK Mariarosaria Taddeo
London, UK
Rome, Italy Ludovica Glorioso

References

Arquilla, J. 1999. Ethics and information warfare. In *Strategic appraisal: The changing role of information in warfare*, ed. Zalmay Khalilzad and John Patrick White, 379–401. Santa Monica: RAND.
Arquilla, J., and Douglas A. Borer. 2007. *Information strategy and warfare: A guide to theory and practice*. New York: Routledge.
Brenner, Susan W. 2009. *Cyber threats the emerging fault lines of the Nation State*. New York [u.a.]: Oxford University Press. http://www.oxfordscholarship.com/oso/public/content/law/9780195385014/toc.html
Chadwick, Andrew, and Philip N. Howard (eds.). 2009. *Routledge handbook of internet politics*, Routledge handbooks. London: Routledge.
Cornish, Paul. 2015. Survival: Global politics and strategy. *Survival: Global Politics and Strategy* 57(3): 153–176.
Denning, D. 2007. The ethics of cyber conflict. In *Information and computer ethics*. Hoboken: Wiley.
Dipert, R. 2010. The ethics of cyberwarfare. *Journal of Military Ethics* 9(4): 384–410.
Dipert, Randall. 2013. "The Essential Features of an Ontology for Cyberwarfare." In Conflict and Cooperation in Cyberspace, edited by Panayotis Yannakogeorgos and Adam Lowther, 35–48. Taylor & Francis. http://www.crcnetbase.com/doi/abs/10.1201/b15253-7.
European Union. 2015. *Cyber diplomacy: Confidence-building measures – Think tank*. Brussels. http://www.europarl.europa.eu/thinktank/en/document.html?reference=EPRS_BRI(2015)571302
Floridi, L. 2014. *The fourth revolution, how the infosphere is reshaping human reality*. Oxford: Oxford University Press.
Floridi, L., and M. Taddeo (eds.). 2014. *The ethics of information warfare*. New York: Springer.
Glorioso, Ludovica. 2015. Cyber conflicts: Addressing the regulatory gap. *Philosophy & Technology* 28(3): 333–338. doi:10.1007/s13347-015-0197-8.
Lin, Herbert. 2012. Cyber conflict and international humanitarian law. *International Review of the Red Cross* 94(886): 515–531. doi:10.1017/S1816383112000811.
MarketsandMarkets. 2015. *Cyber security market by solutions & services – 2020*. http://www.marketsandmarkets.com/Market-Reports/cyber-security-market-505.html?gclid=CNb6w7mt8MgCFQoEwwodZVQD-g
O'Connell, M.E. 2012. Cyber security without cyber war. *Journal of Conflict and Security Law* 17(2): 187–209. doi:10.1093/jcsl/krs017.
Schmitt, M. 2013. Cyberspace and international law: The penumbral mist of uncertainty. *Harvard* 126(176): 176–180.

Steinhoff, Uwe. 2007. *On the ethics of war and terrorism*. Oxford/New York: Oxford University Press.

Taddeo, M. 2012a. An analysis for a just cyber warfare. In *2012 4th international conference on Cyber Conflict* (CYCON 2012), 1–10.

Taddeo, M. 2012b. Information warfare: A philosophical perspective. *Philosophy and Technology* 25(1):105–120.

Taddeo, M. 2014a. Just information warfare. *Topoi*, April, 1–12. doi:10.1007/s11245-014-9245-8.

Taddeo, M. 2014b. The struggle between liberties and authorities in the information age. *Science and Engineering Ethics*, September, 1–14. doi:10.1007/s11948-014-9586-0.

Taddeo, M. 2016. Just information warfare. *Topoi*, April, 1–12. doi:10.1007/s11245-014-9245-8. Reprint *Ethics and Policies for Cyber Operations*, Ed. Mariarosaria Taddeo & Ludovica Glorioso, Philosophical Studies, Book Series, Springer.

Taddeo, M., and Elizabeth Buchanan. 2015. Information societies, ethical enquiries. *Philosophy & Technology* 28(1): 5–10. doi:10.1007/s13347-015-0193-z.

Taddeo, M., and Floridi Luciano. 2014. The ethics of information warfare – An overview. In *The ethics of information warfare*, Law, governance and technology series. New York: Springer.

Chapter 1
Deterrence and the Ethics of Cyber Conflict

Paul Cornish

> We need to furnish a rational basis for our moral thinking both in general and, in particular, in relation to the difficult issues of war and peace. To be able to write [...] about morality and war, it is necessary first to secure the foundations of morality.
>
> (David Fisher, 2011 (Fisher D. 2011. *Morality and War: Can War be Just in the Twenty-first Century?* Oxford University Press, p3.))

Abstract The notion that the resort to and conduct of conflict can, and should be constrained on ethical grounds is well understood. Why then is it proving difficult to apply that understanding to cyber space? In the first place, it is as yet unclear how we might define 'conflict', 'violence' and 'aggression' in cyber space; what the 'cyber domain' might be; and what it might be to be secure within or from that domain. Do we apply existing understandings and simply prefix them with 'cyber', or is there something qualitatively different about cyber security and the conduct it permits or requires?

Then there is a far deeper, more structural problem with which to contend. The ethical problem with cyber space is that it is an artificial environment offering both co-operation on the one hand, and contention and confrontation on the other. And it is difficult to know the identifiers of good and bad behaviour at any given moment. Cyberspace is stubbornly pre-political, pre-strategic and therefore pre-ethical; it is not yet susceptible to the forms of political and strategic organisation with which we are familiar and is thus resistant to the normative constraints with which we have expected to manage and moderate conflict and organised violence.

Cyber security must be conceptualised as an arena for human exchange in which diplomacy, negotiation, bargaining, compromise, concession and the exercise of moral judgement are not only considered proper but are also made possible. This chapter argues that deterrence thinking can assist in this process.

Keywords Cyber conflict • Ethics • Aggression • Anonymity • Deniability • Just war tradtion • Law of war • Harm • Intention • Discrimination • Trust • Deterrence

---

P. Cornish (✉)
Global Cyber Security Capacity Centre, Oxford, UK
e-mail: pncornish@icloud.com

© Springer International Publishing Switzerland 2017
M. Taddeo, L. Glorioso (eds.), *Ethics and Policies for Cyber Operations*, Philosophical Studies Series 124, DOI 10.1007/978-3-319-45360-2_1

1.1 Introduction

What is so difficult, ethically, about the prospect of conflict in cyberspace? Writing in 2010, Randall Dipert referred to 'a virtual policy vacuum' where cyberwarfare is concerned; 'there are no informed, open, public or political discussions of what an ethical or wise policy for the use of such [cyber] weapons would be.' Of particular note for the purposes of this chapter, Dipert went on to argue that 'cyberwarfare appears to be almost entirely unaddressed by the traditional morality and laws of war.'[1] Things have changed somewhat in the 5 years since Dipert's article was published; there is now a very great deal of open, public debate about the ethics of 'cyberwar' or 'cyberwarfare', with scholars and policy analysts addressing increasingly complex and sophisticated aspects of the problem.[2] Nevertheless, there remains an element of scepticism, or at least caution, in much of the new wave of writing on cyber-ethics. Ethical and legal constraints on the recourse to, and conduct of conventional armed conflict are deeply embedded in history, sophisticated and widely observed. Yet their applicability to conflict in the cyber domain is not automatic. As Ryan Jenkins notes, 'The just war tradition has been developed around incontrovertibly *physical* means of waging warfare: shooting bullets, dropping bombs, etc. But sensational attacks like Stuxnet seem to usher in a new era of *non*-physical warfare.' The challenge can neither be ignored nor wished away; 'The virtual genie', he continues, 'has been let out of the bottle and there is no putting him back in.'[3]

I welcome this caution. Jenkins observes, as do others, 'that technological development has outpaced the development of our normative theories.'[4] I take a rather different approach, however. My opening premise is that, far from being laggardly, muted and lacking in ambition, the ethical debate concerning cyberspace has in some respects run away with itself. While the debate is certainly becoming more 'open' and 'public' I am much less confident, as I will explain in this chapter, that the debate is either as 'informed' (and informing) or as 'political' as it should be. Rather than ask how ethical and legal frameworks (such as the just war tradition and international humanitarian law respectively) should be applied to conflict in cyberspace, I believe it would be more pertinent to ask why it has proved, and is proving, so difficult to do so.

This chapter is by way of a first-order inquiry, therefore; if we are not yet at the point that the expression 'the ethics of cyber conflict' makes as much sense as 'the ethics of armed conflict' (or as much sense as we would like, perhaps), then we should ask why that is the case, and what should be done about it. I will argue that

---

[1] Dipert (2010), pp385, 405.

[2] See, for example, Danks and Danks (2013).

[3] Jenkins (2013), p68. Jenkins' response (p75) to the challenge he sets is, nevertheless, optimistic; 'cyber-weapons *can* be more morally sound than conventional weapons ever could.' [Emphases in original]

[4] Jenkins, Is Stuxnet Physical?, p69.

one problem with cyberspace is that it is persistently pre-political, pre-strategic and, therefore, pre-ethical. This is not to say that cyberspace is an ethical and normative desert – there should be as much scope for ethics in cyberspace as in any other environment in which human interchange takes place. But it is to say (and here I am entirely in agreement with Dipert) that it is not yet fit for a sophisticated framework for ethically-guided judgement as in, for example, the just war tradition and its coverage of conventional war and warfare. In Fisher's words, we do not yet have the 'foundations of morality' in respect of cyberspace. I believe, furthermore, that these foundations must lie in an ordered politics for cyberspace. As Hedley Bull once observed, in his seminal *Anarchical Society*, 'Order in social life is desirable because it is the condition of the realisation of other values' (Bull 1977). If my position is a reasonable one, then what should be done; how can cyberspace be made more political and more ordered, such that an ethical framework can have more meaningful purchase?

This chapter is concerned with the relationship between politics, strategy and ethics, and the functioning of that relationship in a given (and rather specialised) environment – i.e. cyberspace. In the first part of the chapter I discuss briefly some of the peculiarities and challenges of the cyber environment. I then examine the challenge of cyber conflict more closely, from two perspectives: in part two of the chapter I ask how it is that cyber conflict has come to represent such a problem for ethics; and in part three I reverse the direction of inquiry to suggest that ethics might itself be seen as a digital problem.

In the fourth and final part of the chapter I return to a discussion of the relationship between politics, strategy and ethics before attempting to answer the question I have set myself above: *how can cyberspace be made more political and more ordered, such that an ethical framework can have more meaningful purchase?* I will suggest that deterrence, most familiar to us as a symptom of a more-or-less co-operative and stable political/strategic relationship during the Cold War, can also be instrumental; deterrence can serve as a mechanism both for ethical awareness-raising and ethical constraint in the cyber era. In other words, I will argue that deterrence can make it possible for cyberspace to be shaped into an environment (albeit virtual) in which human interaction (i.e. politics) and contestation (i.e. conflict) can be made susceptible to credible and effective ethical oversight.

1.2 Cyberspace: The Moral Frontier

Cyberspace seems to be shrouded in layer upon layer of epistemic and analytical uncertainty. One explanation for this might be that our current state of knowledge is simply deficient to the task, but that the problem of cyberspace will clarify as we work harder to understand it. Alternatively, it might be that cyberspace is simply beyond our comprehension, always will be and perhaps always should be. The second explanation is the least acceptable; given all that is involved in cyberspace, a position of passive acquiescence to the possibly far-reaching consequences of this human invention seems irresponsible and wholly lacking in ambition. In any case, while we might not (yet) know everything, it is not as if we know *nothing* about cyberspace.

Cyberspace has a number of features with which we are already familiar and which should provoke us to serious thought about security, order and justice. First, we know that one property of cyberspace (or, at least, one ambition of its proponents) is the rapid and universal diffusion of information and communications technology. Near instant global communications, more or less unfiltered, can place very small amounts of power in the hands of enormous numbers (billions) of people; or, conversely, the same phenomenon can place enormous financial, criminal and even destructive power in the hands of a very small number of technologically skilled people whose motives might seem disconcertingly unknowable and unpredictable. Second, another feature of cyberspace, with which we cannot but be familiar, is surely our accelerating dependency upon information and communications technology (ICT). At every level and in every function of society – the machinery of government; law enforcement; the private sector; arts and culture; research and development; education and the lives of individual people – it is becoming increasingly difficult to imagine what life was like without social media, email, smartphones, broadband and so on. What is more, it is not always the case that our dependency is sufficiently resilient, with built-in redundancy. Unmitigated dependency is a useful idea for understanding vulnerability; the relationship between the two is directly proportionate. And finally anonymity appears to be an enduring and pervasive feature of cyberspace. The potential for users of cyberspace to remain anonymous makes it difficult, among other things, to establish intentions, responsibility and culpability and to ensure a proportionate and discriminate response to those anonymous actors.

Diffusion, dependency and anonymity: if these are indeed core features of cyberspace then it must consequently be difficult to know what it is to be secure in, or from cyberspace. What is at stake? Which vulnerabilities matter more than others? What harm can be done? By whom or what? And to what end? Do familiar terms such as 'attack', 'defence', 'aggression' and even 'conflict' have sufficient explanatory power where cyberspace is concerned?

The next layer of difficulty is then revealed: how can the pursuit of security in cyberspace be evaluated in ethical terms? If, as I suggest above, we are uncertain as to our basic terms of reference and unsure as to what might be at risk in or from cyberspace, then we cannot be confident that we have an appropriate moral framework with which to assess our (and others') actions; what should be within the scope of that moral framework (and what should not); when it should apply (and when it should not)? In illustration of this difficulty, Christopher Eberle argues that 'the vast majority of cyber-attacks can provide no just cause for war. Indeed, because most cyber-conflict would seems far more akin to intense economic conflict than to conventional war, we should assess most cyber-conflict with moral principles tailored to boycotts not bombs' (Eberle 2013, p59). Is it then simplistic to seek to apply established normative, legal and ethical traditions concerned with restraint upon armed conflict in the international system, such as the United Nations Charter,

International Humanitarian Law and the Judaeo-Christian just war tradition, even though these traditions (discussed further below) are predominantly concerned with war as an observable, experiential and so-called 'kinetic' phenomenon between states? Is it acceptable simply to prefix these legal and ethical frameworks with 'cyber', even though we are not as clear as we might wish to be as to the meaning of that term?

As if the layers of this particularly tough fruit did not present enough challenges, at its core there is a yet more complex, more structural problem to be found; concerned with discernment and judgement. Cyberspace is clearly an environment both of opportunity, co-operation and fulfilment on the one hand, and contention, predation and confrontation on the other. Yet for all the 'take-up' of digital means recorded each year around the world,[5] humanity has not yet acquired sufficient familiarity with this relatively new form of communication and information transfer to know, with sufficient confidence, the identifiers of good and bad behaviour at any given moment. It is this which informs my view that the central, paradoxical problem with cyberspace is that it is in a state of arrested development; a pre-political, pre-strategic and therefore pre-ethical 'non-communicative adolescence' (Cornish 2012). Cyberspace is not yet susceptible to the forms of mature political organisation with which we are familiar and, in consequence, it is resistant to the normative constraints with which we have traditionally expected to manage and moderate competition, conflict and organised violence in the international system. Before suggesting how this imbalance might be redressed, I turn to the more specific problem of conflict in/from cyberspace.

1.3 Cyber Conflict as an Ethical Problem

At closer inspection, the prospect of conflict in or from cyberspace poses a number of questions as far as the application of traditional, conflict-focused ethical constraints is concerned. These questions fall into four groups: definitional, quantitative, technological and strategic.

The definitional question is fundamental. Terms of reference are needed in order to structure analysis and to rationalise policy decisions. Yet as I observe above, although long-familiar terms such as 'attack', 'defence', 'aggression' and 'conflict' are now often and widely used in the context of cyberspace, there is little or no consensus as to what these important terms might actually mean. The search for specific terms of reference with which to explain and evaluate coercive confrontation

---

[5] According to the International Telecommunications Union (ITU), in 2015 there are 7 billion mobile cellular subscriptions around the world; a seven-fold increase since 2000. By the end of 2015 the ITU estimate that there could also be as many as 3.2 billion Internet users globally, as against 400 million in 2000. International Telecommunications Union, *ICT Facts and Figures*. ITU, Geneva: http://www.itu.int/en/ITU-D/Statistics/Documents/facts/ICTFactsFigures2015.pdf Accessed 5 June 2015.

is probably as old as conflict itself, and frustration as to the outcome of that search is not new, either. After several decades of deliberation it was not until 1974 that the United Nations General Assembly settled upon a definition of aggression. The authors of the definition were 'convinced that the adoption of a definition of aggression ought to have the effect of deterring a potential aggressor, would simplify the determination of acts of aggression and the implementation of measures to suppress them and would also facilitate the protection of the rights and lawful interests of, and the rendering of assistance to the victim.'[6] Yet although the motivation behind it was impeccable, the UN General Assembly definition remained controversial in some quarters and has in any case has had only limited influence on the deliberations of the UN Security Council. And it is not clear that it has had the deterrent effect expected of it.

Ethical analysis of cyber conflict also generates a number of quantitative questions. Most important among these is the question of harm; what is meant by harm inflicted in or from cyberspace, and how can it be measured? This question points to a deeper uncertainty as to the effects (physical, human, psychological) of cyber conflict and how those effects might be prioritised both internally and relative to the effects of other forms of conflict and coercion (such as state-sanctioned organised violence, terrorism and the effects of crime), and relative to the effects of natural hazards and disasters (such as extreme weather events, volcanic eruptions and earthquakes, for example). Without some understanding of the nature and scale of harm which could result from cyber conflict it cannot be possible to answer the most basic of ethical questions to do with conflict and coercion; where is the threshold at which the harm resulting from a cyber action of some sort moves from the tolerable (e.g. inconvenience, discomfort, disruption) to the unacceptable (e.g. multiple deaths, irreversible physical damage, or even social collapse).

The prospect of cyber conflict is also known to pose several technological challenges, the first of which is generally described as the 'attribution problem'. Given what is known about the openness (by design) of cyberspace and about the wide availability of ICT, how can the source of a cyber intrusion of some sort be established (or 'attributed')? It is essential to know the identity of an assailant in order to enable ethical evaluation of the assailant's intention. The question of intention is central to the just war tradition: without knowing an adversary's intention it is not possible to make a judgement according to the principle of double effect, whereby the otherwise morally unacceptable consequence of an action can be tolerated as long as the action was undertaken in pursuit of a morally good end and as long as the adverse consequence was not intended (even though it might have been foreseeable). Edward Barrett's response to this challenge is to argue that with 'mock-ups of targeted systems … the intended and unintended effects of these [cyber] weapons are foreseeable…' (Barrett 2013, p11). This might be to expect too

---

[6] Definition of Aggression, UN General Assembly Resolution 3314 (XXIX), 14 December 1974: http://daccess-dds-ny.un.org/doc/RESOLUTION/GEN/NR0/739/16/IMG/NR073916.pdf?OpenElement

much, however, given that ICT networks often prove to be more complex and diverse than first thought, with the consequences of an attack upon such networks being largely beyond the knowledge and control of any attacker.

The attribution problem raises another, more straightforward ethical concern. If it is not possible to know, to a reasonable level of certainty, the identity of a cyber attacker then it must be correspondingly difficult to ensure that any response to the attack is discriminate. Who is to blame for the attack and – in many ways more importantly – who is not? Ethical constraint on the conduct of armed conflict requires the capacity to discriminate between legitimate objectives, targets and combatants on the one hand, and innocent non-combatants and civilian objects on the other. But is discrimination to this standard at all possible when, as Mark Duffield noted in 2001, 'network war' does not recognise the existence of civilians? (Duffield 2001, p91).

The extent to which cyber conflict can be prevented and managed is also governed by another technological challenge known as the 'dual-use problem': 'there are no exotic components to cyberweapons […] Any computer is a potential cyberweapon and anyone with advanced knowledge of information systems is a potential cybercombatant.'[7] And the relatively easy availability of ICT, when set against the complexity and structural vulnerability of an ICT-dependent advanced economy, also means that whereas cyber 'offence' might be comparatively cheap and easy, cyber 'defence' is more likely to be difficult and costly; another consideration for governments seeking to ensure a sufficient level of national security with limited financial resources.

Finally, a number of operational/strategic questions present themselves, each of which has implications for ethical oversight of cyber conflict. When, for example, can political leaders and national security decision-makers be confident that a cyber conflict has begun, and how can they be sure that the 'aggression' with which they are confronted is of such severity as to warrant a 'warlike' response? What could such a response reasonably (and legally) entail; should the response be in kind (i.e. making use of cyber defensive and/or offensive means of some sort), or could the response legitimately involve the use of armed force? Would a traditional military response to a 'cyber raid' of some sort be ethically proportionate?

If the timing and level of the response to a cyber attack are problematic, so too is the claim that it is legitimate to pre-empt attacks rather than be required to endure the damage of an attack before responding; as the international lawyer Derek Bowett observed, 'No state can be expected to await an initial attack which, in the present state of armaments, may well destroy the state's capacity for further resistance and so jeopardise its very existence.'[8] Pre-emptive or anticipatory self-defence is possibly one of the most contentious features of the modern laws of armed conflict.

---

[7] Dipert, The Ethics of Cyberwarfare, p385.
[8] Quoted in Shaw (1991), p694.

It can be justified in cases of imminent and manifest danger,[9] yet the necessary level of certainty might be impossible to establish where cyber conflict is concerned. In that case, two options become available, neither of which is attractive: either to endure the incursion or attack before being able to respond legitimately (assuming that the attack has been detected and its source identified); or to embark upon a policy of coercive prevention to ensure that such an attack can never take place. 'Preventive' war is the most controversial extension of the pre-emption debate. As Anthony Coates argues, preventive war 'attempts to justify recourse to war in the absence of an immediate and direct threat' and as such 'appears far too permissive' (Coates 1997, p159). And even if it does prove possible to embark upon and to 'fight' a cyber (or cyber-related) conflict in a morally sound way, one more hurdle remains; it must be asked how (and when, and by what means) cyber conflict could be brought to a definitive (and ethically acceptable) conclusion.

These definitional, quantitative, technological and strategic questions must all affect the prospects for some form of ethical governance of cyber conflict. It might be, of course, that the various challenges and difficulties described above are no more than temporary uncertainties; that cyber conflict is, fundamentally, as conducive to ethical governance as any other form of conflict and that the questions posed will all be answered in the fullness of time. Conversely, it might also be that a fundamental epistemic error has been made in assuming that cyber conflict can and must be containable within a broadly familiar framework of permissions and prohibitions and that 'the ethics of cyber conflict' will require some harder thinking if it is to be at all persuasive. Before continuing that discussion, the next part of the chapter asks whether ethics itself (or, at least, one ethical tradition) might be part of the problem.

1.4 Ethics as a Digital Problem

As I acknowledge below, the Judaeo-Christian just war tradition is by no means the only framework for the ethical constraining of conflict and war, and is in certain respects a hybrid of several approaches. The just war tradition is, however, in the ascendant insofar as it has developed to be the basis for much contemporary jurisprudence and practice in international humanitarian law (otherwise known as the law of armed conflict). On that basis, the just war tradition can be useful as a case study, of sorts, with which to demonstrate that traditional approaches to the ethical constraining of war and warfare can prove to be rather inelastic when it comes to cyber conflict.

---

[9] The usual reference point in this debate is the *Caroline* case of 1837 which established the principle that the pre-emptive use of armed force could be justified only where there was 'a necessity of self-defense … instant, overwhelming, leaving no choice of means, and no moment for deliberation.' US Secretary of State Daniel Webster quoted in Walzer (2000), p74.

The first and most obvious questions concerning the fitness of the just war tradition to deal with cyber conflict are concerned with the tradition's underpinning ambition. As one of many systems for the judgement of human behaviour *in extremis*, the just war tradition is predicated on there being individuals with a direct experience of such dire circumstances – i.e. fighting. As I have discussed above, this requirement might not always be met in a cyber conflict. More importantly, the just war tradition is also about classifying people, actions and, to a lesser extent, physical property; the tradition presupposes not only knowledge of the combatants – on all sides – but also the ability to distinguish between combatants and non-combatants; necessary and disproportionate actions; military objectives and civilian objects. It follows that identity, action and intention should be known and/or knowable. Having made these classifications, the just war tradition then operates as a normative theory of warfare and armed conflict, rather than simply a descriptive critique of those activities. In other words, the just war tradition is not simply about judging behaviour, but is also concerned with improving that behaviour; a task which cannot be undertaken seriously and effectively if the actors in a given conflict, their actions and their intentions, are neither known nor knowable. If it is not possible reliably to establish the identity of the combatants in a given conflict then it must be correspondingly difficult to know that certain people who are close by and who seem to be involved in one way or another, are actually *non*-combatants. It would seem, therefore, that the just war tradition embodies an expectation which simply cannot be met while the 'attribution problem' persists.

The second concern is with respect to trust. The just war tradition is based on the ethic of reciprocity; an organising principle of the tradition is that a level of trust can and must be achieved between adversaries, even when implacably opposed. Other ethical traditions might not set so much store by the so-called 'Golden Rule' yet the very existence of customary international humanitarian law (particularly the Geneva Conventions of 1949) points to there being some element of reciprocity and trust at work. But it is not easy to see how trust and reciprocity could be built in an anonymous environment. The purpose of a transactional, reciprocal ethical framework (such as the just war tradition) is to identify, evaluate and modify human behaviour under certain conditions. Yet that purpose might be impossible to achieve in an anonymous environment in which neither agents, behaviours nor conditions are sufficiently clear.

Finally, the just war tradition could also be severely limited in its technological scope. The just war tradition is concerned with traditional war and warfare which is physically known, witnessed and/or experienced and regarded as a regretfully necessary activity, and with the limitation of the effects of war on human life and liberty. It is testament to the enduring strength of the just war tradition that it continued to inform debate for many centuries after its inception. Yet it might have met its match with the prospect of cyber conflict because it is ill-equipped to deal with conflict in which technology is not merely an accessory to an otherwise familiar human activity, but has become its defining feature. Put in this way, the just war tradition loses its elasticity when it comes to intangibles such as communications networks and information flows:

> Just War theory rests on an anthropocentric ontology, i.e. it is concerned with respect for human rights and disregards non-human entities as part of the moral discourse, and for this reason it does not provide sufficient means for addressing the case for cyber warfare. It is this hiatus that presents the need for philosophical and ethical work to fill it and provide new grounds to ensure just cyber warfare (Taddeo 2012).

This hiatus has to be closed somehow: if cyber conflict is a serious possibility (if not already a serious fact) then it cannot be permitted to be an ethical no-go zone.

1.5 Politics, Strategy, Ethics and Deterrence

I suggest ('guess' would be more accurate) that humankind has been conscious of the relationship between politics, strategy and ethics for as long as our species has been social, warlike and moral. And I also suggest that what we now know as the complex edifice of the 'ethics of warfare' has developed – in very broad terms – through three stages of modernisation which I describe here as 'reflective', 'impulsive' and 'progressive'.

Michael Walzer suggests that 'For as long as men and women have talked about war, they have talked about it in terms of right and wrong.'[10] Although impossible to prove or disprove, Walzer's assertion seems, intuitively at least, to be reasonable enough. But for how long have men and women been talking? There seems to be no consensus on this point but let us accept, for argument's sake, that structured, vocal communication began when *homo sapiens* arrived at behavioural modernity – roughly 50,000 years ago – when, along with the capacity for symbolic thought, culture and creativity, early humans also developed the capacity for intelligent conversation. In that case, at what point could men and women have begun talking about an activity which we might today describe as war? *Homo sapiens* was a skilled hunter, certainly, but when did he also become a warrior? In *A History of Warfare* John Keegan suggests, albeit a little tentatively, that this might have occurred in the Neolithic era, some 10,000 years ago, with the advent of the simple bow and arrow: 'Cave art of the New Stone Age undoubtedly shows us scenes of bowmen apparently opposed in conflict' (Keegan 1993, p119). Perhaps, then, we can now attach an approximate date to Walzer's assertion; for the purposes of this chapter we can suppose that the first stage in the development of a recognisably modern 'ethics of warfare' took place some 10,000 years ago, when men and women could begin talking about a discrete activity, identifiable to them as violent conflict, 'in terms of right and wrong'.

In its second stage of modernisation human reflection on morality in warfare became more critical and more organised. This is perhaps the point at which several

---

[10] Walzer, *Just and Unjust Wars*, p3.

millennia of Neolithic moral reflection began to become more coherent and driven, in the form of an impulse to restrain warfare on moral grounds. The idea of righteous warfare – with restraints both on the resort to war and on conduct within it – can be found in the *Mahabharata*, one of the Sanskrit texts at the heart of Hinduism; possibly written between 3000 and 4500 years ago. The three Abrahamic religions offer cognate ideas dating from a similar era: Judaism (c. 3000 years old); Christianity (c. 2000 years old); and Islam (c. 1400 years since the death of Mohammed[11]). And it was at this time that the body of ideas known today as the just war tradition found its origin. The first traces of just war thinking have been found in Classical Antiquity; in the Greece of Pericles, Thucydides and Plato ('The ancient Greek period [...] saw restraints placed on both recourse to war and conduct of war' (Bellamy 2006, p18)); and in the Rome of Cicero ('one of the first thinkers to insist on the need for developing a legal and normative framework for war' (Reichberg et al. 2006, p51)).

Latterly, in the course of a third stage of modernisation, moral reflection about war and warfare has become yet more structured and, above all, more visibly progressive. This stage began very much more recently; its inception is often attributed to late Renaissance thinkers such as Francisco de Vitoria (1492–1546), Francisco Suárez (1548–1617) and Hugo Grotius (1583–1645). The achievement of these and other thinkers was twofold. In the first place, they encouraged the birth of what might best be described as a complex international institution for the prevention, management and amelioration of armed conflict. This institution has several components, the most significant of which, arguably, are those bodies of international law (particularly International Humanitarian Law; otherwise known as the Law of Armed Conflict or the Law of War) which insist that whenever violence is used between states it must be held tightly within a legal embrace, buttressed by a range of judicial and enforcement mechanisms. The institution has also generated a set of applied ideas such as conflict prevention, crisis management, deterrence, *détente* and peacekeeping. Another component are those international governmental and non-governmental organisations, ranging from no less than the United Nations to the International Committee of the Red Cross, which are dedicated to the avoidance and management of conflict and the remediation of its effects. Other organisations, such as the North Atlantic Treaty Organisation, have taken a more muscular approach to the restraint of organised violence, arguing that it might be necessary and even preferable, but only in cases of self-defence and existential emergency. The second achievement of its architects is that the institution should also be considered a *project*; it is designed, in other words, to persist as a permanent feature of an ever-improving national and international normative discourse on war and warfare.

These three stages are cumulative: the first creates the possibility of the second, while the second establishes the (unending) case for the third and together they constitute an impressive edifice: behavioural modernity gave mankind the beginnings

---

[11] Islam is considered by its adherents to have existed since Adam, held to be the first human.

of a moral vocabulary; the invention of strategy (i.e. the use of organised violence by one community against another) saw that moral vocabulary applied to warfare and its restraint; and in the last few centuries the ethical impulse has driven the creation of the progressive international institution described above. This edifice has also proved to be remarkably durable. Over ten or so millennia of human history, the emerging claim that war should and can be restrained on moral grounds has withstood a series of political and technological challenges: such as theocracy, democracy and totalitarianism on the one hand; and the crossbow, the atomic bomb and laser-blinding weapons on the other. There would seem, *prima facie*, no reason to suppose that cyber conflict should be any less tractable.

Yet for all its impressiveness this outcome was neither historically nor morally determined, and nor was it a straightforward, linear process. It would be simplistic to suppose that having first reflected on war and warfare 'in terms of right and wrong' some 10,000 years ago and having then discovered the impulse to restrain organised violence somewhat more recently, that the task is now complete; that all that remains is the continual implementation of the ethical 'project' through the recently evolved international institution. As well as the reflective, the impulsive and the purposive stages of development, one other ingredient was necessary for the achievement of the ethical edifice I have described, and remains essential for its continued relevance. This is a basic human insight which has accompanied the development of a moral consciousness about warfare and which has made it possible for that consciousness (whether derived from a religious or a secular morality) to have effect; to be instrumental. The insight is that war should only be contemplated to the extent that it serves a social or political logic, rather than some amoral, mechanical logic of its own. Once war was understood in this way, its ethical constraining then became feasible. The ethics of armed conflict do not control war and warfare, any more than business ethics control commercial practices. In both cases, the task of ethics is to infiltrate a moral dimension to the choices and decisions which must be made in a given set of circumstances. A complex, applied ethical framework cannot operate in an uncontrolled space, in a vacuum of some sort. And neither can ethics provide the control which is necessary to its operation. In modern human history it is certainly the case that not all war can be said to have been constrained by ethics; but for war to be so constrained, it must first be captured by politics.

This argument stems from a reading of the work of Carl von Clausewitz, the early nineteenth century Prussian general and a highly influential presence in modern strategic thought; in the late 1970s the English philosopher Philip Windsor described Clausewitz's *On War* as 'the only work of philosophic stature to have been written about war in the modern period' (Windsor (1977), p193). When Clausewitz observed that 'war is not a mere act of policy but a true political instrument, a continuation of political activity by other means' (Clausewitz C von. (1976), p87) he placed war and warfare firmly within a political framework. For the purposes of this chapter, the inference to be drawn from Clausewitz's observation is that violent conflict must first be seen as a political act if we are (a) to understand it and have any sort of a fruitful reflection upon it, and (b) if we are to be able to shape, direct and modify it. Whether he intended this result or not,[12] Clausewitz's insight helped to make possible the most recent phase in the ethical project to govern warfare.[13]

This is not, of course, to argue that Clausewitz invented the ethics of armed conflict. As I suggest above, moral reflection on the use of armed force is probably as old as organised armed force itself. But Clausewitz does make it possible to take the first step towards answering the question at the heart of this chapter: how can moral reflection on war and warfare be instrumentalised and turned into practical ethics? I contend that, following Clausewitz, politics *enables* applied ethics where international conflict, aggression, violence are concerned. I cannot see that a fruitful deliberation on the ethical constraining of armed conflict could be had if we were not accustomed to seeing such behaviour (i.e. warfare) as characteristically and unavoidably political.

The next step is to ask how practical ethics can be applied in the complex, evolving environment of cyber conflict. Following my Clausewitz-derived logic, this step requires cyber space to be turned into a politically malleable – and therefore ethically susceptible – environment; where, as in traditional conflict, known actors knowingly interact, and where negotiations, trade-offs and compromises can be made. A way must be found to make cyber space politically *knowable*, in order for moral language and ideas about cyber space and cyber conflict to develop and for a mature, comprehensive, applied ethical framework to become established. At this point a reformulation of the Clausewitzian logic might yield benefit.

Armed conflict is not simply a relationship between politics and war; it is also, self-evidently, a relationship between adversaries, albeit of a rather peculiar, destructive nature. Paradoxically, whereas this adversarial relationship ostensibly represents the *breakdown* of politics between the warring sides, it also carries with it the *possibility* (if not the inevitability) of something closer to a normal political relationship reasserting itself between the adversaries. This is key, since political and strategic relationships (however arid and conflicted) must first exist before they can be stabilised and improved, and indeed made subject to moral evaluation and constraint. Deterrence is another antiquated and very human idea which arguably reached the zenith of complexity during the Cold War in the form of 'deterrence by denial' (ensuring that the benefits of any military action would be outweighed by the costs incurred in undertaking it) and 'deterrence by punishment' (the threat of punitive retaliation, intended to convince a potential assailant that his attack would result in painful and destructive consequences which would, again, outweigh any expected benefit).

---

[12] Hew Strachan argues that this was not Clausewitz's intention and that 'over the past thirty years western military thought has been hoodwinked by the selective citation of one phrase': Strachan (2013), p13.

[13] I develop this argument in Cornish (2003).

As described, Cold War deterrence expressed, or was symptomatic of a basic political/strategic relationship between adversaries; a relationship which was more-or-less trusting and constructive even at times of deep hostility. In the cyber era I suggest that it is not only possible but also imperative to conceive of deterrence as less symptomatic and more instrumental – perhaps even causative – of a political/strategic relationship between cyber adversaries (actual or potential). By reverse-engineering the millennia-long reflective/impulsive/progressive process which I describe in the first part of the chapter, our contemplation of cyber conflict can in a sense be underpinned with a moral foundation upon which can be constructed a prospective and progressive 'project' for the ethically meaningful and practically effective containment of cyber conflict. Yet earlier in the chapter I also described anonymity as an 'enduring and pervasive feature of cyberspace', making it difficult to ascribe 'intentions, responsibility and culpability.' If it is not possible to know with sufficient confidence who (or what) should be deterred, how can cyber-deterrence work? I suggest the answer to this conundrum might lie in deterring neither the actor nor the action, but anonymity itself; in the form of 'deterrence by association'.

Although cyber space is indeed very largely an opaque environment it is nevertheless possible to make an intelligent assessment as to the motivations behind certain behaviours, the rewards being sought and by whom. In an earlier analysis of nuclear proliferation I suggested that deterrence by association could work in the following way: 'if states and commercial organisations can be exposed for having supplied a nuclear weapon capability to a terrorist group, they can then be subject to sanctions; and the threat of sanctions might have the effect of cutting off supply in the first place' (Cornish 2010). The basics of this idea are not new; it has elsewhere been described as 'triadic', 'third party' or 'indirect' deterrence and in the context of cyber conflict Jeffrey Knopf has advanced the notion of 'deterrence by counter-narrative and delegitimisation' (Knopf 2010, p10).

The value of deterrence by association can be explained by analogy with nuclear deterrence. The premise of nuclear deterrence, in part, was that by identifying an adversary's most valued assets (whether, military, economic, industrial or even social), and then threatening to inflict unacceptable damage on those assets, the adversary would modify his behaviour accordingly. In cyber space generally, the most prized assets are, arguably, anonymity, deniability and uncertainty. Thus, the cyber equivalent of the nuclear threat which an adversary would be unwilling to confront, might be nothing more than being willing (and sufficiently confident and courageous diplomatically) to speak against the cyber adversary's prized assets; using information to turn these assets into liabilities, by making them the 'target' of deterrence.

The purpose of this exercise would be to re-engineer the cyber debate in a subtle way, by developing the norm that *association* with pariah behaviours would be reputationally damaging and should therefore be avoided. As the norm becomes more established so the onus would be placed on governments to demonstrate that they were not involved, as sponsors or beneficiaries, in cyber aggression, terrorism, crime, espionage or in any given event, rather than to argue that their involvement cannot be proved.

The essence of deterrence by association (rather than 'deterrence by *proof*') is that a political, rather than a technological standard of evidence can and should be used. This is a difficult case to make in a field in which we have become so used to technological precision and certainty and in which there is reluctance to publicise information concerning cyber attacks for fear of revealing intelligence capabilities and procedures. However if governments, international organisations, the commercial sector and private individuals can all be encouraged not just to form judgements about adverse and/or deviant behaviours but to discuss those behaviours (and their likely origins) more openly, then we will, in effect, see a normative discourse beginning to create its own foundation; the 'reverse-engineering' to which I refer above. Not all cyber aggression or wickedness will be revealed, certainly, but the fact and the consequences of exposure should influence those who are contemplating similar behaviours in the expectation of anonymity and deniability.

1.6 Conclusion

The central argument of this chapter is that cyberspace must first be made politically malleable in order that cyber conflict can then be subject to well-established ethical principles such as those offered by the just war tradition; last resort, just cause, right authority, proportionality, discrimination and so on. The goal should be for cyber security to become an arena in which diplomacy, negotiation, bargaining, compromise, concession and moral judgement are all considered possible, proper and productive. Cyber security has elements both of the familiar and of the peculiar about it and it must be understood in its own terms if it is to be the subject of serious, balanced public policy discourse and strategic and ethical analysis.

Deterrence can assist in this process. Deterrence can function in several ways: it can be both protective and defensive; it can prevent escalation to more severe levels of conflict; and it can be expressive of a residual politico-strategic relationship between otherwise implacable adversaries – a relationship in which trust can develop, enabling the stabilisation and even the resolution of conflict. I have argued in this chapter that deterrence can also work 'in reverse' – by *association* – to help construct a rudimentary relationship between adversaries and competitors, where such a relationship does not exist. Rather than address adversarial actors or actions, the focus of deterrence by association is instead the presumed advantages of anonymity and deniability. If these presumptions can be challenged successfully then cyber conflict will become increasingly receptive to effective (rather than merely declaratory or aspirational) ethical constraint. Deterrence by association seeks to make cyber space knowable; as an environment that has acquired universal value, access to which is increasingly seen as a right (and in some cases a *human* right); and as an environment that is being encroached upon by miscreants of various sorts. By these means, deterrence by association will make the cyber security debate ethically tractable, bringing politico-strategic reality and ethical aspiration into closer, and more effective alignment.

References

Barrett, E.T. 2013. Warfare in a new domain: The ethics of military cyber-operations. *Journal of Military Ethics* 12(1): 4–17.
Bellamy, A.J. 2006. *Just wars: From Cicero to Iraq*. Cambridge: Polity Press.
Bull, H. 1977. *The anarchical society: A study of order in world politics*, 96–97. London: Macmillan.
Clausewitz C von. 1976. *On War*, ed and trans Howard M and Paret P. Princeton: University Press.
Coates, A.J. 1997. *The ethics of war*. Manchester: Manchester University Press.
Cornish, P. 2003. Clausewitz and the ethics of armed force: Five propositions. *Journal of Military Ethics* 2(3): 213–226.
Cornish, P. 2010. Arms control tomorrow: The challenge of nuclear weapons in the twenty-first century. In *America and a changed world: A question of leadership*, ed. R. Niblett, 232–233. London: Wiley-Blackwell/Chatham House.
Cornish, P. 2012. Digital Détente: Managing the cyber confrontation between China and the West. *Public interest report*. Washington, DC: Federation of American Scientists.
Danks, D., and J.H. Danks. 2013. The moral permissibility of automated responses during cyber warfare. *Journal of Military Ethics* 12(1): 18–33.
Dipert, R.R. 2010. The ethics of cyberwarfare. *Journal of Military Ethics* 9(4): 384–410.
Duffield, M. 2001. *Global governance and the new wars: The merging of development and security*. London: Zed Books.
Eberle, C.J. 2013. Just war and cyberwar. *Journal of Military Ethics* 12(1): 54–67.
Fisher, D. 2011. *Morality and war: Can war be just in the twenty-first century?* 3. Oxford: Oxford University Press.
Jenkins, R. 2013. Is stuxnet physical? Does it matter? *Journal of Military Ethics* 12(1): 68–79.
Keegan, J. 1993. *A history of warfare*. London: Hutchinson.
Knopf, J.W. 2010. The fourth wave in deterrence research. *Contemporary Security Policy* 31(1): 1–33.
Reichberg, G.M., H. Syse, and E. Begby (eds.). 2006. *The ethics of war*. Oxford: Blackwell.
Shaw, M.N. 1991. *International law*. Cambridge: Grotius.
Strachan, H. 2013. *The direction of war: Contemporary strategy in historical perspective*. Cambridge: Cambridge University Press.
Taddeo, M. 2012. *Just war theory and cyber warfare*. http://blog.practicalethics.ox.ac/2012/06/just-war-theory-and-cyber-warfare/. Accessed 6 June 2012.
Walzer, M. 2000. *Just and unjust wars: A moral argument with historical illustrations*, 3rd ed. New York: Basic Books.
Windsor, P. 1977. The clock, the context and clausewitz. *Millennium* 6(2): 190–196.

Paul Cornish was educated at the University of St Andrews and the London School of Economics and completed his PhD at the University of Cambridge. He served in the British Army before joining the Foreign and Commonwealth Office, was senior research fellow at Chatham House, taught strategic studies at the UK Defence Academy and the University of Cambridge and directed the Centre for Defence Studies at King's College London. He has held professorial appointments at Chatham House, the University of Bath and the University of Exeter. His work covers national strategy, cybersecurity, security futures, arms control, the ethics of armed conflict and civil-military relations. He is a fellow of the University of Oxford Global Cyber Security Capacity-Building Centre, a principal research associate in the Department of Computer Science at University College London and a participant in UK-China Track 1.5 discussions on cybersecurity. He is editor of the *Handbook of Cyber Security* (forthcoming 2017).

Chapter 2
Blind Justice? The Role of Distinction in Electronic Attacks

Jack McDonald

Abstract This chapter questions the direct applicability of the just war tradition's principle of distinction to electronic attacks involving computer network exploitation (CNE). It offers three principle challenges to maintaining the norm of distinction in electronic attacks that are rooted in the impossibility of foreknowledge of the object of attack in a computer network. In lay terms, without significant inside assistance it is impossible for a hostile agent seeking to exploit a computer network from knowing the network's architecture and role prior to conducting hostile exploitation of the network. Due to this lack of knowledge, it is impossible for the hostile agent to be certain that initial exploitation will be free of negative consequences. This draws attention to the understanding of hostile action in both CNE and computer network attacks (CNA). This impossibility of foreknowledge leads to three challenges in applying the principle of distinction to cyber attacks: the access problem – where if CNE is considered to be an attack, then our understanding of distinction collapses, the boundary problem – where it may be impossible for an agent to know the boundaries or couplings of the system that they are attacking, and the levels problem – where humans are held accountable for non-human agency inherent in the deployment of autonomous software programmes ('viruses', 'malware', etc). This chapter argues that these problems are surmountable, but not with an understanding of distinction that is directly transposed from human interactions.

Keywords Cyber security • Ethics • Just war theory • International humanitarian law • Distinction • Computer network exploitation • Surveillance

J. McDonald (✉)
The Department of War Studies, King's College London, Strand, London WC2R 2LS, UK
e-mail: jack@jackmcdonald.org

2.1 Introduction

For better or for worse, the three core normative principles of the just war tradition – distinction, proportionality and necessity – form the basis of most moral and legal judgement of violence and destruction in the context of war and armed conflict. The just war tradition itself has, as A. J. Coates writes, "monopolised the moral debate about war, at least in the western world" (Coates 2012, 3). The contemporary law of armed conflict derives from the principles of the just war tradition, but significant differences exist between the role of law and morality in contemporary warfare. Modern western militaries now operate with lawyers integrated at many levels of the command structure, and implement legal training programmes for their service personnel to ensure compliance with the law of armed conflict (Carvin and Williams 2014). Ethics and law are both integral to the notions of military professionalism, but law has engendered an archipelago of regulation and potential punishment. Moral censure for actions in war carries a different weight.

Despite the differences between the two disciplines, the analysis of the moral basis of actions during armed conflict is still important. One reason for this is that the law can be used to render legal judgement, but this is different in character to asserting that an action is right. The ongoing dialogue over the reasoning of, and limits to, justice and injustice in war constitutes the just war tradition. By extension, the public debates over the morality of given actions reflect upon societies at war, or those coming to terms with past actions. Recent examples such as A.C. Grayling's criticism of Allied bombing campaigns in the Second World War focus public attention on the deep moral legacy of the past and its continued effects on the present (Grayling 2007). Attempts to define and redefine justifiable action in war have undeniable political and social consequences. Like law, the just war tradition constitutes the rules of war, and by extension constitutes the phenomenon of war itself. Unlike law, the process of reasoning is paramount in the just war tradition – many of the actual moral judgements are the consequence of competing world views that have no parallel in law. Critical philosophers, for example, feel free to dismiss the notion of just war (Butler 2009), whereas international lawyers cannot dismiss the tradition of the Hague Conventions, still present in the Geneva Conventions of 1949.

Both international lawyers and moral philosophers tend to agree that no matter how the character of warfare changes any new or novel forms of political violence that amount to war or armed conflict should be regulated by the law of armed conflict and/or the just war tradition. Discussions of novel technologies, such as the potential development of autonomous weapon systems, tend to be predicated on the necessity of new weapons conforming with existing standards (Sharkey 2010). Both law and ethics have responded to and accommodated changing technology from the invention and use (or potential use) of aerial warfare to nuclear weapons and the use of computers in armed conflicts. In general, it is normal for novel means of warfare to be considered in light of the law as it exists at the point of invention. International lawyers point to the general acceptance of Article 36 of the 1977 Additional Protocol 1 to the Geneva Conventions, which obliges states to assess whether the weapons that they choose to adopt can be used in a manner consistent with the law of armed conflict prior to adopting them (Boothby 2013). However, new forms of technology also give rise to novel forms of long-standing legal and moral dilemmas.

The rise of the internet and global communications networks challenges common understandings of war and security due to the nature of online communication and cyber space (Betz and Stevens 2013). This has been accompanied by the issue of 'cyber attacks' involving the intentional disruption or sabotage of computer systems and networks (Rid 2013a, 55). Cyber attacks differ from normal types of military attack in a number of important ways. First, they are not directly lethal nor destructive – although the second order effects of altering the operation of a computer system may destroy the system itself, or cause consequential physical harm and damage (Rid 2013a, 142). Whether or not alterations to a computer network constitute an attack or not is a major area of contention in the literature on 'cyber war' (Rid 2012; Stone 2013, 101–108). The lack of direct kinetic or physical effect means that the applicability of concepts associated with physical violence is neither obvious nor immediate. In particular, whether such attacks contravene basic principles of law and morality is an obvious concern.

Cyber attacks do not have to be sophisticated – simplistic methods of flooding a networked computer with requests for information can drive them offline, as demonstrated in the 2007 distributed denial of service attacks on Estonia. That said, the types of attacks that drive fears of 'cyber armageddon' are predicated on gaining access to computer systems and networks. One issue with this is the way in which these systems may need to be accessed prior to understanding their role, function, or – in terms of the just war tradition – whether the systems are civilian or military in nature. Accessing computer systems without the explicit authorisation of their owners (which in this context is unlikely to be given) requires hostile actions taken against the system itself. Some argue that the action of 'hacking' to access a computer network is hostile in and of itself. In 2012 the Chairman of the Joint Chiefs of Staff, General Martin Dempsey, stated that hacking was not automatically hostile, but that hacks on critical infrastructure could be (Levine 2012). This points to an important issue: the interpretation of the initial act of accessing a system by an unauthorised person or agent. The importance of this is that knowledge of a computer system's function might be impossible without first accessing it in a hostile manner.

One understudied element of the justice of cyber attacks is the relationship between the exploitation of a computer system and the knowledge that a transgressor gains of the system itself. The nature of protected computer systems is that a well designed security regime will prevent any attacker from knowing with any degree of certainty either the system's function or internal architecture. In other words, an attacker will have to exploit or penetrate the system itself in order to gain a requisite level of access that would enable the attacker to exploit or disrupt the system's normal functions so as to cause the "damage or destruction to objects" that could constitute a cyber attack according to the Tallinn Manual (Schmitt et al. 2013).

For the purposes of this chapter, I will discuss three related problems that arise from the lack of knowledge about a system's characteristics or functions where interference with the system is characterised as an attack. In the just war tradition hostile actions taken in the absence of evidence distinguishing the subject of hostility are usually held to be immoral. This leads to the first issue, which I term the access problem: actions to identify a system as either a legitimate or illegitimate object of attack could take place prior to knowing or understanding a system's ownership or function. Depending how one defines an attack, the initial activity of gaining access to a system may – in itself – constitute a cyber attack. The inherent problem is that while an attacker might have some idea about a given system that they are trying to exploit, confirmation about the system's nature is likely to occur after initial entry to the system is gained. If one considers initial exploitation to constitute an attack, this means that most "cyber attacks" would breach the core principle of distinction – distinguishing legitimate targets of attack from civilian persons or objects – by definition. The initial stages of any cyber operation will inevitably be waged in the blind. This means that technical aspects of computer network exploitation pose a challenge to the normative principle of distinction because transposing it from prior interpretations could render substantial electronic access immoral.

The second challenge is what I term the boundaries problem: computer networks are not singular entities with readily identifiable enclosures. While some computer networks may be constructed as such, military computer networks are hard to separate from civilian ones. While there has already been considerable discussion regarding the overlap and interplay of military and civilian computer networks (Geiß and Lahmann 2012; Lubell 2013), here I focus again on the knowledge question: how would one know in the course of CNE if one had moved beyond the targeted network? This is fundamentally related to the third problem that I term here to be the levels problem: how we connect human decision making to the operation of software programmes, often operating independently of the user. At one level of abstraction, the human being engaged in CNE is conscious of the design of the software tools that they are using, and therefore the form of agency in question is that of human decision makers, much like any other ethical question such as the classic trolley problem. At a less abstract level, the decision making functions of software are often non-visible to the user, who may be conscious of the general operating parameters of the software programmes that they have introduced to a computer system, but are unable to oversee these autonomous software agents at work. This level of abstraction poses a challenge to thinking about CNE from the perspective of the just war tradition since it is concerns non-human agency and agents that by definition cannot exercise the same type of analytical and ethical behaviour as a human being.

Cyber attacks constitute a problem for both international humanitarian law and the just war tradition. The three problems identified have parallels in the study of attack thresholds and military objectives under international law (Mavropoulou 2015). However international law contains specific definitional issues that differentiate it from the just war tradition. For example, the question of applicable law to cyber attacks in international armed conflicts between states differs to those that occur in non-international armed conflicts (Dinstein 2012). International law defines lawful (and unlawful) targets of attack in a far more precise manner (Lubell 2013). Moreover, questions such as proportionality of attacks are fundamentally tied to legal treaties and their interpretation (Geiß and Lahmann 2012). Nonetheless both international humanitarian law and the just war tradition are predicated on the notion of rational agents that are able to observe a situation and make a judgement about the situation regarding the consequences of any use of force or violence. If the very act of observation requires the hostile exploitation of a computer network, then states might – in theory – be able to adhere to the just war tradition, but only due to otherwise indiscriminate acts performed on networks that could be civilian in nature.

2.2 Distinction, Knowledge and Prediction in the Just War Tradition

As an adversarial activity the conduct of war often requires participants to make decisions in conditions of extreme uncertainty. At the same time, the ethical discussion of war and warfare often proceeds from the point of limited – but iron-clad – certainty. When authors discuss the morality of killing civilians, they often do so by declaring known variables at the outset, or by discussing historical examples. Michael Walzer's classic work on the just war tradition is replete with such cases (Walzer 2006). There are a couple of points to make here. One is that the categories of person are often taken to be stable, even if hidden. Although there is a lively debate about the application of just war theory to fighting terrorists and guerrillas, the second major area of contention is when – if ever – civilians should count as military targets. The circumstances and reasons for attacks that result in harm or damage to civilians is one of the central themes of discussion in the just war tradition.

Identifying objects of attack as legitimate targets is inherent in the principle of distinction. This legal and moral principle underpins western concepts of legitimate warfare. In essence, distinction consists of three interlocking principles. First is that attacks are intentional acts that cause harm and either disrupt or destroy a person or object (a relatively wide definition for the use of force or violence). Second, attacks should be directed at the enemy in order to reduce the ability of their military to operate. Following from this, militaries are required to distinguish between legitimate and illegitimate targets of attack prior to any attack. Without these three components the concept of distinction does not make much sense. After all, militaries distinguish between people and objects with attacks in mind (as a functional activity). Furthermore, attacks involve positive action and intention, a feature that also implies personal and political responsibility (Ramsey 2002, 7). Lastly, it is the intent to cause harm that matters – attacks are still attacks even if they cause no damage.

This last point is vital in understanding the problem of foreknowledge in cyber attacks that involve the exploitation of computer networks.

Surveillance and intelligence collection activities are usually beyond the scope of the just war tradition. As a tradition founded on the moral discussion of violence and its consequences, the collection of information pales in comparison. Surveillance is intrusive, but it is mainly understood as a non-violent activity. Surveillance, like organised military violence, is fundamentally political in nature and a core function of the nation state (Dandeker 1990). Yet when one examines the moral debates of the just war tradition surveillance is usually seen as beneficial – greater intelligence collection leads to better decisions, greater precision and less accidents. In some senses, it could be argued that there is a moral imperative to surveil potential targets and areas of operation. Indeed, it is this idea that leads some to argue that forms of precision warfare, such as the use of drones to conduct targeted killings, are more moral than the alternatives (Strawser 2013). If one considers, as most tend to do, that the act of observation is non-violent, even when it is intrusive, then it is clear that the principle of distinction requires as much surveillance as possible up until the point that it is possible to distinguish between potential targets of attack. This division between processes of knowledge formation and hostile action is quite sensible, particularly since knowledge formation is generally held to be a pre-condition for the use of force. Of course there are exceptions to this. The use of torture to acquire information that might identify people as legitimate targets of attack is one area where most people consider the means of attaining knowledge is an inherent moral wrong. People such as Alan Dershowitz who defended the use of torture after 9/11 faced considerable criticism (Strauss 2003–2004) for their position. Although incomparable to the moral wrong of torture, the similarity here is whether the attainment of knowledge required to satisfy the principle of distinction in cyber attacks will require the commission of wrongful acts.

The problem of applying the principle of distinction to attacks on computer systems is that whereas observing physical targets is usually deemed to be a non-hostile act (or indeed a legal and moral necessity) the actions required to effectively identify a computer system as belonging to a military group or performing a military function are in some cases considered hostile acts in and of themselves, particularly by civilians unconnected to military activity (Mazetti and Sanger 2013). Due to the physical operation of the internet, cyber attacks may inherently violate the law of neutrality (Kelsey 2008). In practice, militaries operate with imperfect information, and inevitably must make judgements with serious legal and moral consequences in these circumstances. The problem with attacks directed at computer networks is that the very act of attempting to identify a system as a permissible or impermissible target may on a physical level be no different to the actions that can be taken to disable it (Roscini 2014, 17). After all, the exploitation of a computer system to gain access involves the manipulation of physical states on storage devices, as does the re-writing of code, or the injection of malicious code, onto the same system.

This gives rise to two interpretations of attack relevant to the just war tradition. The first is that any hostile intrusion is considered to be an attack, the second that an attack only takes place when a system is intentionally manipulated in order to impair its function. These are necessarily simplifications, and, as I will later argue, it is extremely difficult to make such a separation due to the nature of CNE. The possibility of distinction in the context of electronic attacks turns on the possibility of non-intrusive knowledge. When discussing the role of knowledge in target selection, the most often emphasised element is the requirement for combatants to form knowledge about a target through observation or other reliable means prior to taking any decision that involves violent or lethal action. In this sense, our first interpretation of CNE holds that non-intrusive knowledge generation is likely impossible, at least to the degree necessary to satisfy the requirement of distinction. If on the other hand the very act of accessing a system in order to identify it, or to identify methods of disabling it, is considered an attack, then it might be very difficult to apply the principle of distinction in practice.

2.3 Fighting Blind?

The access problem, and the problems of identification and knowledge associated with cyber attacks, are reformations of traditional problems. Every war features problems involving the principle of distinction, and novel forms of warfare invariably affect the type and availability of information. With cyber attacks, certain pieces of knowledge – or paths to them – are unlikely to be available to potential attackers. How should individuals or militaries approach such an environment, and is the just war tradition still applicable?

Computer network attacks (CNA) take different forms. Thomas Rid argues that we should focus attention upon acts of sabotage, rather than theorised 'cyber Pearl Harbors' or similar doomsday scenarios involving significant damage to infrastructure (Rid 2013b). Sabotage, in this sense, is not directly violent. As Rid has commented on the Saudi Aramco incident, this involved the 'bricking' of computers (rendering them inoperable) rather than damage to infrastructure. Rid argued that "It's useful to use the word sabotage, because then we don't talk about violence. These are non-violent attacks, yet they have a huge effect" (BBC 2015). Yet acts of war do not have to be violent in this sense. Whether or not incidents like the attack on Saudi Aramco constitute war or not, these would be covered by both international humanitarian law and the just war tradition were they to be conducted under the auspices of war. Therefore the way in which we think about and conceive of hostile actions towards a computer system matters – at least if we are to conceive of these actions as being justifiable within the just war tradition.

The access problem hinges on whether one interprets both CNE and CNA as attacks in the sense of the just war tradition, or whether only CNAs warrant consideration. A vital aspect of the ongoing debate about cyber attacks relates to the functional and technical means of exploiting computer networks. As the security expert Bruce Schneier (2014) writes: "The problem is that, from the point of view of the object of an attack, CNE and CNA look the same as each other, except for the end result." To attack a computer network in order to disrupt or damage it, one has to first gain access to the system itself. In most national jurisdictions, states have varying degrees of legislation that make unauthorised access to a computer network a criminal act.[1] This is because basic forms of exploitation involve the acquisition of login credentials and passwords in order to access a network as an ordinary user or administrator. Without such information, accessing a network requires the exploitation of flaws in software, and sometimes the manipulation of the interaction between software and hardware – the physical elements of the computer system itself – in order to attain access to the system. Importantly, all computer software is also physical, since it is stored on whatever hardware it operates upon.

Regardless of ultimate intent, both CNE and CNA physically change a system. Uploading harmful software to a system can cause physical damage to connected elements of the system as demonstrated by a recent attack on a German steel mill (Lee et al. 2014). This attack that resulted in physical damage targeted the mill's SCADA systems - the industrial control systems that regulate the behaviour of industrial equipment. By causing these physical systems to operate beyond their normal safe limits, the exploitation of a computer network can result in physical damage. In parallel, manipulating an industrial system so that feedback sent to machine controllers or human operators that can cause either operators to destroy critical industrial systems by accident (Langner 2013). Regardless of consequent effects, or intentions, any intentional action that could be classified as a cyber attack – be it damage, intelligence collection or sabotage – requires access to a system. The subsequent steps necessary to enact an attack (such as understanding a system's function and role, altering the system itself to facilitate access, or extracting information from the system) take place after this initial act of hostile access. If we conceive of cyber attacks in this sequential manner, and consider hostile access as a form of attack, then considerable problems arise for thinking about the maintenance of the principle of distinction.

The point of distinction is that states and their agents must distinguish between military and non-military targets, and refrain from attacking the latter. In the physical world, this involves human beings using their senses, as well as collating and assessing information derived from sensors and intelligence analysis. 'Positive identification' is often a synthesis of a range of different pieces of knowledge by human agents. While certain features of computer systems that are connected to the internet will always be observable (or a requisite for exploitation) such as an IP address, the computer system that is reachable through this address will not be known until it is accessed, either legitimately or illegitimately. The problem in this regard is the same one facing signals intelligence agencies across the world in our digital age: the technical means of accessing systems is in many ways indistinguishable from the means to disable it. After all, unless a computer system can somehow be manipulated to cause physical damage to itself (shutting down cooling systems, for example), in both cases the likely physical action needed to access, observe and disable a system is the same: manipulating states on storage devices to overwrite

---

[1] See, for example, the US Computer Fraud and Abuse Act of 1984, or the UK's Computer Misuse Act of 1990.

programmes or change stored data. What outsiders can know about computer systems prior to accessing them is very little, meaning, therefore, that it is almost impossible to satisfy what we might think of as the criteria for distinction if it necessary to take hostile action against a system in order to access it. Even if a system is unprotected, there is a second-order issue that a user unfamiliar with the system may disrupt its operations by accident while finding their way around the system in order to ascertain its purpose and operation. If illegitimate access (CNE) is considered an 'attack' then it would conceptually be little different from accidentally harming someone when hitting them with a vehicle. If, however, we take attack to be intent, this leads us to the problem that unintentional harm is a constituent aspect of network exploitation, something that the existing law of armed conflict works very hard to prevent.

The problem with dividing cyber attacks into these two (admittedly crude) levels of abstraction is that it relies heavily on the perception of the party attempting to exploit other systems. The organisation, state, or individual, on the receiving end of the attack will not necessarily be able to discern this intent. This is a significant and long-standing problem in the available literature on cyber attacks (Libicki 2012). The wider problem is that the principle of distinction does not readily transfer to domains where the acquisition of knowledge is an inherently hostile act. Beyond the digital realm, it is possible to distinguish between surveillance and violence, but at the physical level of computer networks, such distinctions disappear. The access problem is therefore only resoluble if a common interpretation or distinction between CNE and CNA can be found, and given the fundamental differences of interpretation between states (as well as between security communities such as the EU, business and private individuals) it is unlikely that a common interpretation can be found (Schneier 2015). An alternate solution to the access could be the identification of remote computer networks as legitimate targets by using a range of intelligence sources. Thomas Rid points out that successful attacks are likely to require considerable intelligence efforts on the part of states or non-state actors in order to successfully impair the operation of a computer network, or industrial systems regulated by one (Rid 2013b). Even so, the access problem is likely to arise at some point or another even in the context of wide scale intelligence collection.

2.4 Unknown Boundaries

The problem of blind investigation of a system goes to the heart of the concept of distinction. There is no real equivalent in the existing just war tradition. Soldiers aren't permitted to cause harm or use force without first identifying a person or object as something that can be legitimately harmed. The second problem, that I term the boundary problem, is that in many circumstances it is likely to be impossible to predict the ultimate limits of a given computer system identified as a legitimate target of attack. The relative harm that is likely to result from this is low – particularly if a sharp threshold is drawn separating violent attacks from non-violent ones – but this problem still militates against the direct translation of concepts from the physical world to computer networks.

The boundary problem exists because we cannot know a computer network's structure and architecture purely from a description of its physical elements. The two are related, but the point of contemporary computer networks is that they allow physically separated entities to function as a single system. Even if a computer network is identified as a legitimate target for CNE or CNA, it is quite unlikely that a potential attacker will know the complete architecture of the network, and therefore be able to define its boundaries, prior to interacting with the system itself.

The preparatory exploitation of computer networks relies upon information retrieved from a network itself. This can be quite limited, even when a computer's activity is known. A good parallel is found in the work of computer security researchers who work to defeat botnets, which are networks of computers infected by malicious software that enable people (usually criminals) to repurpose them. The problem of 'hacking back', or intruding onto infected computers in order to disable malicious software - again without the knowledge or authorisation of the computer's owner – is similar in character to the present discussion precisely because it hinges on the problem of incomplete information. Unlike military professionals, security researchers are usually bound by domestic legislation that prevents any unauthorised access to a computer system (Kabay 2009). Still, military professionals and security researchers face the same fundamental problems when interacting with unknown computer systems and networks.

Botnets are an interesting example in the present context because they can be identified from traffic patterns. As these networks are usually generated by malicious self-propagating software, they work by connecting to command and control nodes – computers that can be used to control the behaviour of network members. Traffic analysis and other non-intrusive sources of information can therefore be used to identify members of a network without accessing individual computers (Cooke et al. 2005). These networks also pose ethical problems for researchers – should they attempt to access command and control nodes? If access is gained, is it unethical to then let the network keep running? Attitudes to this are changing among security researchers, as evidenced by growing support for aggressive measures to disable botnets in the past 10 years (Dittrich et al. 2010).

The limiting problem of active response against botnets is the effects of 'hacking back' or the production of malware designed to disable botnets is impossible to predict. Researchers have raised the issue of accidentally crashing systems whilst cleansing them, or interfering with critical functions such as life support systems (Amini 2008). From this we can see that the consequential problem of the effects of systems interference follows from the primary one of identification. The boundary problem here is that a clearly "hostile" system remains unknown to those who seek to alter or disrupt its function, and they may be unable to understand the network's architecture from the modes of access attained. Moreover, without knowledge of a system's architecture or onward connectivity, alterations to the system may result in consequential effects unknown to the person altering the system.

Assessing consequential harm is an important aspect of the principle of distinction. Distinguishing between permissible and impermissible targets of attack is a precondition to the legitimate use of force, at least in the moral framework of the just war tradition. Whereas the concept of distinction applied to computer networks might preclude any attack, applied to the architecture of the internet, it may permit a far greater range of objects to be physically attacked. Early computer networks provided military commanders with command and control systems for the nuclear age (Schlosser 2013). The trans-oceanic cables that carry the bulk of data communications are used by the military in the same way as military vehicles share roads with civilian ones. In many senses, however, metaphors rooted in the physical world collapse when one considers the reality of digital networks. This is even more apparent when we consider the means of using code to cause a computer system (or attached machinery) to malfunction.

The boundary problem could be solved again by codifying a set of norms specifically applicable to computer networks. In particular these norms might consider the degree of understanding required of someone engaged in CNE or CNA to understand the architecture of the system that they are interacting with prior to making any significant alterations to the system. Secondly this may impose a duty to understand the system's coupling with other systems, or its real-world function. Given the relative simplicity of some systems, however, this may not be possible in all cases. This could give rise to perverse effects, for example, designing computer networks such that their architecture and coupling is illegible to an infiltrator so as to protect them from attack on moral grounds. Nonetheless, the boundary problem demonstrates the issues inherent with transposing an anthropocentric moral framework to computer networks.

2.5 Abstraction, Distinction and Agency

The notion of what constitutes an attack highlights the relationship between knowledge and the moral requirements of the just war tradition. Nonetheless the tradition itself is firmly rooted in human agency. After all, it is a creation of humans, who are also the primary moral agents capable of violence. One issue with both CNE and CNA is that humans might not be the ones making the decisions that result in a system malfunctioning. The use of code that traverses computer networks without human oversight means that the code itself, while not sentient, will affect the state of the system independent of human supervision. For a moral tradition rooted in human agency and decision-making, the notion of computer systems and programmes of sufficient complexity or disaggregation to render human beings bit players in their operation is a troubling development. This is what I term here the levels problem: the level of abstraction at which the just war tradition is held to apply.

In order to collect or extract information from a system, network exploitation often involves the use of software programmes that run independent of human supervision. As indicated previously, these programmes may interact with the targeted system in ways that produce unforeseen results. Furthermore, given their modes of operation (self-replication and transfer) they are likely to spread beyond the targeted system. Even though the designers of Stuxnet, responsible for damage to Iran's nuclear programme, attempted to limit its spread, they were unable to do so (Langner 2011). The social imaginary of hacking leads to an expectation of real-time control and access of affected systems, whereas the reality of computer network exploitation often includes self-executing programmes designed to alter a computer network's function, or to prepare a system for exploitation at a later date.

The reality of exploiting networks by means of malicious code is that the exact effects of this code is impossible to predict. Sophisticated attacks on computer systems or networks often manipulate system-specific security flaws or architecture. This means that attacks on specific systems requires exploits that "may be so fine-tuned to one specific target configuration that a generic use may be impracticable" (Rid 2013a, 168). Nonetheless, two points are worth considering. First is that computer network exploitation is likely to have unpredictable results that are not bounded in space and time as are physical acts of violence. The second is that there is a certain loss of control and oversight implicit in cyber attacks that use autonomous sub-routines or programmes. Neither of these are in themselves are prohibitive constraints, but again are worth considering in light of the standards that we tend to judge physical violence by.

Human agency is the implicit moral requirement of adherence to the just war tradition. If we consider the broad span of actions that some define as moral in the context of war, inevitability we deal with choices. Some are forced by another person or political entity – hence the requirement for self defence – but in almost all cases worthy of consideration as moral problems, the element of choice is present. The prospect of non-human agents deciding whether or not to execute acts of physical violence drives opposition to autonomous weapons (Purves et al. 2015). Cyber attacks can't cause direct violence in this fashion, but it is worth considering the connections between the two debates to a certain extent. After all, what morally separates a human overseeing the operation of distributed computer code from a human overseeing the operation of an autonomous weapon system? Although the two have drastically different levels of effect, and I do not suggest that they are one and the same, human responsibility for non-human agents will likely be one of the key moral questions of war in the twenty-first century. If a consensus is derived about human responsibility for autonomous computer programmes, then it will likely form part of the debate about human responsibility for the actions of autonomous weapon systems.

Software is too simple to make judgments on the level of human beings. In particular, the kind of judgements that the autonomous software programmes used in CNE and CNA are simplistic, even if the design of this software demonstrates a considerable combination of skill, art and technical prowess on the part of the developer. These autonomous routines are dumb, consisting of *if...then...* logical conditions as they navigate or replicate through a system. Therefore they cannot be considered to be moral agents, unless we radically reimagine the concept of moral agency. This leads to a key issue with the principle of distinction: can a human being be held accountable for the failure of dumb software to distinguish between a civilian and military target (or anything thus far considered in the course of this chapter)? If one wishes to preserve the principle of distinction for the digital realm, then the only answer to this question can be yes. Even so, we again return to the question of knowledge: to what degree is it possible for a person running such code on an unknown system able to foresee its ultimate effects? If they are to be held responsible for all the consequences, then this gives rise to persons being held to account for the unwitting violation of the principle of distinction in circumstances where they had no way of knowing whether such a violation was possible.

Again, the levels problem militates against the direct translation of the just war tradition to the digital sphere. It is unlikely that states or their service personnel will agree upon the appropriate degree of human responsibility for non-human agency. Competing standards and concepts of cyber security divide the permanent five members of the UN Security Council (Korzak 2011). Despite recent inter-state dialogue, a similar disparity exists over autonomous weapons (The United Nations Office at Geneva 2014). Although it is highly unlikely that code will end up killing human beings in a direct fashion on par with physical acts of violence, the way in which cyber attacks are defined and considered is likely to inform the future of the just war tradition in the context of an increasingly automated future.

2.6 Conclusions

Depending on who you ask, cyber attacks either herald the possibility of near-immediate catastrophic destruction of country-wide infrastructure, or a concept that is non-violent but confused with warfare. The purpose of this chapter has been to point out that a very broad definition of 'cyber attack' would render most forms of hostile computer access and network exploitation immoral by the principles of the just war tradition. Conversely, an understanding of cyber attacks as the consequential effects of network exploitation negates this problem. As always, definitions matter. The purpose of this chapter has not been to determine whether cyber attacks are as 'real' as some argue that they are, but to reflect on the way in which this idea challenges the just war tradition because it doesn't contain elements that have been axiomatic to historical debates regarding the morality of war.

Reasoning about cyber attacks allows us to reflect upon the just war tradition itself. In *Morality and War* David Fisher compares and contrasts the three primary modes of ethical reasoning: virtue ethics, deontology and consequentialism. Fisher ultimately argues that the just war tradition makes sense with a mixture of virtue ethics and consequentialism, which he terms virtuous consequentialism (Fisher 2012). This chapter has argued that the technical issues of computer network exploitation precludes the direct transfer of the just war tradition predicated upon rational single-agency. Nonetheless, we live in a digital age, and the just war tradition – that has survived the transition from agrarian warfare to the nuclear age - will doubtless survive.

In many ways the technical issues inherent in computer network exploitation serve to re-enforce Fisher's arguments against a deontological approach to the just war tradition. If anything, the prospect of computer network exploitation places more emphasis upon virtue than any of the other two branches of moral reasoning: distinction as a virtue, rather than a rule, can survive in the digital age. Anyone seeking to exploit computer networks for military purposes is likely to exploit systems whose 'identity' is indistinct until hostile exploitation, and to do so using computer programmes that are unable to make distinctions of their own. In blunt terms, they are likely to break the rules of the just war tradition by design, and well as accident. Yet if the persons conducting such exploitation do so with the principles of the just war tradition in mind, that is, they take the principles of the just war tradition as virtues or goals, then this problem might be reduced to the smallest possible given the technical features of digital networks.

One point to raise about this is that the just war tradition has a very specific scope - war and violent political conflict. Yet the current context of cyber attacks (or activity by intelligence agencies that might support such attacks in future) is invariably conducted in peace. Therefore what is discussed in this chapter is a relatively minor aspect of the overall issue of hostile attacks on computer networks by states (or their agents) in the twenty-first century. Some might wonder: what use is virtue if it leads to the rules being broken? The real problem is that the just war tradition is insufficient to address the myriad range of problems associated with cyber attacks. In the opinion of the author, one tendency that should be resisted is the wish to apply the moral reasoning of the just war tradition beyond the bounds of war and warfare. The ethics of cyber attacks between otherwise peaceful states might be informed by the moral principles of the just war tradition, but any attempt to apply the tradition itself to actions during peace would be a grave mis-step.

This chapter has made the argument that all attempts at computer network exploitation constitute grounds for thinking about and applying the just war tradition if they are conducted by states or their militaries with hostile political intent. In some respects this expands the application of the just war tradition beyond its traditional boundaries (the use of force) to include activity that is intended to be intelligence collection. Some may bridle at this expansion, but it is a consequence of the changing technological means of violence and warfare, rather than an attempt to reformulate the just war tradition itself. Just as the invention of railways and significant civilian infrastructure led to questions of whether it was just or unjust to destroy these, so, too, does the evolution of digital society give rise to questions of justice in relation to computer network exploitation.

The material difference between computer networks and physical infrastructure results in problems that have no direct parallel in past wars and conflicts. If we wish to preserve the just war tradition (a laudable aim, in the opinion of this author) then the tradition must tackle these issues and evolve to take account of the world in which we now exist. It is impossible to wish away these problems as it is to ask militaries to refrain from exploiting networks that they may need disrupt in the course of a military campaign.

References

Amini, P. 2008. *Kraken Botnet Infiltration*. DVLabs. http://dvlabs.tippingpoint.com/blog/2008/04/28/kraken-botnet-infiltration.
BBC Online. 2015. *Is cyber-warfare really that scary?*. http://www.bbc.co.uk/news/world-32534923.
Betz, D.J., and T. Stevens. 2013. Analogical reasoning and cyber security. *Security Dialogue* 44(2): 147–164.
Boothby, B. 2013. *How will weapons reviews address the challenges posed by new technologies?* Military Law & Law of War Review 52:37.
Butler, J. 2009. *Frames of war: When is life grievable?* London: Verso.
Carvin, S., and M.J. Williams. 2014. *Law, science, liberalism and the American way of warfare: The quest for humanity in conflict*. Cambridge: Cambridge University Press.
Coates, A.J. 2012. *The ethics of war*. Manchester: Manchester University Press.
Cooke, E., F. Jahanian, and D. McPherson. 2005. The zombie roundup: Understanding, detecting, and disrupting botnets. In: Proceedings of the USENIX SRUTI workshop, vol. 39.
Dandeker, C. 1990. *Surveillance, power & modernity*. Cambridge: Polity Press.
Dinstein, Y. 2012. The principle of distinction and cyber war in international armed conflicts. *Journal of Conflict and Security Law* 17(2): 261–277.
Dittrich, D., F. Leder, and W. Tillmann. 2010. *A case study in ethical decision making regarding remote mitigation of botnets*. Financial Cryptography and Data Security, Lecture Notes in Computer Science, vol. 6054, 216–230.
Fisher, David. 2012. *Morality and war: Can war be just in the twenty-first century?* Oxford: Oxford University Press.
Geiß, R., and H. Lahmann. 2012. Cyber warfare: Applying the principle of distinction in an interconnected space. *Israel Law Review* 45(3): 381–399.
Grayling, A.C. 2007. *Among the dead cities: Is the targeting of civilians in war ever justified?* London: Bloomsbury.
Kabay, M.E. 2009. *Computer security handbook*, 5th ed. New York: Wiley.
Kelsey, J.T. 2008. Hacking into international humanitarian law: The principles of distinction and neutrality in the age of cyber warfare. *Michigan Law Review* 106: 1427–1451.
Korzak, E. 2011. Computer network attacks, self-defence and international law. In *International law, security and ethics: Policy challenges in the post-9/11 world*, ed. A. Hehir, N. Kurt, and A. Mumford, 147–163. New York: Routledge.
Langner, R. 2011. Stuxnet: Dissecting a cyberwarfare weapon. *IEEE Security & Privacy* 9(2): 49–51.
Langner, R. 2013. *To kill a centrifuge a technical analysis of what Stuxnet's creators tried to achieve*. The Langner Group, Arlington. Online at http://www.langner.com/en/wp-content/uploads/2013/11/To-kill-a-centrifuge.pdf.
Lee, R.M., M.J. Assante, and T. Conway. 2014. ICS CP/PE (Cyber-to-Physical or Process Effects) case study paper – German steel mill cyber attack. SANS ICS. Online at https://ics.sans.org/media/ICS-CPPE-case-Study-2-German-Steelworks_Facility.pdf.
Levine, A. 2012. *Hacking by China not necessarily a "hostile act"*. CNN. http://security.blogs.cnn.com/2012/02/14/hacking-by-china-not-necessarily-a-hostile-act/.
Libicki, M.C. 2012. *The specter of non-obvious warfare*, 88–101. Fall: Strategic Studies Quarterly.
Lubell, N. 2013. Lawful targets in cyber operations: Does the principle of distinction apply? *International Law Studies* 89: 252.
Mavropoulou, E. 2015. Targeting in the cyber domain: Legal challenges arising from the application of the principle of distinction to cyber attacks. *Journal of Law & Cyber Warfare* 4: 23.
Mazzetti, M., and D.E. Sanger. 2013. Security leader says U.S. would retaliate against cyberattacks. *The New York Times*.

Purves, D., R. Jenkins, and B.J. Strawser. 2015. *Autonomous machines, moral judgment, and acting for the right reasons*. Ethical Theory and Moral Practice, online at http://link.springer.com/article/10.1007/s10677-015-9563-y.
Ramsey, P. 2002. *The just war: Force and political responsibility*. Oxford: Rowan & Littlefield.
Rid, T. 2012. Cyber war will not take place! *Journal of Strategic Studies* 35(1): 5–32.
Rid, T. 2013a. *Cyber war will not take place*. London: Hurst.
Rid, T. 2013b. More attacks less violence. *Journal of Strategic Studies* 36(1): 139–142.
Roscini, M. 2014. *Cyber operations and the use of force in international law*, 17. Oxford: Oxford University Press.
Schlosser, E. 2013. *Command and control*. London: Allen Lane.
Schmitt, M.N., et al. 2013. *Tallinn manual on the international law applicable to cyber warfare*. Cambridge: Cambridge University Press.
Schneier, B. 2014. There's no real difference between online espionage and online attack. *The Atlantic*. http://www.theatlantic.com/technology/archive/2014/03/theres-no-real-difference-between-online-espionage-and-online-attack/284233/.
Schneier, B. 2015. *Data and Goliath: The hidden battles to collect your data and control your world*. New York: WW Norton & Company.
Sharkey, N. 2010. Saying 'No!' to lethal autonomous targeting. *Journal of Military Ethics* 9(4): 369–383.
Stone, J. 2013. Cyber war will take place! *Journal of Strategic Studies* 36(1): 101–108.
Strauss, M. 2003–2004. Torture. 48 *New York Law Review* 201.
Strawser, B.J. 2013. Introduction: The moral landscape of unmanned weapons. In *Killing by remote control: The ethics of an unmanned military*, ed. B.J. Strawser. New York: Oxford University Press.
The United Nations Office at Geneva. 2014. *Report of the 2014 informal meeting of experts on Lethal Autonomous Weapons Systems (LAWS)*. Geneva: UN.
Walzer, M. 2006. *Just and unjust wars: A moral argument with historical illustrations*. London: Basic Books.

Jack McDonald PhD, is a research associate and teaching fellow at the Centre for Science and Security Studies, a research centre in the Department of War Studies, King's College London, where he currently convenes the MA in science and security. Jack's research explores the ethical, legal and political implications of novel technologies and their use in warfare. This chapter is informed by his work on an ESRC-funded research grant *SNT Really Makes Reality: Technological Innovation, Non-Obvious Warfare and the Challenges to International Law* (ES/K011413/1) that investigated professional military attitudes to the relationship between emerging technologies and international humanitarian law.

Chapter 3
Challenges of Civilian Distinction in Cyberwarfare

Neil C. Rowe

Abstract Avoiding attacks on civilian targets during cyberwarfare is more difficult than it seems. We discuss ways in which an ostensibly military cyberattack could accidentally hit a civilian target. Civilian targets are easier to attack than military targets, and an adversary may be tempted to be careless in targeting. Dual-use targets are common in cyberspace since militaries frequently exploit civilian cyber infrastructure such as networks and common software, and hitting that infrastructure necessarily hurts civilians. Civilians can be necessary intermediate objectives to get to an adversary's military, since direct Internet connections between militaries can be easily blocked. Cyberwarfare methods are unreliable, so cyberattacks tend to use many different methods simultaneously, increasing the risk of civilian spillover. Military cyberattacks are often seen by civilian authorities, then quickly analyzed and reported to the public; this enables criminals to quickly exploit the attack methods to harm civilians. Many attacks use automatic propagation methods which have difficulty distinguishing civilians. Finally, many cyberattacks spoof civilians, encouraging counterattacks on civilians; that is close to perfidy, which is outlawed by the laws of armed conflict. We discuss several additional problems, including the public's underestimated dependence on digital technology, their unpreparedness for cyberwarfare, and the indirect lethal effects of cyberattacks. We conclude with proposed principles for ethical conduct of cyberwarfare to minimize unnecessary harm to civilians, and suggest designating cyberspace "safe havens", enforcing reparations, and emphasizing cyber coercion rather than cyberwarfare.

Keywords Cyberwarfare • Civilians • Ethics • Distinction • Cyberattack • Networks • Dual-use • Reporting • Propagation • Perfidy • Infrastructure • Product tampering

N.C. Rowe (✉)
Computer Science Department, U.S. Naval Postgraduate School, Monterey, CA, USA, 93845
e-mail: ncrowe@nps.edu

3.1 Introduction

Article 52 of the Additional Protocol I to the Geneva Conventions (1977) (ICRC 2015) is clear in stating principles regarding collateral damage that have been ratified by the majority of the world's countries:

Article 52 – General protection of civilian objects

1. Civilian objects shall not be the object of attack or of reprisals. Civilian objects are all objects which are not military objectives as defined in paragraph 2.
2. Attacks shall be limited strictly to military objectives. In so far as objects are concerned, military objectives are limited to those objects which by their nature, location, purpose or use make an effective contribution to military action and whose total or partial destruction, capture or neutralization, in the circumstances ruling at the time, offers a definite military advantage.
3. In case of doubt whether an object which is normally dedicated to civilian purposes, such as a place of worship, a house or other dwelling, or a school is being used to make an effective contribution to military action, it shall be presumed not to be so used.

These principles have been insufficiently respected for cyberspace as the world sees increasing planning for use of cyberspace by militaries (Geers et al. 2013; Dinniss 2012). The Stuxnet cyberattacks on Iran (Gross 2012) provide an example of a sloppy operation that insufficiently considered collateral damage to civilians. Around 10 million civilian machines were infected worldwide by a worm-based propagation that eventually found its way to targets in nuclear-processing facilities in Iran. The damage to the civilian machines was initially unclear, so quick removal was important. It costs at least $100,000 in U.S. dollars for the world to recognize, analyze, and find countermeasures for a new attack method, since it requires around 1000 h total by well-trained specialized personnel. Stuxnet was sufficiently novel that it probably cost $1,000,000 to analyze it, design signatures to recognize it, and develop methods to remove infected files and processes. Then deployment of the countermeasures in the form of antivirus software required additional downloads by users, which could however be bundled with other security updates so that the extra time for each user was about a second, for a total of 10,000,000 * (1/3600) hours * $100 per hour = $277,000.

Secondary costs were attempts to attribute the Stuxnet attacks, around $100,000 since this was not a high priority. More importantly, the reuse of Stuxnet attack methods in subsequent criminal cyberattacks (Kaplan 2011) probably resulted in 10,000 incidents worldwide probably requiring around $100 per incident to address, for an extra cost of $1,000,000 total.

Thus the total collateral damage of Stuxnet was at least $2.4 million. The international standard for insurance purposes is $50,000 per year of human life, so Stuxnet's collateral damage to civilians was equivalent to the taking of one average human life. The lesson of Stuxnet is that collateral costs, despite initial claims, can be significant with cyber operations.

Enforcing the distinction between military and civilian targets in warfare has a long history (Kinsella 2011). We agree with much of the legal analysis of (Brenner and Clarke 2011) but will focus more on the technical methods that lead to collateral damage in cyberwarfare. Technical threats can also have technical solutions.

3.2 Methods by Which Cyberwarfare Can Hit Civilians

We consider here the kinds of mechanisms of cyberwarfare spread to civilians. We shall use "civilian" in the informal sense of people not employed by militaries, realizing that there are many borderline cases (Kaurin 2007). For instance, people contracted to work for a military are not generally considered civilians. We use the term "cyberwarfare" to refer to any military operations accomplished primarily by the use of computers, networks, software, and digital data (Clarke and Knake 2010; Shakarian et al. 2013).

Nearly all methods proposed for cyberwarfare exploit flaws in software, and most methods are similar to those of cybercrime using malware, rootkits, and bot networks (Elisan 2012). We will use the term "cyberattack" to refer to all these methods.

3.2.1 Civilian Cyberspace Is Ubiquitous

Civilian objects, both hardware and software, are all over cyberspace. The vast majority of Internet traffic is civilian. All the hardware we depend on – desktop computers, laptop computers, tablets, mobile devices, and storage devices – is fundamentally civilian. Similarly, all the software we depend on – operating systems, network protocols, Web browsers, document processing, and security management – is also fundamentally civilian. When military organization use cyberspace, they predominantly build on top of this existing infrastructure with their own data, using methods like encryption to prevent their data from being read or interfered with by civilians and civilian software. That means that, for the most part, there are not many distinctively military targets in cyberspace. In fact, it is very difficult to restrict attacks to only military targets because they must circumvent so much civilian infrastructure.

To be sure, some military activities are critical enough to need special handling in the form of exclusively military hardware and software. Examples are weapons systems, command-and-control systems, military-vehicle controls, and weapons-production systems. But simply because they are critical to militaries, they are well-protected. They are hard to reach on the Internet, or they may be disconnected from it. So if a cyberattack goes astray, the odds are good that it will hit a civilian rather than a military target in cyberspace.

3.2.2 Civilians Are Easy Targets

Besides the difficulty of avoiding civilians in cyberspace, civilian targets are often easier to damage than military ones in cyberspace. Military organizations well understand the importance of maintaining their operations to keep their cyberspace access, communications, and data safe. So they provide many layers of security for those systems in the form of access controls, cryptography, real-time monitoring for suspicious behavior, and deceptions to fool attackers.

Civilians have considerably lower standards of security. Commercial pressures encourage vendors of popular software (e.g. Microsoft Windows, Adobe Reader, Web browsers, and mail systems) to make their products unnecessarily complex. The rate of flaws in software is roughly proportional to the square of its size, so overly complex software runs high rates of bugs. For instance, the size of the minimum Microsoft Windows operating system on desktop computers, according to Microsoft, has gone from 18 megabytes in 1992 to 720 megabytes in 2000 and 20,000 megabytes in 2012. Little of this additional code is necessary for operation of the computer. The more bugs in software, the more vulnerabilities that can provide the basis for cyberattacks.

In addition, civilian targets are not prepared for cyberwarfare. The world has not seen a major cyberwar yet. Many civilians confuse cyberwar with cybercrime and expect that it will play out similarly. They expect, as in the case of cybercrime bank fraud, that someone will quickly and cheerfully refund their damage costs after a cyberwar and everything will be fine. However, cyberwar tends to target important assets, and tries to thoroughly disable them, so recovery from a cyberwar may be very slow.

This means that there are considerably greater opportunities for attacks on civilian targets than military targets, and the attacks can be simpler. Deliberate attacks on civilians are a violation of Article 52. However, when a country is greedy or desperate, they will be sorely tempted to attack civilian sites in cyberspace regardless of Article 52.

3.2.3 Civilians Can Be Desirable Targets

Another appealing thing about civilian targets is that such attacks can send political, social, or cultural messages that an attacker wishes to convey. By attacking U.S. banks, for instance, Islamic militants are making a statement about their advocacy of non-usurious banking under Sharia law. The attacking of military targets often does not send as clear a message, particularly the targets in big military organizations like those of the U.S. which engage in a large variety of activities all over the world. If war is just an extension of politics by other means, its message needs to be clear.

3.2.4 Dual-Use Targets Are Hard to Avoid

Because of the ubiquity of civilians in cyberspace, many military systems and artifacts are "dual-use" resources, or resources intended for both civilians and militaries. Dual-use resources can be legitimate military targets if they are justified as per Article 52. An example would be a civilian mail server hosting a military command-and-control network, which could be attacked to prevent communications during a military operation. Another example would be the Global Positioning System (GPS) used to measure precise locations on the surface of the earth and whose disablement could greatly impede military operations, but could hurt civilian entities such as aircraft and emergency services. However, key issues are how much of the civilian system is of military use and how critical is that military use. Since civilian traffic on the Internet is so much larger than military traffic, the Internet must be described as almost entirely civilian. If the military use is small, it is hard to justify it as a military target according to the standards of Article 52.

It may be possible to attack only the military parts of a dual-use target to satisfy Article 52. For instance, one could modify a mail system by a cyberattack to lose military mail exclusively. But such attacks require detailed knowledge of the software target and are difficult to implement. Most cyberattacks, like most munitions, will engage in undiscriminating destruction because that is the easiest effect to get.

An ethical justification for attacking dual-use targets and harming civilians is that citizens often bear some responsibility for their government's actions. If a government has committed crimes with the support of its citizens, a cyberattack with broad international support against those citizens may be justified although it violates Article 52. However, as we discuss below, cyberattacks have peculiar side effects of being able to harm civilians in countries unrelated to a conflict. Stuxnet was an example with its widespread (albeit mild) damage, but any cyberattack that employs new methods will like cause some harm to the entire international community.

Side effects of disabling even small parts of the Internet can be significant. (Anonymous 2012) reports that the Chinese government's disabling of Domain Name Service (DNS) servers, to prevent Chinese citizens from reaching non-Chinese Web sites, led to failures all over the Internet since DNS servers are essential to Internet routing. If mere acts of censorship can hurt the Internet everywhere, a cyberwar could be much worse.

Dual-use targets can be deliberately constructed to be problematic to attack. For instance, a state can put their hospitals on the same network used by their military for command-and-control as a way to provoke international outcry if the network is attacked. This is an appealing tactic for weak states, although if it can be shown to be deliberate, they get no immunity under international law for their civilians being attacked. Still, it looks bad for the attacker.

3.2.5 *Attacks Can Damage the Environment*

Even if a target is exclusively military, side effects of a cyberattack may hurt the civilian environment. This is most likely with cyberattacks that cause physical damage to a target. Causing an explosion in a nuclear power station used by a submarine, for instance, can release nuclear materials into the environment. A precedent is a cyberattack on an Australian sewage plant that caused a release of large amounts of sewage (Slay and Miller 2008). Matters are exacerbated by the tendency of military planners to think only in terms of military objectives, something that led during the Vietnam War in the 1960s to overuse of herbicides to reduce insurgent cover, destroying forests and causing health problems for the Vietnamese (War Legacies Project 2010).

3.2.6 Civilians Can Be Desirable Intermediate Steps

Stuxnet used many intermediate computers to get to the eventual target of Iranian nuclear facilities, and it is likely that future cyberattacks will be similar. That is because direct military-on-military cyberattacks will likely be blocked because militaries know most of the Internet sites of their possible adversaries already. So it is essential to get to a military cyber target indirectly through Internet sites that the target considers friendly or neutral. Civilian sites would generally be friendly. Unfortunately, this violates the Hague Convention Article V on neutrality:

CHAPTER I: The Rights and Duties of Neutral Powers
Article 1. The territory of neutral Powers is inviolable.
Article 2. Belligerents are forbidden to move troops or convoys of either munitions of war or supplies across the territory of a neutral Power.
Article 3. Belligerents are likewise forbidden to:

(a) Erect on the territory of a neutral Power a wireless telegraphy station or other apparatus for the purpose of communicating with belligerent forces on land or sea;

(b) Use any installation of this kind established by them before the war on the territory of a neutral Power for purely military purposes, and which has not been opened for the service of public messages.

Cyberattacks designed to damage an adversary are a form of munition that sent to a target and then is triggered. Using intermediate Internet sites in neutral countries to convey such cyberattacks is moving munitions across the territory of a neutral power. Even cyberattacks to facilitate intelligence gathering would violate Article 3 on establishing installations on neutral territory not available for public messages (they cannot be public because then they could be found and removed by anti-malware software), and they would also likely function analogously to wireless telegraphy stations. (von Heinegg 2012) argues against this analysis, claiming that Internet sites that forward packets without processing them are like a global public service, and malicious packets cannot hurt these sites. This argument does not refute the danger of denial-of-service attacks where the volume of traffic is a weapon, and the volume could hurt the forwarding site too. In addition, packets get stored in many places on forwarding sites, and malware could conceivably get out and attack the forwarding site if it is vulnerable. More importantly, the only fast way stop an attack of unknown ultimate origin is to stop neutral sites from forwarding the attack by attacking them in turn, which could draw neutral countries unjustifiably into cyber warfare, just what the Hague Convention article is trying to prevent.

Victims could first try to contact the owners of the neutral site to stop the attack, but this is not always possible due to the required time and the possible lack of expertise at the neutral site.

3.2.7 The Unreliability of Cyberwarfare Encourages Overkill

Cyberwarfare methods tend to be unreliable because they depend on flaws and bugs in software and hardware. Flaws and bugs can disappear suddenly when their vendors find them. This does not bother cybercriminals, who if an attack fails, can just try another method or another target because they often do not care how or who they are attacking. But it is an issue for nation-states because they want to achieve more precise and certain effects on a few important targets. This means that cyberwarfare must use simultaneously several methods of rather different types, as Stuxnet did, to have a good chance of an effect. The methods must be of rather different types to reduce the chances that a failure of one significantly increases the chances of a failure of another. But having many methods of attack increases the chances of hitting civilians, because there are more possibilities for targeting mistakes.

3.2.8 Side Effects of Reporting the Attack Can Hurt Civilians

A serious form of collateral damage with cyberattacks is in the potential reuse of the attack in subsequent criminal attacks. Effective cyberwarfare generally requires surprise, achieved by finding and exploiting previously unrecognized flaws and bugs in software. It is especially important to find novel ones because known ones get fixed quickly. It is also especially important to find novel flaws and bugs because they are more likely to work, and failed attack attempts warn an adversary to harden their defenses and give them good clues as to how. So an adequate cyberwarfare attack requires a good number of novel methods to provide a good degree of success on a first strike.

These requirements mean that cyberweapons will be a good source of ideas for cybercriminals as well as the cyberwar units of other states. Certainly cybercriminals do prefer attacks that have been tested and shown to be effective. It was not long before some of the six attack methods of Stuxnet were being reused for cybercrime (Kaplan 2011).

Cybercriminals learn about new cyberattack methods from threat-alerting sites such as www.us-cert.gov, vulnerability databases like nvd.nist.gov, cybersecurity-related newsgroups like those at www.securityfocus.com, and attack-testing software like www.metasploit.com. While these sites are for defense and tend not to give many attack details, there are plenty of fee-based commercial sites that will give more details and even will sell you attack code. The monitoring that provides data for these sites is accomplished by a variety of automated tools (Hashim et al 2013), and vendors compete fiercely to offer the most up-to-date notices of cyberattack methods. Why is such information posted if criminals can exploit it? The consensus of the information-security community is that it is more important to share information freely to enable finding countermeasures quickly than it is to conceal information to prevent a few additional attacks over a few days (TechRepublic 2005). Analysis can stop most cyberattack methods within days with a software modification or "patch" if a wide range of experts can contribute. However, not everyone gets the patch quickly since not everyone uses their systems everyday, and not all vulnerable systems are configured properly. Thus, any new cyberattack method will cause damage for several days to a good number of civilian systems, then continue causing damage at a gradually decreasing rate over a long period of time, and this will be the case regardless of the source of the attack.

Cybercriminals can also learn new cyberattack methods by monitoring the Internet directly. Tools called "sniffers" can look for particular kinds of suspicious traffic, and tools called "honeypots" can serve as decoy sites for collecting attacks. Even Twitter feeds can provide early warning of cyberattacks (Al-Qasem et al. 2013). So observant criminals can learn new attack methods even if no one else notices them.

3.2.9 Automatic Propagation of Cyberattacks

Some cyberattacks like Stuxnet reach their targets by propagating autonomously from one computer or device to another. Viruses (propagation of file infections) and worms (propagation of running processes) are the major examples. Autonomous propagation tries to circumvent normal controls on site access, often through vulnerabilities in software. Automatic propagation is appealing for cyberattacks because the attack can grow fast: The more sites and files are attacked, the more launching pads for further attacks, and the more subsequent attacks. This multiplies the effect of the initial attack quickly, and overwhelming force applied quickly is a key goal of military operations. Even if there are a limited number of ultimate targets as with Stuxnet, propagation to many sites increases the chances of reaching a target and the speed of getting there.

Civilian sites are good places from which to autonomously propagate an attack because civilian systems have fewer controls than military systems. But even if an ethical military planner tries hard to confine the propagation to military systems, this may fail because automatic attacks cannot easily distinguish what they are attacking. Cyberattack code needs to be small to sneak past defenses, and does not have much room to carefully analyze what it is attacking. Typically viruses and worms just scan systems for neighbor systems and go after all of them. Matters can get ugly if military systems have "backdoor" connections to civilian systems for purposes such as software updates. Civilian sites can also be connected to military sites because someone on the military site did not know what they were; sites rarely describe themselves internally, and even when they do, it is not placed consistently. Furthermore, just because a site has many military neighbors does not mean that it does warfighting, since many military hospitals and public-relations sites have such connections, and conversely, many contractors with ".com" sites in the U.S. directly support the military. So one cannot judge whether a site is military or civilian easily, certainly not solely by its IP address or site-owner registration in the Regional Internet Registries such as ARIN.

Another serious danger of automatic propagation of viruses and worms is the difficulty of turning them off. They are like land mines that are committed to actions independent of the context, and they usually have no respect for ceasefires or surrenders since there is little room in their small packages for a communications receiver (and having such a receiver would make them easier to detect anyway). Continuing hostilities after a ceasefire or surrender are explicitly prohibited by the laws of war, so it will be important to stop viruses and worms then.

3.2.10 Spoofing of Civilians by Militaries

One more way in which civilians can be hit by cyberwarfare is when adversaries "spoof" (impersonate) to get past defenses. Since military sites block direct connections from adversaries, it can be effective for an adversary to pretend they are civilian by just modifying their source address rather than going through intermediate sites. Standard network protocols do not allow address modification, but an adversary can design their own protocols. Spoofing is useful with denial-of-service attacks such as those against Georgia in 2008 (USCCU 2009)

If a victim of a spoofed attack counterattacks, their counterattack will likely go to the spoofed address, causing civilian damage if a civilian was spoofed. Counterattacking is a natural human impulse that is hard for many victims to resist even if they are not sure who they are counterattacking. But careless counterattacking can easily do more harm than the original attack.

Spoofing of civilians by militaries is specifically prohibited by the laws of war under the name "perfidy". That is because spoofing of civilians increases disbelief in civilian status and increases the risk of legitimate civilians being harmed. Here is the relevant part of the Additional Protocol I of the Geneva Conventions. Note that perfidy need not risk killing someone by these conventions, just that it "injure" the adversary.

Article 37 – Prohibition of perfidy

1. It is prohibited to kill, injure or capture an adversary by resort to perfidy. Acts inviting the confidence of an adversary to lead him to believe that he is entitled to, or is obliged to accord, protection under the rules of international law applicable in armed conflict, with intent to betray that confidence, shall constitute perfidy. The following acts are examples of perfidy:

   (a) the feigning of an intent to negotiate under a flag of truce or of a surrender;
   (b) the feigning of an incapacitation by wounds or sickness;
   (c) the feigning of civilian, non-combatant status; and
   (d) the feigning of protected status by the use of signs, emblems or uniforms of the United Nations or of neutral or other States not Parties to the conflict.

2. Ruses of war are not prohibited. Such ruses are acts which are intended to mislead an adversary or to induce him to act recklessly but which infringe no rule of international law applicable in armed conflict and which are not perfidious because they do not invite the confidence of an adversary with respect to protection under that law. The following are examples of such ruses: the use of camouflage, decoys, mock operations and misinformation.

Most cyberattacks also rely on a special kind of spoofing, impersonation of routine software by malicious software. That is because there are many defenses against attempts to subvert computers and devices: security kernels, hash and parity values computed on digital objects, anti-malware scanners, intrusion-detection systems, and software-based security-policy implementations. These countermeasures make it difficult to attack machines and software directly. So the only good hope is to subvert the civilian software of those machines. But when done to achieve military objectives, the software is then masquerading as a neutral party when it is in fact a tool of a military cyberattack. So subversion of civilian software is a form of perfidy and is outlawed by international law (Rowe 2013). Some cases are more obvious than others, such as modifying air-operations software to confuse locations of hospitals with locations of military units and thereby cause targeting of hospitals.

Subversion of software may be easier to understand as a form of product tampering. Tampering with commercial products by third parties is illegal in nearly all countries because a modified product can harm a consumer. In the U.S. this is a form of "malicious mischief" and there are serious penalties. Software, as an easily modifiable product, needs especially to trusted to be free of tampering. Most software vendors make customers sign "end-user license agreements" to agree not to modify the software because of its dangers as well as their own interest in controlling variations on the software. So the necessary modifications of software to accomplish cyberattacks violate domestic law in most countries as well as international law.

3.2.11 Psycholfogical Damage

Psychological consequences on civilians of their military being cyberattacked can be significant because the technology is mysterious and provides grounds for irrational fear. If major systems stop working, civilians will wonder what other systems will also stop working soon. This irrational fear can also affect the cyberattacking country because citizens will think their military is cyberattacking some serious threat. Many have written about the irrational fear of terrorism that has gripped the U.S. in recent years (Kimmel and Stout 2006) which has led to abuses of privacy in cyberspace (Angwin 2014).

3.3 Intensifiers for Collateral Damage

In this section we discuss some additional factors at play in civilian collateral damage of cyberattacks.

Military organizations expect that their technology may be damaged during conflict. For their cyber assets, they have extensive backup plans including both hardware and software replacements, including backup sites from which copies can be downloaded. Military organizations also have well-developed contingency plans for when they lose assets including communications. Civilians, on the other hand, are inadequately prepared for the collateral damage that can occur with cyberwarfare. Businesses have plans, but depend too much on legal remedies designed for cybercrime (such as suing someone) instead of hardening their systems, and this will be little help if they are hurt during major sabotage activities in cyberwarfare by countries with which their country does not share tort law. Home-computer and mobile-device users have little protection against cyberattacks since many backup sporadically if at all. They depend extensively on a narrow set of options for finding out about the world (like television and the Internet) that could easily be disabled during cyberconflict. That suggests that collateral damage to civilians will be more serious and long-lasting than the damage to military systems during cyberwarfare.

A related factor is that it is often harder for civilians to repair cyberattack damage than it is for militaries. Civilians often lack training to respond to cyber problems adequately since the technology is changing rapidly and few people, even the developed world, can keep up to date with it. In the less-developed world, fewer people still understand the technology, and a cyberattack on a less-developed country may leave it damaged for years unless it gets extensive outside assistance.

Some military apologists have suggested that cyberspace is a new isolated domain of conflict much like outer space and the depths of oceans, so that cyberwarfare is unlikely to have many consequences for civilians. This may have been true 20 or more years ago, but is less true today due to the increasing ubiquity of digital technology. Food, shelter, jobs, and other basic necessities are heavily dependent on digital technology in most countries. Our social infrastructure of power, transportation, financial services, commerce, medicine, and communications is heavily dependent on it as well, and everything is interconnected. Use of digital technology and cyberspace is no longer optional, and thus collateral damage can easily have consequences for everyone.

Another claim often made by military apologists is that cyberwarfare will be bloodless. However, all effective weapons can hurt and kill people, and cyberweapons are no exception; explosions are not the only way to kill people. Analysis of the U.S. invasion and occupation of Iraq 2003–2013 showed surprising numbers of violent civilian deaths, estimated at 600,000 in the first 3 years (Burnham et al 2006) due to the increased lawlessness in the country in that time. In addition, the crippling of the civilian infrastructure resulted in at least 100,000 additional deaths (Hagopian et al 2013). This was despite a swift military victory in the initial weeks. Cyberwarfare could be even more likely to damage civilian infrastructure.

3.4 Towards Ethical Principles for Cyberwarfare That Minimize Collateral Damage

Despite all these dangers, cyberwarfare can be conducted in ways that greatly minimize the collateral damage to civilians.

3.4.1 General Principles

To provide guidance in designing policies, and eventually laws, that could help reduce the danger of collateral damage, we propose the following principles.

- Avoid deliberate attacks on preponderantly civilian targets under any circumstances, no matter what the incentive. Military attack and defense should involve only military personnel.
- Avoid dual-use targets as much as possible, and proportionately to the degree to which they are civilian. A rule of thumb is that anything whose proportion of military use is less than that of the domestic economy of the victim state (4 % for the U.S.) can be treated as entirely civilian.
- Minimize propagation of cyberattacks through civilian cyberspace during attack setup and control, since propagation alone causes damage and can violate neutrality of nation-states.
- Avoid autonomous propagation of the cyberattacks by methods such as viruses and worms, since they are difficult to control and stop.
- Design cyberattacks so their methods cannot be easily reused by cybercriminals, as by obfuscating (deliberately complicating) the code.
- Prefer to attack specialized military hardware and software that is not used by civilian systems.
- Acknowledge responsibility for the attack and make its purpose clear, to achieve desired effects and avoid scapegoating innocent civilians.
- Either make the attack highly effective so it cannot be blamed on civilian incompetence, or conceal it well so criminals won't find it.
- Minimize the number of cyberattack methods to reduce the chances of reuse by cybercriminals.
- Attack only countries that have the resources to investigate it.
- Avoid perfidious attacks that subvert civilian infrastructure and could encourage mistrust of civilians and civilian artifacts.

3.4.2 Partitioning of Cyberspace

Another principle that will help reduce collateral damage is to separate the arena of cyberwarfare better from civilian activities. It is important to designate and respect cyber "safe havens" analogous to those for refugees in conventional conflicts (Geiss and Lahmann 2012). These would be designated unacceptable targets for cyberwarfare such as medical systems, power systems, banking systems, Google servers, Microsoft Update, and personal Web pages. Since these are almost exclusively civilian, it is hard anyway to justify them as military targets. However, dual-use entities shared by military and civilian users such as mail systems and databases could be legitimate military targets under occasional and carefully justified circumstances, and they should not be included in the "safe havens". We are starting to see some ideas about how to plan cyberattacks to limit collateral damage by trying to carefully identify characteristics of targets (Raymond et al 2013), though one can be skeptical of the ideas that do not take into account possible deception by an adversary, an essential part of military operations.

Partitioning of military cyberspace from civilian cyberspace is technically feasible in large part though there have not been strong incentives for it previously. Segregation need not be physical (accomplished by separate hardware). Separation can be "logical", meaning that military data and network communications are carried through different software mechanisms. Recent work has developed extensive technology for "virtual machines" and "cloud computing" that can allow software to execute in an environment well separated from a host environment so that viruses and worms cannot get out to the host environment (Pearce et al. 2013). Military systems have often pioneered the necessary technology.

3.4.3 Reparations

Cyberwarfare can cause significant damage. If cyberattacks are unprovoked, the laws of war should apply and enforce reparations for the damage. Reparations for cyberattacks can be assistance, perhaps through a third party, in repairing the damaged hardware, software, and data. The assistance of the attacker will often be required since often only the attacker knows exactly what was attacked and damage can be hard to see. An important justification for reparations is the deterrent effect they have on future cyberattacks, and deterrence is often the primary reason for having a military. For example, reparations should be due to Iran for the unprovoked cyberattack of Stuxnet, particularly since Iran was not at war with any country at the time.

3.4.4 A Role for Cyber Coercion

Since cyberwarfare methods are flexible, it is reasonable to consider more limited forms of cyberconflict as alternatives. Cyberconflict short of warfare has been termed "cyber coercion" (Flemming and Rowe 2015) and may suffice to resolve many conflicts. An example could be when an aggressor state prepares for a regional military dispute by sending ships to the area, but discovers that the command-and-control systems for those ships no longer function, and receives a message from an adversary telling them to back off; the induced system malfunction would function as coercion with the threat of further consequences for the aggressor state. Cyber coercion does not need to significantly impact a state's ability to wage war, as does conventional warfare; it suffices to provide a demonstration of capabilities since cyberattacks can often be scaled up.

Future warfare is likely to see many forms of cyber coercion. However, it has some disadvantages compared to conventional conflict. It may not be noticed by the victim unless it is strong enough, as it may be confused with normal system problems. At the other extreme, the victim may escalate the conflict after cyber coercion to demonstrate their own resolve, leading to the cyberwarfare that coercion was intended to avoid. Cyber coercion could unfairly target civilians just as much as full cyberwarfare unless the principles given above are followed. Nonetheless, in many cases cyber coercion may be a more focused and less problematic method of cyber influence short of cyberwarfare.

3.5 Conclusions

States wishing to go to war often provide incomplete justifications that insufficiently consider the costs to civilians involved in the conflict. Citizens should be made more aware of what the likely consequences are. Modern warfare has increasingly emphasized high-technology and infrastructure targets (Smith 2002). But for the new arena of cyberspace, incomplete arguments for offensive cyber operations are especially common, and possible consequences have been insufficiently understood and appreciated. This chapter has argued there are many ways, both overt and subtle, in which civilians can be hurt by cyberconflict, but there are ways to reduce such damages. Civilian distinction is only one of several ethical problems that need to be addressed in cyberwarfare, however (Rowe 2015).

Acknowledgements The views expressed are those of the author and do not represent the U.S. Government. This work was supported by the U.S. National Science Foundation under the Secure and Trustworthy Cyberspace program.

References

Al-Qasem, I., S. Al-Qasem, and A. Al-Hammouri, 2013. *Leveraging online social networks for a real-time malware alerting system*. In: Proceedings of the 38th IEEE conference on local computer networks, Sydney, AU, October, 272–275.

Angwin, J. 2014. *Dragnet nation: A quest for privacy, security, and freedom in a world of relentless surveillance*. New York: Times Books.

Anonymous. 2012, July. The collateral damage of Internet censorship by DNS injection. *ACM SIGCOMM Computer Communications Review*, 42(3):22–27.

Brenner, S., and L. Clarke. 2011. *Civilians in cyberwarfare: Casualties*. http://works.bepress.com/susan_brenner/3. Accessed 1 Nov 2011.

Burnham, G., R. Lafta, S. Doocy, and L. Roberts. 2006, October 11. Mortality after the 2003 invasion of Iraq: A cross-sectional cluster sample. *The Lancet*, 368(9545):1421–1428.

Clarke, R., and R. Knake. 2010. *Cyber war: The next threat to national security and what to do about it*. New York: HarperCollins.

Dinniss, H. 2012. *Cyber warfare and the laws of war*. Cambridge: Cambridge University Press.

Elisan, C. 2012. *Malware, rootkits, and botnets: A beginner's guide*. New York: McGraw-Hill Osborne.

Flemming, D., and N. Rowe. 2015. *Cyber coercion: Cyber operations short of cyberwar*. In: Proceedings of the 10th international conference on cyber warfare and security, Skukuza, South Africa, March.

Geers, K., D. Kindlund, N. Moran, and Rachwald. 2013. *World War C: Understanding nation-state motives behind today's advanced cyber attacks*. http://www.FireEye.com. Accessed 7 Apr 2013.

Geiss, R., and H. Lahmann. 2012, November. Cyber warfare: Applying the principle of distinction in an interconnected space. *Israel Law Review* 45(3):381–399.

Gross, M. 2012. A declaration of cyber-war. *Vanity Fair*, April 2011. Retrieved May 12, 2012, from www.vanityfair.com/culture/features/2011/04/stuxnet-201104.

Hagopian, A., A. Flaxman, T. Takaro, E. Shatari, A. Sahar, J. Rajaratnam, S. Becker, A. Levin-Rector, L. Galway, H. Al-Yasseri, J. Berq, W. Weiss, C. Murray, G. Burnham, and E. Mills. 2013, October 15. Mortality in Iraq associated with the 2003–2011 war and occupation: Findings from a national cluster sample survey by the University Collaborative Iraq Mortality Study. *PLoS Medicine* 10(10). http://www.plosmedicine.org/article/info%3Adoi%2F10.1371%2Fjournal.pmed.1001533. Accessed 9 Nov 2013.

Hashim, S., A. Ramli, F. Hashim, K. Samsudin, R. Abdulla, R. Azmir, L. Barakat, A. Osamah, I. Ahmed, and M. Al_Habshi. 2013, September. Scarecrow: Scalable malware reporting, detection, and analysis. *Journal of Convergence Information Technology* 8(14): 9–19.

International Committee of the Red Cross (ICRC). 2015. *Treaties and customary law*. http://www.icrc.org/en/war-and-law/treaties-customary-law. Accessed 11 Jan 2015.

Kaplan, D. 2011, October 18. New malware appears carrying Stuxnet code. *SC Magazine*. http://www.scmagazine.com/new-malware-appears-carrying-stuxnet-code/article/214707. Accessed 1 Aug 2012.

Kaurin, P. 2007. When less is more: expanding the combatant/noncombatant distinction. In *Rethinking the just war tradition*, ed. M. Brough., J. Lango and H. van der Linden, Chapter 6. New York: SUNY Press.

Kimmel, P., and C. Stout (eds.). 2006. *Collateral damage: The psychological consequences of America's war on terrorism*. Westport: Praeger.

Kinsella, H. 2011. *The image before the weapon: a critical history of the distinction between combatant and civilian*. Ithaca: Cornell University Press.

Pearce, M., S. Zeadally, and R. Hunt. 2013, February. Virtualization: Issues, security threats, and solutions. *ACM Computing Surveys* 45(2):17.

Raymond, D., G. Conti, T. Cross, and R. Fanelli. 2013. *A control measure framework to limit collateral damage and propagation of cyber weapons*. In: Proceedings of fifth international conference on cyber conflict, Tallinn, Estonia.

Rowe, N. 2013. Cyber perfidy. In *The Routledge handbook of war and ethics*, ed. F. Allhoff, N. Evans and A. Henschke, Chapter 29, 394–404. New York: Routledge.

Rowe, N. 2015. Distinctive ethical challenges of cyberweapons. In *The research handbook on cyber security*, ed. N. Tsagourias and R. Buchan, Chapter 14, 307–325. Cheltenham: Edward Elgar Publishing.

Shakarian, P., J. Shakarian, and A. Ruef. 2013. *Introduction to cyber-warfare: A multidisciplinary approach*. Amsterdam: Syngress.

Slay, J., and M. Miller. 2008. Lessons learned from the Maroochy water breach. In: *Critical infrastructure protection*, ed. E. Goetz and S. Shenoi, Chapter 6. New York: Springer.

Smith, T. 2002. The new law of war: Legitimizing hi-tech and infrastructural violence. *International Studies Quarterly* 46: 355–374.

TechRepublic. 2005. *Flaw finders go their own way*. http://www.techrepublic.com/forum/discussions/9-167221, dated January 26, 2005. Accessed 1 Aug 2012.

USCCU (United States Cyber Consequences Unit). 2009, August. *Overview by the US-CCU of the cyber campaign against Georgia in August of 2008*. US-CCU special report. http://www.usccu.org. Accessed 2 Nov 2009.

von Heinegg, W. 2012. *Neutrality in cyberspace*. In: Proceedings of the 4th international conference on cyber conflict, Tallinn, Estonia.

War Legacies Project. 2010. *Agent orange record*. http://www.agentorangerecord.com. Accessed 2 Mar 2015.

Neil C. Rowe is professor of computer science at the US Naval Postgraduate School where he has been since 1983. He has a PhD in computer science from Stanford University (1983). His main research interests are in data mining, digital forensics, modelling of deception and cyberwarfare.

Chapter 4
Towards a Richer Account of Cyberharm: The Value of Self-Determination in the Context of Cyberwarfare

Patrick Taylor Smith

Abstract Cyberharm is an increasingly used and useful concept for the ethical analysis of actions in cyberspace. At the moment, two accounts of cyberharm dominate the discussion: the instrumentalist view where only harm to material human interests is morally relevant and the intrinsic view where information systems have independent moral status. I reject the latter as ontologically implausible and the former as normatively impoverished. I then describe a richer account of the human interests that are affected and constituted by our relationships with information systems. Relying upon legitimate human entitlements to the social bases for friendship, proper pride, and theoretical reasoning, I show that the richer account can incorporate the best elements of both, rival views. I then apply this richer account to the issue of when cyberharm constitutes a casus belli, arguing that the value of political self-determination is essential to understanding when cyberattacks generate unilateral rights to self-defense.

Keywords Cyberwarfare • Just war theory • Information ethics • Sovereignty

4.1 Introduction

Randall Dipert, in his "The Ethics of Cyberwarfare," argues that *intentional cyberharm* is an increasingly useful category of normative analysis for actions in cyberspace. He defines cyberharm in the following way:

> The broadest useful notion in the discussion of cyber-ethics is *intentional cyberharm*: this is the intentional harm caused by an agent, *via* an informatics network such as the Internet, in which the functioning of a system (a person, a machine, a software or an economy) impaired or degraded. (Dipert 2010, 397)

P.T. Smith (✉)
National University Singapore, Singapore, Singapore
e-mail: patrick.taylor.smith@gmail.com

This definition both reflects and expresses a key tension in our analysis of cyberethics and cyberwarfare. That is, is an action that degrades or impairs a computer system or network doing something that is directly harmful or is that degradation or impairment only instrumentally relevant insofar as it does real harm to material human interests? So, should our moral ontology be expanded to include information systems as bearers of value such that harming those systems is seen as intrinsically bad? Or, conversely, should we view these systems as mere tools that can be used to harm but cannot be harmed themselves? Let us call the view that we should say 'yes' to the former question the *intrinsic cyberham view* (Floridi 2003; Taddeo 2012, 2016 are exemplars) and the view we should say 'yes' to the latter question the *instrumental cyberharm view* (shortened to the 'intrinsic view' and the 'instrumental view', respectively).

I argue that the intrinsic view is implausible but that it is responding to a genuine theoretical need left unfulfilled by the instrumentalist view. Namely, standard instrumentalist views operate with an impoverished understanding of the human interests that are relevant to the proper functioning of information technology. Instrumentalist views are often committed to the idea that only harms to material human interests—death, injury, and serious economic damage—count as the morally relevant cyberharms (Schmitt 1999; Cook 2010; Brownlie 1963; Eberle 2013 and Kahn 2013 are just a few examples). Once we incorporate a richer sense of the human interests that are relevant to cyberharm, then we can see that information systems not only serve as instruments for material human interests but also, sometimes, partly constitute those interests. That is, our relationship with information systems is more intimate than the instrumentalist view allows. However, the richer account raises a significant worry about when cyberharms require or justify a coercive, military action in self-defense. In response, I argue that only certain kinds of cyberharm—those related to the self-determination of one's political community—justify a unilateral right to self-defense.

4.2 Harm and Cyberharm

Before analyzing cyberharm, it is important to first understand the nature of harm. In this chapter, I will rely on a moralized conception of harm. It is a necessary condition of harming something or someone that the harmed party be made worse off. Yet, this naturally raises the question, "worse off when compared to what?" Answering that question requires specifying the relevant baseline. A moralized conception of harm uses a baseline that makes essential reference to what the harmed agents are entitled. On this view, an agent or agents harm someone when they intentionally act to undermine the legitimate moral entitlements of those being harmed. Consider the following scenario:

Bingley owns Wickham and routinely subjects Wickham to abuse. In fact, Bingley flogs Wickham three times a day. Darcy purchases Wickham and treats him somewhat better, only flogging him once a day.[1]

Does Darcy harm Wickham by enslaving him and only flogging him once a day? The answer seems clearly, "Yes." Yet, non-moralized baselines have difficulty justifying that claim. Subjunctively, Wickham is better off under Darcy than had Darcy not acted at all; he would be flogged three times rather than once. Historically, he is better off than he was under Bingley; he only is flogged once compared to before the purchase. Of course, Darcy could have made Wickham even better off—in comparison to Bingley's ownership—by setting Wickham free. But surely it cannot be the case that we harm everyone that we fail to maximally benefit. The more plausible view, then, is that Wickham is morally entitled to—or may legitimately demand— his manumission and to be free of floggings (Pogge 2005). So, Darcy harms Wickham—even though Wickham is better off under Darcy than he was or that he would have been—because Darcy acts in a way that makes Wickham worse off than he may *legitimately expect* or demand. So, the moral entitlements to non-enslavement and to physical integrity represent a normatively-laden baseline by which we can judge whether someone has been made worse off for the purpose of making judgments about harm.

Cyberharm focuses specifically on the disruption, damaging, impairing, or sabotaging of computer-information systems. This focus complicates the question of how to understand the moralized baseline. When we theorized about Wickham, we were considering a case where the moral status of the harmed entity—Wickham— and the relevant entitlement to non-enslavement were relatively uncontroversial. Discussions of cyberharm are quite different. In fact, the clearest and best way to distinguish between different accounts of cyberharm concerns which agents can be harmed and the moral considerations that set the baseline. According to the instrumental view, human beings can be directly harmed and information systems are only indirectly relevant. That is, disruptions to computer systems are morally important when they impact human interests, including and especially material damage to persons and property. So, on the instrumentalist view, cyberharm occurs when the disruption of a computer system negatively affects the legitimate, material interests of a person, either by harming their property entitlements or—in extreme cases— causing them unjustified pain, ill health, or death.

Conversely, the intrinsic view is that the intentional disruption of computer systems, making them operate in ways contrary to their design, is intrinsically harmful to those systems regardless of those effects on human beings. To see how the views diverge, consider the following case:

---

[1] This example is borrowed and adapted from Nozick (1969). For a more in-depth discussion of how harm depends on a baseline—and why a moralized baseline—is most plausible, see Risse (2005) and the Pogge chapter in *Ethics and International Affairs: A Reader* edited by Rosenthal and Barry (2009).

LONELY: A person finds an unused, obsolete server in the computer network of a corporation. This server is forgotten and will never be used by the corporation again, and the hacker uses her time to crack the software in order to see what happens when particular manipulations and degradations of the code occur. It is a purely intellectual enterprise.

On the instrumentalist view, no cyberharm has occurred because no human entitlement has been negatively affected. The intrinsic view, on the other hand, would suggest that there is cyberharm: the server itself has been harmed, and deliberate action to make a computer system malfunction is intrinsically bad.

The intrinsic view gains some initial plausibility from two sources. First, there has been a widespread movement away from what some have taken to be the excessively anthropocentric nature of mainstream ethics. One could argue that a consistent feature of contemporary ethics has been the expansion of the class of entities that have intrinsic moral status. This expansion began with sentient animals (Singer 1975) and others have argued that we should include environmental entities—such as species, topographical formations, or ecological systems—that are not alive and yet nonetheless (purportedly) have interests that are intrinsically valuable (Naess 1973; Devall and Sessions 1985). Once we accept that non-living systems can have intrinsic moral status in the environmental context, it seems no more problematic to ascribe them to information systems. We can support this claim further by referring to (alleged) facts about the ontological commonalities between both human beings and computer systems as both—metaphysically, at their most fundamental level—are information systems (Floridi 2003). The intrinsic view is then predicated upon the idea that it is arbitrary to grant moral status to one information system while denying it to another.

Second, and more importantly, there is *negative* support for the intrinsic view; the instrumentalist view is inadequate as an exhaustive account of possible cyberharms. Human agents are more inextricably linked to their information systems than the instrumentalist view is willing to countenance. The intrinsic view is partly predicated on the idea that there is no sharp distinction between humans and the information systems within which they find themselves and that the wrongness of undermining those systems represents a wrong beyond that which is commensurable to the values in the instrumental view. That is, we can imagine cases where cyberattacks are wrong or harmful even if the physical or economic damage is minimal. Instances of cyber-bullying or online harassment often have this structure; inflicting stress and anxiety through those sorts of attacks can be wrongful even if they do not result in economic damage or ill-health, though they often do. We can readily imagine—or even ourselves have experienced—the stress and anxiety associated with the sense of disconnect and frustration that accompanies the failure of our information systems.

More broadly, human beings have become altered by their interaction with modern information systems. These alterations include changes in social equilibrium, where leaving a voice mail can change from the standard method of communicating to an act of discourtesy. Or social gatherings, which can have important effects on people's life chances and happiness, that come to depend on quick communications that, in turn, depend on the effective operation of wireless networks. In fact, our brain chemistry and capabilities have altered in light of the ready availability of online information. For example, our memories get worse with greater smartphone use.[2] On other hand, playing certain computer games can increase our attentional focus and speed effective decision-making.[3] On the extended mind hypothesis (Clark and Chalmers 1998), these information systems actually become *part* of our mind in virtue of their integration into our mental and social lives. As a result, their availability quickly becomes necessary. There is an irrevocable alteration akin to the effects of literacy that comes with using and participating in them. As a consequence, their use can become non-optional and what may have been understood as a luxury a few years before becomes a key element of one's ability to pursue a rational plan of life.

What's more, one's participation in social media becomes a constitutive part of one's identity such that assaults upon one's public, internet presence represent an assault upon one's own self-understanding. So, there is a sense in which your public persona—represented by your use of particular information networks—becomes something that is intrinsically valuable, much like how an individual might reasonably be interested in their honor—understood as a person's public standing—for its own sake. We might think that a kind of *proper pride* in one's standing in one's community commits oneself to valuing that public persona as *constituted* by their profiles in various social media networks. An individual might take an assault on their public honor as a serious harm even if that assault does not inflict economic or physical harms; similarly, one might think there is a direct harm in others creating false profiles in your name, taking down social media sites, or doxing even if steps are taken to minimize psychological or economic damage.

The intrinsic account is meant to capture these various ways in which the interaction between human agents and their computer networks is not entirely unidirectional or instrumental. While it is undeniably true that we use information systems instrumentally to further our material interests as described by the instrumentalist view, this is an impoverished understanding of how we relate to information systems. They alter our own capacities, shape our expectations, make possible our public identities, and come to partly constitute our identities in ways that the standard view fails to capture. The intrinsic view, which take the computer systems to have their own independent moral status, is an attempt to capture the bidirectional and interdependent relationship between informational systems and human agents.

While I grant that the instrumentalist view is inadequate, the intrinsic view is a problematic response to its limitations. First, the argument that it is a category mistake to grant moral status to non-sentient entities is difficult to resist. It is hard to understand what it would mean for an information system to be *interested* in or to *care* about how it operated. At the very least, it seems plausible that ascribing that sort of status to an entity requires that the world appear to it a certain way and that

---

[2] There are many studies that suggest this; much of the research is ably summarized in Nicholas Carr's article in *Wired* "The Web Shatters Focus, Rewires Brains."

[3] The positive effects of playing video games are described in Mark Griffiths' November 11th, 2014 article in *The Washington Post* "Playing Video Games is Good for Your Brain."

the entity be somehow motivated to make the world one way rather than another. But environmental and informational systems do not care about things, are not conscious, and are not motivated to act in response to the things they care about. As a result, it is unclear what it would mean to harm those kinds of entities. We can see this in LONELY. If a particular computer system is genuinely isolated from anything related to the interests of any sentient being, then it is very hard to see what is harmful about the hacker's experimentations.[4] What's more, the 'ontological' argument about the common metaphysical status of both computer systems and human beings is unconvincing. Even if we granted the underlying metaphysical picture, it may very well be that the particular arrangement or unique, emergent properties of human or sentient beings give rise to their particular moral status. After all, we have known for quite some time that both human beings and the planet Jupiter are both made up of atoms but that puts very little pressure on us to grant moral status to Jupiter. I have no intention of resolving this debate decisively in this chapter. But, at the very least, these objections show that an account of cyberharm that did not commit itself to controversial metaethical and metaphysical theses would be preferable as long as it captured what made the intrinsic view plausible.

I believe that such a view is available. Once we grant that cyberharm has a moralized baseline and we understand both the intrinsic and instrumentalist views as being defined by their respective moral commitments, then the answer seems clear. The problem with the instrumentalist view is not that it fails to grant moral status to information systems but rather that it is based on an impoverished understanding of the ways that human beings can, do, and should care about information systems. What is needed is a more pluralistic and flexible understanding of the relevant human interests and not an expansion of our moral ontology. I will call this account the rich human interests view, and I suggest that it accounts for the counterexamples that seem to motivate the intrinsic view. That is, attacks that disrupt social media sites or undermine a person's ability to use the internet—yet generate no or *de minimis* economic loss or physical injury—are harmful because they undermine that person's entitlement to manage their public persona or to deploy their mental capacities in ways they have good reason to expect or demand. On the other hand, the less plausible implications of the intrinsic view are avoided. The hacker disrupting the isolated server for her own amusement—where the disruption affects no human interest no matter how broadly construed—is not engaging in cyberharm. Nor does a person who deliberately disrupts a computer system in order to make it operate more conveniently for human beings engage in cyberharm. The reason is simple: cyberharm is indexed to a theory of human (or perhaps, sentient beings) interests and entitlements; if no human interest or entitlement is affected, then no harm has occurred. So, the rich human interests view—by acknowledging that computer systems are important to human values beyond economic production—captures our considered intuitions about cases without depending on the controversial claim that

---

[4] Perhaps one could argue that increased energy use or wear and tear on components in LONELY would be harmful, but those claims would depend upon the property entitlements of the system's owners.

computer systems have intrinsic moral status. However, computer systems gain *some* standing in virtue of the fact that they have become extensions of ourselves and are at least partly constitutive our theoretical capabilities and public identities.

The relevant human interests are likely to be highly contextualized. If we are analyzing a cyberattack on a corporation, then we will probably emphasize the economic losses. But even in that context, we might think that the human interests involved are a bit more nuanced. For example, consider the cyberattacks that resulted in the leaking of many internal, Sony emails. As a consequence, high ranking managers at Sony suffered reversals, including and especially Amy Pascal.[5] Now, I would suggest that the public humiliation that Pascal suffered at the hands of the leak was an instance of cyberharm independent of the economic injury that either Sony or Pascal suffered. Perhaps Pascal receives a publicity boost for her production company, everyone purchases tickets for *The Interview*, and Sony learns to avoid even more devastating cyberattacks in the future. If those events come to pass, then it might turn out that these attacks were a net economic benefit to those involved. Yet, it is false that these purported economic benefits would necessarily resolve the question of whether the cyberattacks and subsequent leak were harmful. Similarly, the not-uncommon release of the private photos of celebrities is unlikely to generate lasting economic harm for those celebrities' respective careers, but it seems true that many were harmed by their release. Similarly, we could readily imagine a celebrity who was not particularly upset or stressed about the release of the photos and yet felt that their release was harmful. In other words, there is an interest the celebrity has—beyond their own psychological discomfort or economic well-being—in their own privacy, in controlling their public persona. Of course, one might point to the *legal* rights that those individuals had to their photos, but this begs the question. After all, we would want to know what justifies the provision of legal entitlements to one's private, digital photos. What's more, we would likely claim that a legal system that did not include those protections would be, in an important way, normatively defective. So, at the very least, an individual is entitled to some degree of control over their public persona as represented on social media and through the dissemination of private information, photos, and writings. Broadly speaking, we might call this an interest in privacy. And insofar as privacy is an important human interest, then one can engage in cyberharm by attacking or disrupting computer systems in such a way that undermines it.

And beyond privacy, we have an interest in controlling how the public parts of our identity are perceived and constructed.[6] What's more, these public elements of our identity are partly constitutive of our own self-conception. Insofar as information systems are necessary for the construction, maintenance, and control of those public identities, they become part of us. As a consequence, an attack on those systems—or an attack on our public identities through those information

---

[5] See "Amy Pascal Lands in Sony's Outbox" in *The New York Times*, Feb. 5th, 2015.
[6] We might think of this as our interest in *amour propre* or "proper pride" that Rousseau describes in the *Discourses on Inequality*.

systems—can translate into a direct attack upon ourselves. Which is, perhaps, why those attacks can give rise to such devastating psychological consequences.

Similarly, we have a fundamental human interest in our ability to effectively negotiate our social world and relationships. Our friendships and romances are not important simply because they contribute to our psychological well-being and economic security; friendships simply *are* constitutive features of a good human life. Similar things can be said for opportunities to engage in imaginative play and leisure time. Of course, such things are pleasant and often contribute to economic productivity, but those are not the only reasons why they are valuable. Our ability and opportunity to pursue and participate in a web of social relationships are essential to human flourishing. For example, Martha Nussbaum includes both the ability to form friendships and participate in the social world as fundamental capacities needed to ground human rights. She describes one of the basic capacities thusly:

> Being able to live for and to others, to recognize and show concern for other human beings, to engage in various forms of social interaction; to be able to imagine the situation of another and to have compassion for that situation; to have the capacity for both justice and friendship. (Nussbaum 2008, 516–517)

And insofar as information systems have become the essential mediators for our social lives, we have come to be dependent on them. They enable and structure our relationships and leisure time, and our social lives become significantly impoverished when we cannot participate. So, if we think these endeavors and relationships represent key human interests, then cyber disruptions that impose unacceptable costs on our ability to form, pursue, and maintain friendships are examples of cyberharm regardless of their economic or psychological consequences.

Finally, we might also agree that with Nussbaum that our capability to explore and make reliable, theoretical judgments about the world is a key element of our agency and our flourishing. She writes:

> Being able to use the senses; being able to imagine, to think, and to reason—and to do these things in a way informed and cultivated by an adequate education including, but by no means limited to, literary and basic mathematical and scientific training... (Nussbaum 2008, 516–517)

Exercising theoretical reason is both pleasant (at least occasionally) and economically productive, but that is hardly the only reason for engaging in it. We might, again, think that working through one's view of the world is vital for its own sake and for the sake of democratic and social engagement. What's more, information technology has a much deeper and more fundamental relationship with our theoretical reason than the instrumentalist view countenances. After all, many explorations of the world and theoretical judgments are essentially impossible without technology. Our theoretical capacities are quite literally constructed out of devices meant to deploy our cognitive capacities more efficiently. And as was mentioned above, these new technologies have the effect of structuring, channeling, and reshaping our own capacities as we use them. The relationship is bilateral and dynamic: we construct and revise our technologies to serve us and those technologies in turn shape our mental habits and capacities. Removing those technological capacities thus has the effect of handicapping us.

So, the rich human interests view relies on the idea that we are generally entitled to a broader set of values in the process of living a decent, flourishing life than is normally granted in the standard view. These values include but are not limited to (I also include some standard threats to those values and interests):

(1) Physical integrity: *threatened by* death and injury
(2) Economic security and productivity: financial damage
(3) Control over one's public persona: lack of privacy or participation
(4) Reasonable opportunity to engage in social life: inability to connect
(5) An ability to pursue and revise one's theoretical and practical conceptions of the world: lack of information, restructured cognitive capacities

These all represent important human interests that can be harmed by cyberattacks. As such, cyberharm occurs when the normal functioning of computer information systems is disrupted in such a way as to drop an agent below the moralized baseline as set by an individual's entitlement to an adequate set of values necessary for flourishing life. If other values can be shown to be important to decent human lives, then they should be included as part of the baseline that makes up the conditions for cyberharm. At any rate, once we recognize that harm is predicated on a moralized baseline and that that baseline can include a richer set of human values, then it is easier to see how we can expand the concept of cyberharm to cover a more plausible set of cases without excessively expanding our moral ontology.

4.3 Self-Defense and Cyberharm[7]

It might fairly be stated, at this point, that I have misunderstood the instrumentalist view. One might think that the instrumentalist view of cyberharm is not a *full* account of when the disruption of computer, information systems is harmful. Rather, the instrumentalist view is an account of cyberharm that has a particular political salience. That is, the instrumentalist view describes when cyberharm reaches the level of seriousness and urgency so as to justify actions in self-defense. The instrumentalist view, then, argues that a cyberattack becomes an 'armed attack' in international law—thus, justifying unilateral self-defense—when the cyberharm produced by that attack is sufficiently serious. Namely, self-defense is justified when the cyberattack produces damage, injury, or death equivalent to a physical, military attack. While the instrumentalist view may or may not be correct as an account of all cyberharm, the idea that cyberattacks need to be especially harmful before they justify military action has substantial *prima facie* plausibility. Furthermore, the defender of the instrumentalist view would also say that the rich human interests view is problematically inflationary if applied to the context of

---
[7] These latter sections depend substantially on Smith (2016).

self-defense. Again, while it may be correct that cyberattacks that undermine one's control over one's public persona are harmful, it is grossly implausible that the United States ought to go to war over North Korea's violation of Amy Pascal's privacy. The objection is twofold. First, it looks like many of the values described by the rich human interests view are insufficiently urgent to justify military action. Second, many of the values seem insufficiently political to justify military action. That is, some of the values are best understood as representing ideals of interpersonal relationships (such as friendship) that would be inappropriate if made part of the evaluation and justification of public policy. So, while it may be true that information networks are becoming essential for maintaining robust friendships and that disrupting those networks constitutes a cyberharm for that reason, we generally do not think that "unreasonably increasing the costs of making and maintaining robust friendships" is a good reason to go to war. The challenge to the rich human interests view is to characterize a proper subset of normative concerns that justifies unilateral military in self-defense. Let's call this proper subset 'militarily enforceable values.'

The instrumental view seems much better as an account of this proper subset of values. So, instead of asserting—implausibly—that death, injury, and serious economic damage are the only harms that cyberattacks can cause, the instrumental view may merely be asserting that only those specific harms justify a military response. This view has several beneficial features. First, it avoids the apparent arbitrariness in the claim that *physical* or kinetic actions that lead to death and destruction are attacks while *non-physical* or information-based actions that cause the same damage are not attacks. Ryan Jenkins writes:

> Suppose Stuxnet [a computer worm aimed at Iran's nuclear program] is not a physical thing...the bare nonphysicality of the attack is not morally relevant. Imagine a hypothetical case of physical aggression that parallels Stuxnet as closely as possible: a commando raid on Iran's uranium enrichment facilities at Natanz that successfully destroys several hundred uranium enrichment centrifuges...Such a raid clearly falls under *jus ad bellum*...there seems to be differences [between the raid and Stuxnet]...What seems most relevant in these two cases is not whether the means were physical but whether an action causes damage to a state's tangible assets. (Jenkins 2013, 8–14)

There does not appear to be an important difference between damage caused by physical or non-physical means based on *intrinsic* features of the actions, and the standard view captures that. What's more, the instrumental view can easily show why other sorts of cyberattacks—such as economic espionage or surveillance—do not justify a right to a unilateral military response. While these cyberattacks may cyberharm, they do not produce the right kind of physical, kinetic consequences to generate a claim to engage in self-defense.

Despite these benefits, the instrumentalist view suffers from two significant flaws. First, the view has a problematically narrow view of the interests that justify a military response in self-defense. Consider the following two scenarios where military action seems justified:

> INVASION: Oceania invades Eurasia and seizes a slice of undefended territory in a *fait accompli*. The territory is lightly populated and local defense forces retreat or surrender almost immediately. Eurasia continues as an independent polity and engages in military action

ANNEXATION: Eastasia occupies Eurasia at the end of a just war. Their *post bellum* reconciliation strategy is to simply occupy Eurasia, granting full political liberties to Eurasians, and integrating Eurasia into its polity. A resistance movement develops with the purpose of fighting the annexation.

The common feature of these scenarios is that a particular interest of the individuals of Eurasia is undermined by an enemy action *even if* there is no particular complaint that can be leveled in terms of the instrumentalist view. In INVASION, no damage or deaths are caused by the actions of the invading country, but the interest that the people of Eurasia have in administering their territory as they see fit is undermined. In ANNEXATION, the death and damage is justified by the fact that Eastasia was engaged in a just war, so it cannot be used to justify Eurasian resistance to the Eastasian occoupation. The wrong that Eastasia commits in ANNEXATION is undermining the *political autonomy* of Eurasia by forcing it to become a permanent part of Eastasia. In other words, both of these scenarios justify military action by appealing to the need to protect the right that a political community has to *self-determination*, especially understood in terms of territorial integrity and political autonomy.[8] The key point is that these values—independent of death and destruction—can justify a military response. The instrumentalist view cannot easily capture these values

Second, the instrumentalist view is problematically inflationary concerning the right to engage in self-defense. There are many ways, which we generally consider to be peaceful, to generate damaging effects in a targeted state. Trade sanctions, embargos, and protectionist policies all have the potential to generate serious damage and even death. Yet, if we decouple our concept of an attack from the use of physical, kinetic means to accomplish a political goal, it becomes very difficult to see how the instrumentalist view can avoid the worrisome implication that an embargo from one country can be met with military force by another country as long as that targeted state can show that that embargo—via the economic disruption it causes—causes death or destruction. Consider the following state action:

RARE EARTH MINERALS: X and Y are engaged in a territorial dispute. X, either coincidentally or intentionally, has developed a monopoly on rare earth minerals that play an important constituent role in electronics. The Y government, fearing that this monopoly might be used to X's geopolitical advantage, funds a research project to create ersatz replacements for those minerals. The research program succeeds, causing large-scale economic dislocations in X when Y's demand for rare earth minerals plummets. As a further consequence, some citizens of X die, suffer injuries, and generally see their interests ill-served.

Here we have an intentional strategic action that is designed to weaken the target nation, causes significant death and destruction, and yet I would submit that we do not think that X can bomb the research facility in order to prevent the research from

---

[8] The idea that a right to self-determination can give rise to a right to engage in coercive action is commonplace. See Levitov (Forthcoming), Walzer (1980), Altman and Wellman (2009), Rawls (1999), and Lea Ypi (2013).

causing that destruction.[9] The main difference between RARE EARTH MINERALS and a typical military action, then, lies in the means employed. The instrumentalist view has considerable difficulty conceptualizing this difference as it purports to treat all means for producing death and destruction equivalently.

Proponents of the instrumentalist view are aware of this problem, and they attempt to sidestep it by arguing that only cyberattacks that create death and destruction through *direct* and *proximate* means justify a military response (Schmitt 1999). So, if a cyberattack (or any attack) generates the morally relevant effects in a way that takes many causal steps or a considerable length of time, this undermines the claim that the attack justifies defensive military action. Yet, it is hard to see why this should be so. If I create an elaborate machine that requires many steps or a long time to detonate a bomb, then I have still detonated the bomb and am responsible for the attendant effects. In other words, it is unclear what the *normative* weight of causal directness or temporal proximity would be if we were independently confident that the causal chain would generate the relevant consequences. If an indirect or distant cause was nonetheless certain to produce the relevant consequences, then directness and proximity appear to be irrelevant for evaluating the action.

This exploration of the instrumentalist view and its weaknesses provides some desiderata for our theory of cyberharm as an account of when cyberattacks justify a military response. First, the value of political self-determination should be central to any account of when cyberattacks become sufficiently severe, urgent, or serious that we can use physical force to stop them. Second, we need to show why cyberattacks like Stuxnet are much more akin to normal military attacks than RARE EARTH MINERALS in terms of how they produce their effects. That is, we need to create an account that both includes a richer account of militarily enforceable values and that shows how we can distinguish between various strategic actions in terms of their means and not merely in terms of their effects. It is to this view that I now, somewhat briefly turn.

4.4 Cyberwarfare, Self-Determination, and the Imposition of Will

Using a moralized baseline to provide content for the concept of cyberharm makes it a comparatively easy theoretical matter to incorporate additional values or interests. That is, we take cyberharm to be indexed to what individuals are *entitled* to expect in terms of the performance of information systems. So, if we believe that individuals have a right to self-determination, and a particular action in cyberspace undermines or violates that right, then that cyberattack represents an instance of cyberharm. And if threats to self-determination more generally justify military

---

[9] This is not to say that RARE EARTH MINERALS is perfectly acceptable as a matter of international or global justice. The targeted nation may very well be justified in availing itself of retaliatory economic actions or legal action in multilateral organizations.

action in response, then this particular kind of cyberharm can similarly be responded to with force as long as all other conditions for just war are met.[10] The key point is that founding cyberharm on a moralized baseline does not undermine our commitment to the view that differences that are predicated *merely* on the difference between the physical and the cyber are arbitrary. If a cyber-actions can undermine self-determination, then it is in principle open to the same sort of response as a physical attack.

One important feature of the rich human interests view is that it is essentially pluralistic about how we respond to different kinds of cyberharm. If an action in cyberspace is indeed a cyberharm, then we know that some sort of political response is likely warranted because harm is indexed to individual entitlements. Yet, the claim that any cyberharm requires some kind of social or political response in order to protect the relevant interest does not commit us to the less plausible view that all cyberharms demand the same policy response. Rather, the specific dynamics of any particular interest determine the appropriate response. In this case, we want to characterize the interests and entitlements that can, in principle, justify unilateral military action in self-defense. We have some good reason to think that the value of self-determination is a human interest that can—in other circumstances—ground precisely those sorts of responses. So, what is needed is to demonstrate that cyberattacks can, in fact, undermine the value of self-determination. What's more, the view needs to show that, unlike the standard view, the inclusion of self-determination does not generate an excessively inflationary understanding of when states can go to war.

The key deflationary feature of my view is that the value of self-determination is not undermined by the simple imposition of costs on an agent's choices. This is ubiquitous. Suppose I wish to purchase a particular item that has been made costlier because others want it as well. Or perhaps I want to go see my doctor but no appointments are available for a few days. A country wants to develop its own products for export but is undercut by the cheaper production in a competitor on the other side of the world. I wish to purchase a product, message a friend, email my congressman, or coordinate a business meeting but the computer programs I need to accomplish these goals has been taken down through a denial of service attack. It is an unavoidable consequence of being a member of a social world that others can impose costs on different ways of acting. If the view is that we can go to war whenever the action of an agent imposes a cost on our ability to do what we want, then the self-determination view would be inflationary to the point of absurdity.

Fortunately, an individual's entitlement to self-determination is not undermined merely because the actions of others influence the cost of acting in one way or another. Rather, it is undermined by the way in which one's choices are structured (Blake 2001). That is, one's self-determination is undermined when another agent imposes their will upon you through coercion or manipulation. Consider the classic case of the highwayman who threatens a bystander, "Your money or your life!"

---

[10] These include, amongst others, the requirement of necessity, proper authority, and proportionality. See Walzer (2000).

One way to understand this threat is an imposition of a cost: walking around alive requires that you give this person your money. Yet, this is not really an accurate understanding of the situation because the option of *keeping your money* is not really a live one. After all, if you refuse, the highwayman will murder you and then take your money. And if you move to avoid the highwayman, then he or she will attempt to overcome your resistance. The highwayman has a goal and that person will *do what it takes* to accomplish it. Your judgment or will, with regards to your money, are—in an important sense—irrelevant. Whatever you decide, you will no longer be in possession of your wallet or purse. To describe the highwayman as merely imposing a cost is to present the misleading picture of the mugger as offering a consideration whereby you can judge whether it is worth it to continue. Rather, the highwayman is trying to *replace* your will with their own, trying to make your agency irrelevant. This represents an especially egregious threat to your ability to determine your own life and, as a result, you can respond with physical force in order to protect yourself from the mugger. There is a distinction, then, between actions that achieve their goals operating through one's agency and actions that achieve their goals by bypassing, eliminating, or making irrelevant one's agency. It is only actions of the latter sort that undermine the value of self-determination sufficiently in order to warrant military action. The issue at hand, then, becomes whether cyberattacks can be impositions of will.

Before I turn to the implications of this view for cyberwarfare, two questions immediately present themselves about this account and I wish to answer them briefly. First, one might wonder whether there is a sufficiently sharp distinction between impositions of costs and impositions of will. It is clear that impositions of cost can become or fade into impositions of will. Yet, the mechanism by which this happens is not merely the result of the magnitude of the costs that are inflicted. Rather, an imposition of costs can become an imposition of will *if* the costs are sufficiently high as to undermine the preconditions for exercising either individual or collective agency. So, if a small country is utterly dependent on a larger country's trade and an embargo strips that country of the resources it needs to engage in collective political action, then the imposition of those costs through an embargo might eventuate in an imposition of will. Yet, an economic cost of similar extent and magnitude imposed on a larger and more well-resourced country might not—if the consequences of those costs did not result in the fatal undermining of that country's ability to determine its own policies—be an imposition of will. So, we now have a second pathway by which cyberattacks may warrant a military action in self-defense. They may fundamentally undermine the ability of the state operate in a politically autonomous fashion.

Second, how might we know when an action ought to be understood as an imposition of will? I think that one key epistemic factor in understanding whether an action is an imposition of will is the response to resistance. This is easy to see in the highwayman case: the mugger is willing to do whatever it takes, no matter how you resist, to achieve his or her goals. This is what provides the sense that your agency is irrelevant. Now, I do not want to suggest that if one fails to resist or if one eventually gives way to resistance, then no imposition of will occurs. Rather, I want to say that if one gives way in the face of any resistance no matter how slight, then the agent who gives way cannot be genuinely attempting to impose one's will. I call this type of action—one with the surface logic of an imposition but where the attack gives way in the face of any resistance—*probing*. My view is that probing actions do not warrant military response unless it satisfies the second condition.[11]

To sum up, the rich human interest view resolves questions about cyberwarfare in the following way. Unilateral military action is justified primarily as an attempt to protect both entitlements to self-determination for both collectivities and individuals. Yet, not all actions that affect our choice situation undermine the value of self-determination to the extent that military action is justified in response. Actions that impose costs, operating through the agency and judgment of the target agency, do not undermine the value of self-determination sufficiently as to justify a military action in self-defense. Only actions that represent impositions of will—thereby *bypassing* individual and collective agencies and judgment—justify that kind of response. Yet, actions that impose costs can nonetheless become impositions of will if the costs are both accurately directed and sufficiently serious as to undermine the ability of polities to collectively self-determine. So, cyberattacks that undermine or attack the value of self-determination represent the kinds of cyberharms that are potentially just causes for going to war. One test that can be used to ascertain whether an action is attempted imposition of will or an imposition of cost is how the attack—or the agent engaged in the attack—responds to attempts to foil his or her plan. That is, we can use the response to resistance as an indicator of the type of imposition a particular strategic action justifies a military response.

4.5 Conclusion: Implications for the Characterization of Cyberattacks

In this final section, I will consider what my view implies about various questions related to cyberwarfare. First, let us take the most general question: can cyber-operations that aim at achieving a political objective cause the kind of cyberharms that may warrant a military response? The answer, I think, is clearly yes. Consider the following scenario

> LOGIC BOMB: A maritime border dispute occupies Eurasia and Oceania. Eurasian agents defeat an Oceanian firewall. They plant a program in the navigational software of the Oceanian flagship, causing it to run aground.

---

[11] This result is consistent with international law concerning *de minimis* armed attacks. See *Nicaragua vs. United States*.

LOGIC BOMB satisfies the conditions of a cyberharm that justifies a military response. It purports to create a political state of affairs opposed to the collective judgment of Oceania. And it achieves this by overcoming resistance and bypassing the decision-making process of the targeted state. Other ways of making the flagship unavailable would not necessarily have this feature. For example, an embargo that caused the Oceanian flagship to be unavailable because the Oceanians *decided* to skimp on maintenance would not count as a potential justification for military action because the embargo overcomes no resistance and does not attempt to replace, bypass, or destroy the collective capacity to self-determine for the target state. Yet, LOGIC BOMB *does* overcome resistance and attempts to make the judgments of the target state irrelevant and does so in order to achieve a political objective contrary to those judgments. So, LOGIC BOMB straightforwardly satisfies the account of cyberharms as just causes for war.

Surprisingly, this view offers theoretical reasons for thinking that distributed denial of service attacks—the most common form of cyberattack—do not cause the appropriate kind of cyberharm. DDoS attacks undermine the effectiveness of websites and other applications by using botnets of computers to ping those websites for information at rates that cannot be easily processed. As a consequence, the attacked sites lose functionality, even shutting down completely. Yet, DDoS's do not attempt to overcome resistance; there is no sense in which DDoS's attempt to bypass any attempt to block a particular effect. Rather, these kinds of attacks are much more akin to probes. That is, they impose a cost for operation, just as a probing movement into a territory imposes costs by requiring the target state to respond. The probing attack into the territory then immediately retreats in the face of resistance. Yet, it is true that DDoS's can *amount* to an imposition of will if those costs come to have a certain character. A key example of this might be the large-scale DDoS attack on key institutions in Estonia in 2007 if that attack had been somewhat more sustained and serious. If the attacks shut down vital transportation and communication infrastructure necessary for the people of Estonia to collectively deliberate about how to respond to Russian pressure, then those attacks could graduate from an imposition of cost to an imposition of will. In that case, then the cyberharm produced would be the right kind to potentially justify a military response in self-defense. So, whether a DDoS attack represents the right kind of cyberharm is a complicated manner, but it will usually be the case that they do not represent the kind of threat that normally justifies military retaliation even if they are fairly serious.

The rich human interests view is a flexible and pluralistic account that avoids the inflation of the standard view when it comes potential casus belli. By limiting potential just causes for war to only those actions which represent the kind of imposition of will that undermines the value of self-determination, the rich human interests view can offer a superior alternative to the standard view and significantly constrain when military responses are justified in response to cyberattacks. As such, the rich human interests view can offer a moralized baseline to guide our understanding of cyberharm while acknowledging that different human interests make appropriate different public policy requirements.

References

Altman, Andrew, and Christopher Wellman. 2009. *A liberal theory of international justice*. Oxford: Oxford University Press.
Blake, Michael. 2001. Distributive justice, state coercion, and autonomy. *Philosophy and Public Affairs* 30: 257–296.
Brownlie, Ian. 1963. *International law and the use of force by states*. Oxford: Oxford University Press.
Clark, Andy, and David Chalmers. 1998. The extended mind. *Analysis* 58: 7–19.
Cook, James. 2010. 'Cyberation' and just war doctrine: A response to Dipert. *The Journal of Military Ethics* 9: 411–423.
Devall, Bill, and George Sessions. 1985. *Deep ecology: Living as if nature matter*. Salt Lake City: Peregrine Smith.
Dipert, Randall. 2010. The ethics of cyberwarfare. *Journal of Military Ethics* 9: 384–2010.
Eberle, Christopher. 2013. Just war and cyberwar. *The Journal of Military Ethics* 12: 54–56.
Floridi, Luciano. 2003. On the intrinsic value of information objects and the infosphere. *Ethics and Information Technology* 4: 287–304.
Jenkins, Ryan. 2013. Is stuxnet physical? Does it matter? *The Journal of Military Ethics* 12: 68–79.
Kahn, Leonard. 2013. Understanding just cause in Cyberwar. In: *Routledge handbook of war and ethics: Just war*, ed. Fritz Alhoff, Evans, Nicholas G., and Henschke, Adam, 382–393. London: Routledge.
Levitov, Alex. 2015. Human rights, self-determination, and external legitimacy. *Politics, Philosophy, and Economics* 14(3): 291–315.
Naess, Arne. 1973. The shallow and the deep, long-range ecology movement. *Inquiry* 16: 95–100.
Nozick, Robert. 1969. Coercion. In *Philosophy, science, and method: Essays in honor of Ernest Nagel*, ed. White Morgenbesser, 440–772. New York: St Martin's Press.
Nussbaum, Martha. 2008. Human capabilities, female human beings. In *Global justice: Seminal essays*, ed. T. Pogge and D. Mollendorf, 495–552. St. Paul: Paragon Press.
Pogge, Thomas. 2005. Baselines for determining harm. In *Ethics and international affairs*, ed. Joel Rosenthal and Christian Barry, 329–334. Washington, DC: Georgetown University Press.
Rawls, John. 1999. *The law of peoples*. Cambridge: Harvard University Press.
Risse, Mathias. 2005. How does the global order harm the poor? *Philosophy and Public Affairs* 33: 2349–2376.
Schmitt, Michael. 1999. Computer network attack and the use of force in international law: Thoughts on a normative framework. *Columbia Journal of Transnational Law* 37: 885–937.
Singer, Peter. 1975. *Animal liberation*. New York: Random House.
Smith, Patrick Taylor. 2016. Cyberattacks as Casus Belli: A sovereignty-based account. *Journal of Applied Philosophy*: doi: 10.1111/japp.12169.
Taddeo, Mariarosaria. 2012. Information warfare: A philosophical perspective. *Philosophy and Technology* 25: 105–120.
Taddeo, Mariarosaria. 2016. Just information warfare. *Topoi* 35(1): 213–224.
Walzer, Michael. 1980. The moral standing of states: A response to four critics. *Philosophy and Public Affairs* 9: 209–229.
Walzer, Michael. 2000. *Just and unjust wars: A moral argument with historical illustrations*, 3rd ed. New York: Basic Books.
Ypi, Lea. 2013. What's wrong with colonialism. *Philosophy and Public Affairs* 41: 158–191.

Patrick Taylor Smith is currently assistant professor of political science in the Global Studies Programme of the National University of Singapore. After receiving his PhD from the University of Washington-Seattle in 2013, he was a postdoctoral fellow at the McCoy Family Center for Ethics in Society at Stanford University. While he works primarily on issues of global justice, he also has considerable interests in justice and emerging technology. As a result, he has published or presented papers on geoengineering, cyberwarfare and autonomous weapons systems. His work has appeared in such journals as *Transnational Legal Theory*, *Philosophy and Public Issues* and *Journal of Applied Philosophy*.

Chapter 5
Just Information Warfare

Mariarosaria Taddeo

Abstract In this chapter I propose an ethical analysis of information warfare, the warfare waged in the cyber domain. The goal is twofold, filling the theoretical vacuum surrounding this phenomenon and providing the conceptual grounding for the definition of new ethical regulations for information warfare. I argue that Just War Theory is a necessary but not sufficient instrument for considering the ethical implications of information warfare and that a suitable ethical analysis of this kind of warfare is developed when Just War Theory is merged with Information Ethics. In the initial part of the chapter, I describe information warfare and its main features and highlight the problems that arise when Just War Theory is endorsed as a means of addressing ethical problems engendered by this kind of warfare. In the final part, I introduce the main aspects of Information Ethics and define three principles for a just information warfare resulting from the integration of Just War Theory and Information Ethics.

Keywords Cyber Conflicts • Entropy • Information Ethics • Information War • Just War Theory • War

5.1 Introduction

Since 2010, the cyberspace has been officially listed among the domains in which war may be waged these days. It comes fifth along land, sea, air and space, for the ability to control, disrupt or manipulate the enemy's informational infrastructure has become as decisive with respect to the outcome of conflicts as weapon superiority. Information and communication technologies (ICTs) have proved to be a useful and convenient technology for waging war, the military deployment of ICTs has radically changed the way wars are declared and waged nowadays. It has actually determined the latest revolution in military affairs, i.e. the informational turn in

M. Taddeo (✉)
Oxford Internet Institute, University of Oxford, Oxford, UK
e-mail: mariarosaria.taddeo@oii.ox.ac.uk

military affairs (Toffler and Toffler 1997) (reference removed for double-blind review).[1] Such a revolution is not the exclusive concern of the military; it has also a bearing on ethicists and policymakers, since existing ethical theories of war and national and international regulations struggle to address the novelties of this phenomenon.

In this chapter I propose an ethical analysis of information warfare (IW) with the twofold goal of filling the theoretical vacuum surrounding this phenomenon and of providing the conceptual grounding for the definition of new ethical regulations for IW. The proposed analysis rests on the conceptual investigation of IW that I provided in (reference removed for blind review), where I highlight the informational nature of this phenomenon and maintain that IW represents a profound novelty, which is reshaping the very concept of war and raises the need for new ethical guidelines.

Following on that analysis, in this chapter I argue that considering IW through the lens of Just War Theory (JWT) allows for the unveiling of fundamental ethical issues that this phenomenon brings to the fore, yet that attempting to address these issues solely on the basis of this theory will leave them unsolved. I then suggest that problems encountered when addressing IW through JWT are overcome if the latter is merged with Information Ethics (Floridi 2013). This is a macro-ethical theory, which is particularly suitable for taking into account the features and the ethical implications of *informational phenomena*, like for example internet neutrality (Turilli et al. 2011), online trust, peer-to-peer (Taddeo and Vaccaro 2011) and IW. Merging the principles of JWT with the macro-ethical framework provided by Information Ethics has two advantages: it allows the development of an ethical analysis of IW capable of taking into account the peculiarities and the novelty of this phenomenon; it also extends the validity of JWT to a new kind of warfare, which at first glance seemed to fall outside its scope (reference removed for blind review).

In the initial part of this chapter, I describe IW and its main features, I will then focus on JWT and on the problems that arise when this theory is endorsed as a means of addressing the case for IW. Information Ethics will then be introduced, its four principles will provide the grounds for the analysis proposed in the final part of this chapter, where I describe the principles for a just IW and discuss how JWT can be applied to IW without leading to ethical conundrums. Having delineated the path ahead of us, we should now begin our analysis by considering in more detail the nature of IW.

5.2 Information Warfare

The expression 'information warfare' has already been used in the extant literature to refer solely to the uses of ICTs devoted to breaching the opponent's informational infrastructure in order to either disrupt it or acquire relevant data and information

---

[1] For an analysis of revolution in military affairs considering both the history of such revolutions and the effects of the development of the most recent technologies on warfare see (Benbow 2004; Blackmore 2011).

5 Just Information Warfare

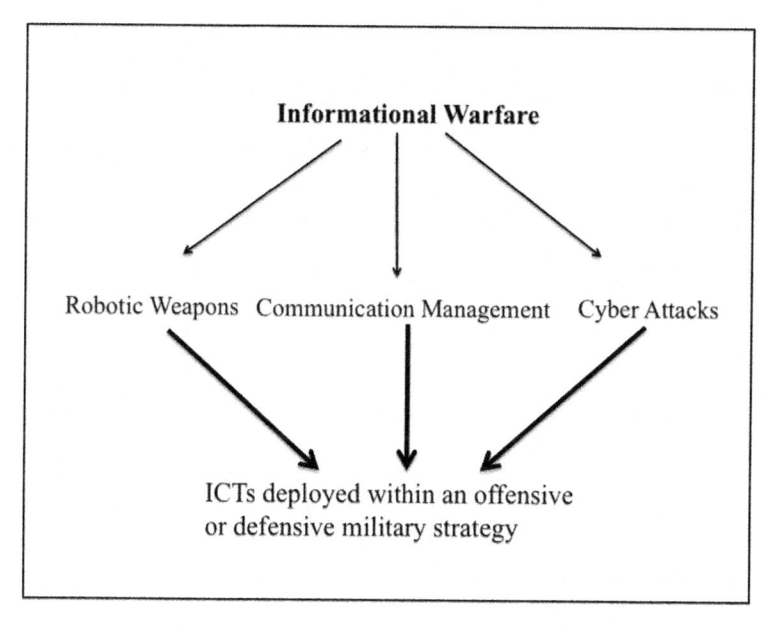

```
                    Informational Warfare
                   /        |        \
                  /         |         \
                 ↓          ↓          ↓
        Robotic Weapons  Communication Management  Cyber Attacks
                  \        |        /
                   \       |       /
                    ↓      ↓      ↓
              ICTs deployed within an offensive
                or defensive military strategy
```

Fig. 5.1 The different uses of ICTs in military strategies (reference removed for blind review, p.)

about the opponent's resources, military strategies and so on; see for example Libicki (1996), Waltz (1998), Schwartau (1994).

Distributed denials of service (DDoS) attacks, like the ones launched in Burma during the 2010 elections,[2] the injection of Stuxnet in the Iranian nuclear facilities of Bushehr,[3] as well as 'Red October' discovered in 2013 are all famous examples of how ICTs can be used to conduct the so-called cyber attacks.[4] Nonetheless, such attacks are only one of the instances of IW. In the rest of this chapter, I will use IW to refer to a wide spectrum of phenomena, encompassing cyber-attacks as well as the deployment of robotic-weapons and ICT-based communication protocols (see Fig. 5.1).

Endorsing a wide spectrum definition of IW offers important advantages, both conceptual and methodological. The conceptual advantage revolves around the identification of the informational nature of this phenomenon. In all the three cases, information plays a crucial role, it is either the target, the source or the medium for the accomplishment of a given goal. Now, while this is evident for the cases of communication management and cyber attacks, further explanation may be needed to highlight the informational nature of the deployment of (semi)autonomous robotic weapons, which may be less obvious. Such weapons are usually deployed (or

---

[2] http://www.bbc.co.uk/news/technology-11693214   http://news.bbc.co.uk/2/hi/europe/6665145.stm

[3] http://www.cbsnews.com/stories/2010/11/29/world/main7100197.shtml

[4] For an annotated time line of cyber attacks see NATO's website http://www.nato.int/docu/review/2013/Cyber/timeline/EN/index.htm

designed to be deployed) to participate in traditional military actions and usually have destructive purposes. See for example Israel's Harpi[5] or Taranis.[6]

Nonetheless, while (semi) autonomous weapon may be used to perform tasks and achieve goals not dissimilar to the ones pursued in traditional warfare, their modes of operations are quite different from traditional ones as they rely extensively on the collection and elaboration of information. The ability and the way in which a machine collects, manipulates and checks information against the requirements for an action to be performed are crucial for the accomplishment of the given task. Information is in this case the means for the achievement of the goal and it shows to be common aspect to all these three cases. In the rest of this chapter I endorse an informational level of abstraction (LoA) to focus on such a common factor.

A brief digression from the analysis of IW is in order here to introduce LoAs. Any given systems, for example a car, can be observed focusing on certain aspects and disregarding others, the choice of the aspects on which one focuses, i.e. the observables, depends on the purpose of the observer. An engineer interested in maximising the aerodynamics would focus on the shape of car's parts, their weight and possibly the materials of which the parts are made. A costumer interested in the aesthetics of the car will focus on its colour and on the overall look of the car. The engineer and the costumer observe the car endorsing different LoAs. A LoA is a finite but non-empty set of observables accompanied by a statement of what feature of the system under consideration such a LoA stands for. A collection of LoAs constitutes an interface. An interface is used when analysing a system from various points of view, that is, at varying LoAs. It is important to stress that a single LoA does not reduce a car to merely the aerodynamics of its parts or to its overall look. A LoA is a tool that helps to make explicit the observation perspective and constrain it to only those elements that are relevant in a particular observation.[7]

Endorsing an informational LoA to analyse cyber attacks, the deployment of robotic-weapons and ICT-based communication protocols allows for unveiling the common factor to these three phenomenon rather than their differences. A different (lower) LoA can be endorsed in a second moment in order to analyse the specific occurrences of these three phenomena and their ethical implications. This approach neither undermines the differences between the use of a computer virus, ICT-based communication protocols and robotic weapons nor denies that such different uses generate different ethical issues. Rather, it aims at focusing first on the aspects that are common among the military uses of ICTs, since the analysis of these aspects provides the groundwork for addressing specific ethical problems brought to the fore by the different modes of military deployment of ICTs.

The methodological advantage of endorsing a wide spectrum definition concerns the scope of the analysis, by considering indiscriminately the different uses of ICTs

---

[5] This is an autonomous weapon system designed to detect and destroy radar emitters http://www.israeli-weapons.com/weapons/aircraft/uav/harpy/harpy.html

[6] This is a UK drone which can autonomously search, identify and locate enemies although it should be stressed that it can only engage with a target upon the authorization of mission command http://en.wikipedia.org/wiki/BAE_Systems_Taranis

[7] For more in details analysis of LoA see (Floridi 2008).

in warfare, the analysis will address the totality of the cases of IW rather than focusing solely on some of its specific occurrences.

IW is thus defined as follows:

Information Warfare is the use of ICTs within an offensive or defensive military strategy endorsed by a [political authority] and aiming at the immediate disruption or control of the enemy's resources, and which is waged within the informational environment, with agents and targets ranging both on the physical and non-physical domains and whose level of violence may vary upon circumstances (reference removed for blind review).

This definition highlights two important aspects of IW, its *informational nature* and its *transversality*, which put it in relation with the so-called information revolution (Floridi 2014) (reference removed for peer review). The information revolution is a complex phenomenon. It rests on the development and the ubiquitous dissemination of the use of ICTs, which have a wide impact on many of our daily practises: from our social and professional lives to our interactions with the environment surrounding us. ICTs allow for developing and acting in a new domain, the digital or informational one. This is a virtual, non-physical domain, which has grown important and hosts a considerable relevant part of our lives. With the information revolution we witness a shift, which has brought the *non-physical domain* to the fore and made it as important and valuable as the physical one (reference removed for blind review).

IW is one of the most compelling instances of such a shift. It shows that there is a new environment, where physical and non-physical entities coexist and are equally valuable, and in which states have to prove their authority and new modes of warfare are being specifically developed.[8] The shift toward the non-physical domain provides the ground for the transversality of IW. This is a complex aspect that can be better understood when IW is compared with traditional forms of warfare. Traditionally, war entails the use of a state's *violence* through the state *military* forces to determine the conditions of governance over a determined territory (Gelven 1994). It is a necessarily violent phenomenon, which implies the sacrifice of human lives and damage to both military and civilian infrastructures. The problem to be faced when waging traditional warfare is how to minimise damage and losses while ensuring the enemy is overpowered.

IW is different from traditional warfare in several respects, mainly because it is not a necessarily violent and destructive phenomenon (Arquilla 1998; Dipert 2010; Barrett 2013). For example, IW may involve a computer virus capable of disrupting or denying access to the enemy's database, and in so doing it may cause severe damage to the opponent without exerting *physical* force or violence. In the same way, IW does not necessarily involve human beings. An action of war in this context can be conducted by an autonomous robot, such as, for example, the EADS Barracuda,

---

[8] The USA only spent $400 million in developing technologies for cyber conflicts: http://www.wired.com/dangerroom/2010/05/cyberwar-cassandras-get-400-million-in-conflict-cash/

The UK devoted £650 million to the same purpose: http://www.theinquirer.net/inquirer/news/1896098/british-military-spend-gbp650-million-cyber-warfare

and the Northrop Grumman X-47B,[9] or by an autonomous cruising computer virus (Abiola et al. 2004), targeting other artificial agents or informational infrastructures, like a database or a website. IW can be waged exclusively in a digital context without ever involving physical targets, nevertheless it may escalate to more violent forms (Arquilla 2013; Waltz 1998; Clarke 2012; Brenner 2011; Bowden 2011).

Consider for example, the data diffused for GridExII.[10] This is a simulation that has been conducted in the US in November 2013. More than two hundred utility companies collaborated with US government to simulate a massive cyber attack on US basic infrastructure. Had the attack been real, estimates mention hundreds of injuries and tens of deaths, while millions of US Citizens would have been left in the darkness.

As remarked above, the transversality of IW is the key feature of this phenomenon; it is the aspect that differentiates it the most from traditional warfare. Transversality is also the feature that engenders the ethical problems posed by IW. The potential bloodless and non-destructive nature of IW (Denning 2007; Arquilla 2013) makes it desirable from both an ethical and a political perspective, since at first glance, it seems to avoid bloodshed and it liberates political authority from the burden of justifying military actions to the public. However, the disruptive outcomes of IW can inflict serious damage to contemporary information societies at the same time, IW has the potential to lead to highly violent and destructive consequences, which would be dangerous for both military forces and civil society.

The need for strict regulations for declaring and waging IW in fairness is now compelling. To this end an analysis that discloses the ethical issues related to IW while pointing at the direction for their solution is a preliminary and necessary step. This will be the task of the next section.

5.3 IW and Just War Theory

Ethical analyses of war are developed following three main paradigms: JWT, Pacifism or Realism. In the rest of this paper, the analysis will focus only on JWT. Two reasons support this choice: the ethical problems with which JWT is concerned are generated by the very same decision to declare and to wage war, be it a traditional or an informational war. Therefore JWT sheds light on the analysis of the ethical issues posed by possible declaration of IW. More in general, the criteria for a *just* war proposed by this theory remain valid when considering IW, the justification to resort to war and the criteria for *jus in bello* and *post bellum* proposed are desirable also in case of IW and there is no doubt that just war principles and their preservation hold in the case of traditional warfare as well as in the case of IW.

---

[9] Note that MQ-1 Predators and EADS Barracuda, and the Northrop Grumman X-47B are Unmanned Combat Aerial Vehicles used for combat actions and they are different from Unmanned Air Vehicles, like for example Northrop Grumman MQ-8 Fire Scout, which are used for patrolling and recognition purposes only.

[10] http://www.nytimes.com/2013/11/15/us/coast-to-coast-simulating-onslaught-against-power-grid.html

Nevertheless, it would be mistaken to consider JWT both the necessary and sufficient ethical framework for the analysis of IW, for addressing this new form of warfare solely on the basis of JWT generates more ethical conundrums than it solves. The problem arises because JWT mainly focuses on the use of force in international contexts and surmises sanguinary and violent warfare occurring in the physical domain. As the cyber domain is virtual and IW mainly involves abstract entities, the application of JWT becomes less direct and intuitive. The struggle encountered when applying JWT to the cases of IW becomes even more evident if one considers how pivotal concepts such as the ones of harm, target, attack have been reshaped by the dissemination of IW.[11] The very notion of harm for example, which is at the basis of JWT, struggle to apply to the case of. This a problem has been already highlighted in the extant literature, see for example (Dipert 2010) who argues that any moral analysis of this kind of warfare needs to be able to account for a notion of harm "[focusing] away from strictly injury to human beings and physical objects toward a notion of the (mal-) functioning of information systems, and the other systems (economic, communication, industrial production) that depend on them" (p. 386).

Particularly relevant to shed some light on the novelty posed by IW, is the transversality of the ontological status of the entities involved in the latter. Traditional warfare concerns human beings and physical objects, while IW involves artificial and non-physical entities alongside human beings and physical objects. Therefore, there is a *hiatus* between the ontology of the entities involved in traditional warfare and of those involved in IW. Such a hiatus affects the ethical analysis, for JWT rests on an anthropocentric ontology, i.e. it is concerned with respect for human rights and disregards all non-human entities as part of the moral discourse, and for this reason it does not provide sufficient means for addressing the case for IW (more details on this aspect presently).

The gap between the ontology assumed by JWT and the one of IW has also been described by Dipert, who stresses that "[s]ince cyber warfare is by its very nature information warfare, an ontology of cyber warfare would necessarily include way of specifying *information objects* […], *the disruption and the corruption of data and the nature and the properties of malware*. This would be in addition to what would be required of a domain-neutral upper-level ontology, which addresses this type of characteristics of the most basic categories of entity that are used virtually in sciences and domain: material entity, event, quality of an object, physical object. A cyber warfare ontology would also go beyond […] of a military ontology, such as agents, intentional actions, unintended effects, organizations, artefacts', commands, attacks and so on." (emphasis added) (Dipert 2013 p. 36)

The case of the autonomous cruising computer virus will help in clarifying the problems at stake (Abiola et al. 2004). These viruses are able to navigate through the web and identify autonomously their targets and attack them without requiring

---

[11] The need to define concepts such as those of harm, target and violence is stressed both by scholar who argue in favor of the ontological difference of the cyber warfare (Dipert 2013) and exploit this point to claim that JWT is not an adequate framework to address IW and by those who actually maintain that JWT provides sufficient element to address the case of IW LUCAS.

any supervision. The targets are chosen on the basis of parameters that the designers encode in the virus, so there is a boundary to the autonomy of these agents. Still, once the target has been identified the virus attacks without having to receive 'authorisation' from the designer or any human agent.

In considering the moral scenario in which the virus is launched three main questions arise. The first question revolves around the identification of the moral agents, for it is unclear whether the virus itself should be considered the moral agent, or whether such a role should be attributed to the designer or to the agency that decided to deploy the virus, or even to the person who actually launched it. The second question focuses on moral patients. The issue arises as to whether the attacked computer system itself should be considered the moral receiver of the action, or whether the computer system and its users should be considered the moral patients. Finally, the third questions concerns the rights that should be defended in the case of a cyber attack. In this case, the problem is whether any rights should be attributed to the informational infrastructures or to the system compounded by the informational infrastructure and the users.

The case of the autonomous computer virus indicates that IW includes informational infrastructures, computer systems, and databases. In doing so, it brings new objects, some of which are intangible, into the moral discourse. The first step toward an ethical analysis of IW is to determine the moral status of such (informational) objects and their rights. Help in this respect is provided by Information Ethics, which will be introduced in the Sect. 5.4. Before focusing on Information Ethics, we shall first consider in detail some of the problems encountered when applying JWT to IW.

5.3.1 The Tenets of JWT and IW

Let me begin this section by stressing that the proposed analysis does not claim that JWT does not adequately respond to contemporary global politics or to new methods for waging violent warfare.[12] In the rest of this section I shall analyse the tenets of *last resort*, *more good than harm*, and *non-combatants immunity* to consider the problems that arise when these principles, which are desirable also in case of IW, are applied to the occurrences of a war in the cyber (non-physical) domain. I argue that the nexus of the ethical problems posed by IW rest on the ontological hiatus between IW and JWT, for the latter focuses on violent warfare, bloodshed and physical damages, and these aspects are peculiar of kinetic warfare but are not peculiar of IW.

The principle of 'war as last resort' prescribes that a state may resort to war only if it has exhausted all plausible, peaceful alternatives to resolve the conflict in question, in particular diplomatic negotiations. This principle rests on the assumption that war is a violent and sanguinary phenomenon and as such it has to be avoided until it remains the only reasonable way for a state to defend itself. The application

---

[12] See (Withman 2013) for an analysis of validity of JWT with respect to contemporary violent warfare.

of this principle is shaken when IW is taken in consideration, because in this case war may be bloodless and may not involve physical violence at all. In these circumstances, the use of the principle of war as last resort becomes less immediate.

Imagine, for example, the case of tense relations between two states and that the tension could be resolved if one of the states decide to launch a cyber attack on the other state's informational infrastructure. The attack would be bloodless as it would affect only the informational grid of the other state and there would be no casualties. The attack could also lead to resolution of the tension and avert the possibility of kinetic war in the foreseeable future. Nevertheless, according to JWT, the attack would be an act of war, and as such it is forbidden as a first strike move.

The impasse is quite dramatic, for if the state decides not to launch the cyber attack it will be probably forced to engage in a sanguinary war in the future, but if the state authorises the cyber attack it will breach the principle of war as last resort and commit an unethical action. This example is emblematic of the problems encountered in the attempt to establish ethical guidelines for IW. In this case, the main problem is due to the transversality of the modes of combat described in Sect. 5.2, which makes it difficult to define unequivocal ethical guidelines.

In the light of the principle of last resort, soft and non-violent cases of IW can be approved as means for avoiding traditional war (Perry 2006), as they can be considered a viable alternative to bloodshed, which may be justly endorsed to avoid traditional warfare (Bok 1999). At the same time, even the soft cases of IW have a disruptive purpose – disrupting the enemy's (informational) (Arquilla and Ronfeldt 1997; Arquilla 2013). Such a disruptive intent, even when it is not achieved through violent and sanguinary means, must be taken in consideration by any analysis aiming at providing ethical guidelines for IW.[13]

Another problem arises when considering the principle of 'more good than harm'. According to this principle, before declaring war a state must consider the *universal* goods expected to follow from the decision to wage war, against the *universal* evils expected to result, namely the casualties that the war is likely to determine. The state is justified in declaring war only when the goods are proportional to the evils. This is a fine balance, which is straightforwardly assessed in the case of traditional warfare, where evil is mainly considered in terms of casualties and physi-

---

[13] It is worthwhile noticing that the problem engendered by the application of the principle of last resort to the soft-cases of IW may also be addressed by stressing that these cases do not fall within the scope of JWT as they may be considered cases of espionage rather than cases of war, and as such they do not represent a 'first strike' and the principle of last resort should not be applied to them. One consequence of this approach is that JWT would address war scenarios by focusing on traditional cases of warfare, such as physical attacks, and on the deployment of robotic weapons, disregarding the use of cyber attacks. This would be quite a problematic consequence because, despite the academic distinction between IW and traditional warfare, the two phenomena are actually not so distinct in reality. Robotic weapons fight on the battlefield side by side with human soldiers, and military strategies comprise both physical and cyber attacks. By disregarding cyber attacks, JWT would be able to address only partially contemporary warfare, while it should take into consideration the whole range of phenomena related to war waging in order to address the ethical issues posed by it (for a more in depth analysis of this aspect see (reference removed for blind review)).

cal damages that may result from a war. The equilibrium between the goods and the evils becomes more problematic to calculate when IW is taken into consideration.

As the reader may recall, IW is transversal with respect to the level of violence. If strictly applied to the non-violent instances of IW, the principle of more good than harm leads to problematic consequences. For it may be argued that, since IW can lead to the victory over the enemy without determining casualties, it is a kind of warfare (or at least the soft, non-violent instances of IW) that is always morally justified, as the good to be achieved will always be greater than the evil that could potentially be caused.

Nonetheless, IW may result in unethical actions –destroying a database with rare and important historical information, for example. If the only criteria for the assessment of harm in warfare scenarios remain the consideration of the physical damage caused by war, then an unwelcome consequence follows, for all the non-violent cases of IW comply by default to this principle. Therefore, destroying a digital resource containing important records is deemed to be an ethical action tout court, as it does not constitute physical damage *per se*.

The problem that arose with the application of this principle to the case of IW does not concern the validity *in se* of the principle. It is rather the framework in which the principle has been provided that becomes problematic. In this case, it is not the prescription that the goods should be greater than the harm in order to justify the decision to conduct a war, but rather is the set of criteria endorsed to assess the good and the harm that shows its inadequacy when considering IW.

A similar problem arises when considering the principle of 'discrimination and non-combatant immunity'. This principle refers to a classic war scenario and aims at reducing bloodshed, prohibiting any form of violence against non-combatants, like civilians. It is part of the *jus in bello* criteria and states that soldiers can use their weapons to target exclusively those who are "engaged in harm" (Walzer 2006, p. 82). Casualties inflicted on non-combatants are excused only if they are a consequence of a non-deliberate act. This principle is of paramount importance, as it prevents massacres of individuals not actively involved in the conflict. Its correctness is not questionable yet its application is quite difficult in the context of IW.

In classic warfare, the distinction between combatants and non-combatants reflects the distinction between military and civil society. In the last century, the spread of terrorism and guerrilla warfare weakened the association between non-combatants and civilians. In the case of IW such association becomes even feebler, due to the blurring between civil society and military organisations (Schmitt 1999; Shulman 1999)(reference removed for blind review).

The blurring of the distinction between military and civil society leads to the involvement of civilians in war actions and raises a problem concerning the discrimination itself: in the IW scenario it is difficult to distinguish combatants from non-combatants. Wearing a uniform or being deployed on the battlefield are no longer sufficient criteria to identify someone's social status. Civilians may take part in a combat action from the comfort of their homes, while carrying on with their civilian life and hiding their status as informational warriors.

This case provides also a good example of the policy gap surrounding IW, for one of the most important aspects of the distinction between military and civilian concerns the identification of the so-called civilian objects, i.e. buildings, places and objects that should not be considered military targets. Chapter III of the Protocol I of the Geneva Convention[14] defines civilian objects as material tokens, which are further categorised of cultural or religious type, environmental or necessary to the survival of the population. This chapter shows to be ontologically limited as it considers as 'objects' only physical, tangible entities.[15] Furthermore, civilian objects are distinguished from military one, as the latter are deemed to be objects that "make an effective contribution to military action and whose total or partial destruction, capture or neutralization, in the circumstances ruling at the time, offers a definite military advantage". The reader may easily see how such a definition may be used to qualify a civilian informational infrastructure in time of IW, making the line between civilian and military even less evident and making even more compelling the need for policies able to accommodate a more inclusive definition of objects, and more in general able to address the conceptual changes posed by this new kind of warfare.

Before introducing Information Ethics, I shall remark that several analyses have been proposed claiming that the existing apparatus of laws resting on JWT is adequate and sufficient to address the cases of IW (see for example Lucas, Sminth). This is an interesting and also useful approach, so far it allowed for applying current international laws to IW, avoiding that the cyber sphere would become an unregulated domain. However, the approach encounters a major flaw, for especially law practitioners and policy-makers identify JWT with the body of international and national laws regulating warfare and overlook the conceptual roots on which this theory rests. In doing so, the universal nature of JWT is missed and so is the possibility of expanding the scope of this theory, the effect is that rather than revising the conceptual roots of the theory to address this new phenomenon, the latter is 'forced' to fit the parameters set for kinetic warfare.

The approach hence misses to consider and to account for the conceptual changes prompted by IW (see the ones discussed in Sects. 5.2 and 5.3) and risks to confuse the remedy for the solution and, in the long run, to pose conceptual limitations to the laws and regulation for IW. See for example the principle of just cause applied to IW. As Barrett (2013) put it "[s]ince damage to property may constitute a just cause, can temporary losses of computer functionality also qualify as a *casus belli*? Like kinetic weapons, cyber-weapons can physically destroy or damage computers. But offensive computer operations, because of their potential to be transitory or reversible, can also merely compromise functionality. While permanent loss of functionality create the same effect as physical destruction, temporary functionality losses are unique to cyber-operations and require additional analysis" (p. 6).

The issue is not whether the case of IW can be considered in such a way to fit the parameters of kinetic warfare and hence to fall in the domain of JWT as we know it.

---

[14] "ICRC Databases on International Humanitarian Law"

[15] On this point see also (Dipert 2010, p. 400).

This result is easily achieved once the focus is restricted to physical damage and tangible objects. The problem lays at a deeper level and questions the vey conceptual framework on which JWT rests and it ability to *satisfactory* and *fairly* accommodate the changes brought to the fore by the information revolution, which are affecting the way we wage war, but also the way in which we conduct our lives, perceive ourselves and the very concepts of harm, warfare, property, state even personal identity.

It would be misleading to consider the problems described in this sections as reasons for dismissing JWT when analysing IW. These problems rather point to a more fundamental issue; namely the need to consider more carefully the case of IW, and to take into account its peculiarities.

5.4 Information Ethics

Information Ethics is a macro-ethics, which is concerned with the whole realm of reality and provides an analysis of ethical issues by endorsing an informational perspective. Such an approach rests on the consideration that "ICTs, by radically changing the informational context in which moral issues arise, not only add interesting new dimensions to old problems, but lead us to rethink, methodologically, the very grounds on which our ethical positions are based" (Floridi 2006) (p. 23).

In one sentence Information Ethics is defined as a *patient-oriented*, *ontocentric*, and *ecological* macroethics. Information Ethics is patient-oriented because it considers the morality of an action with respect to its effects on the receiver of the action. It is ontocentric, for it endorses a non-anthropocentric approach for the ethical analysis. It attributes a moral value to all the existing entities (both physical and non-physical) by applying the principle of ontological equality: "This ontological equality principle means that any form of reality [...], simply for the fact of being what it is, enjoys a minimal, initial, *overridable*, equal right to exist and develop in a way which is appropriate to its nature" (Floridi 2013). The principle of ontological equality is grounded on an information-based ontology,[16] according to which all existing things can be considered from an informational standpoint and are understood as informational entities, all sharing the same informational nature.

The principle of ontological equality shifts the standing point for the assessment of the moral value of entities, including technological artefacts. At first glance, an artefact, a computer, a book or the Colosseum, seems to enjoy only an instrumental value. This is because in considering them one endorses an anthropocentric LoA, in other words one considers these objects as a user, a reader, a tourist. In all these cases the moral value of the observed entities depends on the agent interacting with them and on her purpose in doing so.

---

[16] The reader may recall the informational LoA mentioned in Sect. 5.2. Information Ethics endorses an informational LoA, as such it focuses on the informational nature as a common ground of all existing things.

The claim put forward by Information Ethics is that, these LoAs are not adequate to support an effective analysis of the moral scenario in which the artefacts may be involved. The anthropocentric, or even the biocentric, LoA prevent to properly consider the nature and the role of such artefacts in the reality in which we live. The argument is suggested that all existing things have an informational nature, which is shared across the entire spectrum – from abstract to physical and tangible entities, from rocks and books to robots and human beings, and that all entities enjoy some minimal initial moral value *qua informational* entities.

Information Ethics argues that universal moral analyses can be developed by focusing on the common nature of all existing things and by defining good and evil with respect to such a nature. The focus of the ethical analysis is shifted, the initial moral value of an entity does not depend on the observer, but is defined in absolute terms and depends on the (informational) nature of the entities. Following the principle of ontological equality, minimal and overridable rights to exist and flourish pertain to all existing things and not just to human or living things. The Colosseum, Jane Austin's writings, a human being and computer software all share initial right to exist and flourish, as they are all informational entities.[17]

A clarification is now necessary. Information Ethics endorses a minimalist approach, it considers informational nature as the minimal common denominator among all existing things. Such a minimalist approach should not be mistaken for reductionism, as Information Ethics does not claim that the informational one is the unique LoA from which moral discourse is addressed. Rather it maintains that the informational LoA provides a *minimal starting point*, which can then be enriched by considering other moral perspectives.

Lest the reader be mislead, it is worthwhile emphasising that the principle of ontological equality does not imply that all entities have the same moral value. The rights attributed to the entities are *initial*, they are overridden whenever they conflict with the rights of other (more morally valuable) entities. Furthermore, the moral value of an entity is determined according to its potential contribution to the enrichment and the flourishing of the informational environment. Such an environment, the *Infosphere*, includes all existing things, be they digital or analogical, physical or non-physical and the relations occurring among them, and between them and the environment. The blooming of the Infosphere is the ultimate good, while its corruption, or destruction, is the ultimate evil.

In particular, any form of corruption, depletion and destruction of informational entities or of the Infosphere is referred to as *entropy*. In this case entropy refers to "any kind of *destruction* or *corruption* of informational objects (mind, not of information), that is, any form of impoverishment of *being*, including *nothingness*, to phrase it more metaphysically", (Floridi 2013) and has nothing to do with the concept developed in physics or in information theory (Floridi 2007).

---

[17] For more details on the information-based ontology see (Floridi 2002).The reader interested in the debate on the Informational ontology and the principles of Information Ethics may whish to see (Floridi 2007).

Information Ethics considers the duty of any moral agent with respect to its contribution to the informational environment, and considers any action that affects the environment by corrupting or damaging it, or by damaging the informational objects existing in it, as an occurrence of entropy, and therefore as an instance of evil (Floridi and Sanders 2001). On the basis of this approach Information Ethics provides four principles to identify right and wrong and the moral duties of an agent. The four moral principles are:

0. entropy ought not to be caused in the infosphere (null law);
1. entropy ought to be prevented in the infosphere;
2. entropy ought to be removed from the infosphere;
3. the flourishing of informational entities as well as of the whole infosphere ought to be promoted by preserving, cultivating and enriching their properties.

These four principles together with the theoretical framework of Information Ethics will provide the ground to proceed further in our analysis, and define the principles for a just IW.

5.5 Just IW

The first step toward the definition of the principles for a just IW is to understand the moral scenario determined by this phenomenon. The framework provided by Information Ethics proves to be useful in this regard, for we can now answer the questions posed in Sect. 5.3 concerning the identification of moral agents, moral patients and the rights that have to be respected in the case of IW. The remainder of this chapter will not focus on the problems regarding moral patients and their rights. The issue concerning the identification of moral agents in IW requires an in-depth analysis (see for example (Asaro 2008)) which falls outside the scope of this chapter. I shall clarify a few aspects concerning morality of artificial agents relevant to the scope of this analysis, before setting this issue aside.

The debate on morality of artificial agents is usually associated to the issues of ascribing to artificial agents moral responsibility for their actions. Floridi and Sanders (2004) provide a different approach to this problem decoupling the moral *accountability* of an artificial agent, i.e. its ability to perform morally qualifiable actions, from the moral *responsibility* for the actions that such an agent may perform.

The authors argue that an action is morally qualifiable when it has morally qualifiable effects, and that every entity that qualifies as an interactive, autonomous and adaptable (transition) system and which performs a morally qualifiable action is (independently from its ontological nature) considered a morally accountable agent. So when considering the case for IW, a robotic weapon and a computer virus are considered moral agents as long as they show some degree of autonomy in interacting and adapting to the environment and perform actions that may cause either moral good or moral evil.

As argued by Floridi and Sanders, attributing moral accountability to artificial agents extends the scope of ethical analysis to include the actions performed by the agents and permits to determine moral principles to regulate such actions. This approach particularly suits the purpose of the present analysis, for the reader may accept suspending judgment on the moral responsibility for the actions that artificial agents may perform in case of IW, and agree that such actions are nevertheless morally qualifiable, and that as such they should be the objects of a prescriptive analysis.

Once we have put aside the issue concerning the morality of artificial agents, we are left with questions concerning the moral stance of the receivers of the actions performed by such agents and of the rights that ought to be respected in the case of IW. The principle of ontological equality states that all (informational) entities enjoy some minimal initial rights to exist and flourish in the Infosphere, and therefore every entity deserves some minimal respect, in the sense of a "disinterested, appreciative and careful attention" (Hepburn 1984; Floridi 2013).

When applied to IW, this principle allows for considering all entities that may be affected by an action of war as moral patients. A human being, who enjoys the consequences of a cyber attack and an informational infrastructure that is disrupted by a cyber attack are both to be held moral patients, as they are both the receivers of the moral action. Following Information Ethics, the moral value of such an action is to be assessed on the basis of its effects on the patients' rights to exist and flourish, and ultimately on the flourishing of the Infosphere.

The issue then arises concerning which and whose rights should be preserved in case of IW. The answer to this question follows from the rationale of Information Ethics, according to which an entity may lose its rights to exist and flourish when it comes into conflict (causes entropy) with the rights of other entities or with the well-being of the Infosphere. It is a moral duty of the other inhabitants of the Infosphere to *remove* such a malicious entity from the environment or at least to impede it from perpetrating more evil.

This framework lays the ground for the first principle for just IW. The principle prescribes the condition under which the choice to resort to IW is morally justified.

I. IW ought to be waged only against those entities that endanger or disrupt the well-being of the Infosphere.

Two more principles regulate just IW, they are:

II. IW ought to be waged to preserve the well-being of the Infosphere.
III. IW ought not to be waged to promote the well-being of the Infosphere.

The second principle limits the task of IW to restoring the *status quo* in the Infosphere before the malicious entity began increasing the entropy within it. IW is just as long its goal is to *repair* the Infosphere from the damage caused by the malicious entity.

The second principle can be described using an analogy; namely, IW should fulfil the same role as police forces in a democratic state. It should act only when a crime has been, or is about to be, perpetrated. Police forces do not act in order to ameliorate the aesthetics of cities or the fairness of a state's laws; they only focus on reducing or preventing crimes from being committed. Likewise, IW ought to be endorsed as an *active* measure in response to increasing of evil and not as proactive strategy to foster the flourishing of the Infosphere. Indeed, this is explicitly forbidden by the third principle, which prescribes the promotion of the well-being of the Infosphere as an activity that falls beyond the scope of a just IW.

These three principles rest on the identification of the moral good with the flourishing of the Infosphere and the moral evil with the increasing of entropy in it. They endorse an informational ontology, which allows for including in the moral discourse both non-living and non-physical entities. The principles also prescribe respect for the (minimal and overridable) rights of such entities along with those of human beings and other living things, and respect for the rights of the Infosphere as the most fundamental requirement for declaring and waging a just IW.

In doing so the three principles overcome the ontological hiatus described in Sect. 5.3, and provide the framework for applying JWT to the case of IW without leading to the ethical conundrums analysed in Sect. 5.3.1. The description of how JWT is merged with Information Ethics is the task of the next section.

5.6 Three Principles for a Just IW

The application of the principle of 'last resort' provides the first instance of the merging of JWT and Information Ethics. The reader may recall that the principles forbids embracing IW as an 'early move' even in those circumstances in which IW may avert the possibility of waging a traditional war. The principle takes into account traditional (violent) forms of warfare, and it is coupled with the principle of 'right cause', which justifies resort to war only in case of 'self-defence'. However right this approach may be when applied to traditional (violent) forms of warfare, it proves inadequate when IW is taken into consideration. The impasse is overcome when considering the principles for just IW.

The first principle prescribes that any entity that endangers or disrupts the well-being of the Infosphere loses its basic rights and becomes a licit target. The second principle prescribes that a state is within its rights to wage IW to re-establish the *status quo* in the Infosphere and to repair the damage caused by a malicious entity. These two principles allow for breaking the deadlock described in Sect. 5.3.1, because a state can rightly endorse IW as an early move to avoid the possibility of a traditional warfare, as the latter threatens greater disruption of the Infosphere, and as such it is deemed to be a greater evil (source of entropy) than IW.

A caveat must be stressed in this case: the waging of IW must comply with the principles of 'proportionality' and 'more good than harm'. In waging IW, the endorsed means must be sufficient to stop the malicious entity, and in doing so the means ought not to generate more entropy than a state is aiming to remove from the Infosphere in the first place. This leads us to consider in more detail the principle of 'more good than harm'.

The issues that arose in the case of IW are due to the definition of the criteria for the assessment of the 'good' and the 'harm' that a warfare may cause. As described in Sect. 5.3.1, endorsing traditional criteria leads to a serious ethical conundrum, since all (the majority of) the cases of IW that do not target physical infrastructures or human life comply by default to this principle regardless of their consequences.

Such a problem is avoided if damage to non-physical entities in considered as well as physical damage. More precisely, the assessment of the good and the harm should be determined by considering the general condition of the Infosphere 'before and after' waging the war. A just war never determines greater entropy than that in the Infosphere before it was waged. Once considered from this perspective, the principle of more good than harm acts as corollary of the second principle for just IW. It ensures that a just IW is waged to restore the *status quo* and does not increase the level of entropy in the Infosphere.

Increasing entropy in the Infosphere also provides a criterion for reconsidering the application of the principle of 'discrimination and non-combatants' immunity' to IW. As it has been argued in Sect. 5.3.1, IW blurs the distinction between militaries and civilians, as it neither requires military skills nor does it require a military status of the combatants to be waged. This makes problematic the application of this principle to IW; nevertheless the principle has to be maintained as it prescribes the distinction between licit and illicit war targets.

Help in applying this principle to IW comes from the first principle for just IW, which allows for dispensing with the distinction between militaries and civilians, and for substituting it with the distinction between licit targets and illicit ones. The former are those malicious entities who endanger or disrupt the well-being of the Infosphere. According to the principle, IW rightfully targets only malicious entities, be they military or civilian. The social status ceases to be significant in this context, because any entity that contributes to increasing the evil in the Infosphere loses its initial rights to exist and flourish and therefore becomes a licit target. More explicitly, it becomes a moral duty for the other entities in the Infosphere to prevent such entity from causing more evil.

Before concluding this chapter, I shall briefly clarify an aspect of the proposed analysis, lest the reader be tempted to consider it warmongering.

The third principle provided in Sect. 5.5 stresses that IW is never justly waged when the goal is improving the well-being of the Infosphere. This principle rests on the very same rationale that inspires Information Ethics, according to which the flourishing of the Infosphere is determined by the blooming of informational entities, of their relations and by their well-being. IW is understood as a form of disruption and as such, by definition, it can never be a vehicle for fostering the prosperity of the Infosphere nor is it deemed to be desirable *per se*. IW is rather considered a necessary evil, the bitter medicine, which one needs to take to fight something even more undesirable, i.e. the uncontrolled increasing of the entropy in the environment. With this clarification in mind we can now pull together the threads of the analysis proposed in this chapter.

5.7 Conclusion

The goal of this chapter is to fill the conceptual vacuum surrounding IW and of providing the ethical principles for a just IW. It has been argued that to this purpose JWT provides the necessary but not sufficient tools. For, although its ideal of just warfare grounded on respect for basic human rights in the theatre of war holds also in the case of IW, it does not take into account the moral stance of non-human and non-physical entities which are involved and mainly affected by IW. This is the ontological hiatus, which I identified as the nexus of the ethical problems encountered by IW.

This chapter defends the thesis that in order to be applied to the case for IW, JWT needs to extend the scope of the moral scenario to include non-physical and non-human agents and patients. Information Ethics has been introduced as a suitable ethical framework capable of considering human and artificial, physical and non-physical entities in the moral discourse. It has been argued that the ethical analysis of IW is possible when JWT is merged with Information Ethics. In other words, JWT *per se* is too large a sieve to filter the issues posed by IW. Yet, when combined with Information Ethics, JWT acquires the necessary granularity to address the issues posed by this form of warfare.

The first part of this chapter introduces IW and analyses its relation to the information revolution and its main feature, namely its transversality. It then describes the reasons why JWT is an insufficient tool with which to address the ethical problems engendered by IW and continues by introducing Information Ethics. The second part of the chapter defends the thesis according to which once the ontological hiatus between the JWT and IW it is bridged, JWT can be endorsed to address the ethical problems posed by IW.

The argument is made that such a hiatus is filled when JWT encounters Information Ethics, since its ontocentric approach and informational ontology allow for ascribing a moral status to any existing entity. In doing so, Information Ethics extends the scope of the moral discourse to all entities involved in IW and provides a new ground for JWT, allowing it to be extended to the case for IW.

In concluding this chapter I should like to remark that the proposed ethical analysis should in no way be understood as a way of advocating warfare or IW. Rather it is devoted to prescribing ethical principles such that if IW has to be waged then it will at least be a just warfare.

References

Abiola, A., J. Munoz, and W. Buchanan. 2004. *Analysis and detection of cruising computer viruses*. In In: Proceedings of 3rd EIWC.

Arquilla, J. 1998. Can information warfare ever be just? *Ethics and Information Technology* 1(3): 203–212.

Arquilla, J. 2013. Twenty years of cyberwar. *Journal of Military Ethics* 12(1): 80–87. doi:10.1080/15027570.2013.782632.

Arquilla, J., and D.F. Ronfeldt (eds.). 1997. *In Athena's camp: Preparing for conflict in the information age*. Santa Monica: Rand.

Asaro, P. 2008. How just could a robot war be? In *Current issues in computing and philosophy*, ed. P. Brey, A. Briggle, and K. Waelbers, 50–64. Amsterdam: IOS Press.

Barrett, E.T. 2013. Warfare in a new domain: The ethics of military cyber-operations. *Journal of Military Ethics* 12(1): 4–17. doi:10.1080/15027570.2013.782633.

Benbow, T. 2004. *The magic bullet?: Understanding the "Revolution in Military Affairs"*. London: Brassey's.

Blackmore, T. 2011. *War X*. Toronto: University of Toronto Press.

Bok, S. 1999. *Lying: Moral choice in public and private life*. 2nd Vintage Books ed. New York: Vintage Books.

Bowden, M. 2011. *Worm: The first digital world war*. New York: Atlantic Monthly Press.

Brenner, J. 2011. *America the vulnerable: New technology and the next threat to national security*. New York: Penguin Press.

Clarke, R.A. 2012. *Cyber war: The next threat to national security and what to do about it*. 1st Ecco pbk. edn. New York: Ecco.

Denning, D. 2007. The ethics of cyber conflict. In *Information and computer ethics*. Hoboken: Wiley.

Dipert, R. 2010. The ethics of cyberwarfare. *Journal of Military Ethics* 9(4): 384–410.

Dipert, R. 2013. The essential features of an ontology for cyberwarfare. In *Conflict and cooperation in cyberspace*, ed. Panayotis Yannakogeorgos and Adam Lowther, 35–48. Boca Raton: Taylor & Francis. http://www.crcnetbase.com/doi/abs/10.1201/b15253-7.

Floridi, L. 2002. On the intrinsic value of information objects and the infosphere. *Ethics and Information Technology* 4(4): 287–304.

Floridi, L. 2006. Information ethics, its nature and scope. *SIGCAS Comput. Soc.* 36(3): 21–36. doi:10.1145/1195716.1195719.

Floridi, L. 2007. Understanding information ethics. *APA Newsletter on Philosophy and Computers* 7(1): 3–12.

Floridi, L. 2008. The method of levels of abstraction. *Minds and Machines* 18(3):303–329. doi:10.1007/s11023-008-9113-7.

Floridi, L. 2013. *Ethics of information*. [S.l.]: Oxford University Press.

Floridi, L. 2014. *The fourth revolution, how the infosphere is reshaping human reality*. Oxford: Oxford University Press.

Floridi, L., and J. Sanders. 2001. Artificial evil and the foundation of computer ethics. *Ethics and Information Technology* 3(1): 55–66.

Floridi, L., and J.W. Sanders. 2004. On the morality of artificial agents. *Minds and Machines* 14(3): 349–379. doi:10.1023/B:MIND.0000035461.63578.9d.

Gelven, M. 1994. *War and existence: A philosophical inquiry*. University Park: Pennsylvania State University Press.

Hepburn, R.W. 1984. *"Wonder" and other essays: Eight studies in aesthetics and neighbouring fields*. Edinburgh: University Press.

"ICRC Databases on International Humanitarian Law." 00:00:00.0. http://www.icrc.org/ihl/INTRO/470.

Libicki, M. 1996. *What is information warfare?* Washington, D.C.: National Defense University Press.

Perry, D. 2006. 'Repugnant Philosophy': Ethics, espionage, and covert action. In *Ethics of spying: A reader for the intelligence professional*, ed. J Goldman. Lanham: Scarecrow Press.

Schmitt, M.N. 1999. The principle of discrimination in 21st century warfare. SSRN Scholarly Paper ID 1600631. Rochester: Social Science Research Network. http://papers.ssrn.com/abstract=1600631.

Schwartau, W. 1994. *Information warfare: Chaos on the electronic superhighway.* 1st ed. New York/Emeryville: Thunder's Mouth Press /Distributed by Publishers Group West.

Shulman, M. R. 1999. Discrimination in the laws of information warfare. SSRN Scholarly Paper ID 1287181. Rochester: Social Science Research Network. http://papers.ssrn.com/abstract=1287181

Taddeo, M., and A. Vaccaro. 2011. Analyzing peer-to-peer technology using information ethics. *The Information Society* 27(2): 105–112. doi:10.1080/01972243.2011.548698.

Toffler, A., and A. Toffler. 1997. Foreword: The new intangibles. In *In Athena's camp preparing for conflict in the information age*, ed. Arquilla John and David F. Ronfeld. Santa Monica: Rand.

Turilli, M., A. Vaccaro, and M. Taddeo. 2011. Internet neutrality: Ethical issues in the internet environment. *Philosophy & Technology* 25(2): 133–151. doi:10.1007/s13347-011-0039-2.

Waltz, E. 1998. *Information warfare: Principles and operations.* Boston: Artech House.

Walzer, M. 2006. *Just and unjust wars: A moral argument with historical illustrations*, 4th ed. New York: Basic Books.

Withman, J. 2013. Is just war theory obsolete?". In *Routledge handbook of ethics and war: Just war theory in the 21st century*, ed. Allhoff Fritz, Nicholas G. Evans, and Henschke Adam, 23–34. New York: Routledge.

Mariarosaria Taddeo works at the Oxford Internet Institute, University of Oxford and Faculty Fellow at the Alan Turing Institute. Her recent work focuses mainly on the ethical analysis of cyber security practices and information conflicts. Her area of expertise is Information and Computer Ethics, although she has worked on issues concerning Philosophy of Information, Epistemology, and Philosophy of AI. She published several papers focusing on online trust, cyber security and cyber warfare and guest-edited a number of special issues of peer-reviewed international journals: Ethics and Information Technology, Knowledge, Technology and Policy, Philosophy & Technology. She also edited (with L. Floridi) a volume on 'The Ethics of Information Warfare' (Springer, 2014) and is currently writing a book on 'The Ethics of Cyber Conflicts' under contract for Routledge. Dr. Taddeo is the 2010 recipient of the Simon Award for Outstanding Research in Computing and Philosophy and of the 2013 World Technology Award for Ethics. She serves editor-in-chief of Minds & Machines, in the executive editorial board of Philosophy & Technology. Since 2016, Dr Taddeo is Global Future Council Fellow for the Council on the Future of Cybersecurity of the World Economic Forum.

Chapter 6
Regulating Cyber Operations Through International Law: In, Out or Against the Box?

Matthew Hoisington

Abstract A great deal has been written about the international legal regulation of cyber operations with respect to the use of force (*jus ad* bellum), in situations of armed conflict (*jus in bello*) and during times of peace. International lawyers have offered their expertise on how and in what ways international law should respond to the challenges and opportunities raised. The approaches have generally, although not exclusively, fallen into two separate categories. Either the underlying issues presented by cyber operations can be addressed by existing rules and international legal structures, or cyber operations present something so fundamentally new that a whole new set of rules and structures is required. A third approach suggests that existing structures may in fact be up to the challenge of cyber operations, but that in order to be effective, the discipline must reject those parts of itself that are incompatible with this new subject matter—in effect, going *against* existing doctrine in an effort to address the regulatory demands presented. Which of these three approaches—"in," "out" or "against" the box—is the right one? A critical examination of existing law, policy and practice in this area yields interesting, if not entirely satisfying, answers.

Keywords Cyber Operations • Jus ad bellum • Jus in bello • Use of force • Regulation • Innovation • Technology • Public international law

---

This article is written in my personal capacity and does not necessarily reflect the views of the United Nations.

M. Hoisington (✉)
United Nations, Office of Legal Affairs, Office of the Legal Counsel, New York, USA
e-mail: hoisington@gmail.com

© Springer International Publishing Switzerland 2017
M. Taddeo, L. Glorioso (eds.), *Ethics and Policies for Cyber Operations*, Philosophical Studies Series 124, DOI 10.1007/978-3-319-45300-2_6

6.1 Introduction

Technological advancements challenge legal rules in unique ways.[1] By their very nature, such advancements are often unforeseen, and the arc of their development is, in some cases, unknowable. Where a particular rule or set of rules seems anachronistic or otherwise ill-suited to integrate the technological advancement, the rules themselves, as well as the broader structures underlying them—for instance, systems of recognition, obligation, enforcement, and adjudication, among others—are called into doubt. The trial for legal regulation becomes balancing the competing prerogatives of stability and change. To the extent that existing rules can be applied or reinterpreted to cover the technological advancement, the system as a whole maintains its general shape. This can elucidate expectations among the subjects of the legal regulation. Unfortunately, it can also be misguided. Sometimes the existing rules and structures are not enough and the situation requires something fundamentally new. Evolution must be replaced by revolution.

The seemingly endless possibilities of cyberspace, both for governments and for their citizens, epitomize this challenge. What can conceivably occur through the use of cyber technologies continues to defy expectations. In some cases, the possibilities are invigorating, while others create significant apprehension and fear. The law must account for both ends of the spectrum, enabling innovation and progressive development, while also affording protections through prohibiting certain acts and punishing those who violate the applicable rules.

While domestic regulation covers much of the relevant ground, and governments are rightfully seized with promulgating necessary legislation,[2] the pervasive nature of cyberspace and the instantaneous character of cyber operations, as well as the link between such operations and issues of international peace and security and human rights, among other subjects of international concern, necessitates international regulation. Accordingly, a great deal has recently been written about how international law, in particular, may regulate cyber operations. These proposals have focused not

---

[1] For a review of the response of law to technological advances with particular reference to cyberspace *see generally* Joel Trachtman, *Cyberspace, Sovereignty, Jurisdiction and Modernism*, 5 IND. GLOBAL LEGAL STUD. 571 (1997–1998); David Friedman, *Does Technology Require New Law?* 25 HARV. J. L. & PUB. POL'Y 71 (2001–2002); Noel Cox, *The Relationship between Law, Government, Business and Technology*, 8 DUQ. BUS. L. J. 31 (2006); Daniel J. Gifford, *Law and Technology: Intersections and Relationships*, 8 MINN. J. L. SCI. & TECH. 571 (2007); Sen. Patrick Leahy, *New Laws for New Technologies: Current Issues Facing the Subcommittee on Technology and the Law*, 5. HARV. J. L. & TECH. 1 (1991–1992); Joel Reidenberg, *Lex Informatica: The Formation of Information Policy Rules through Technology*, 76 TEX. L. R. 554 (1997); Kieran Tranter, *Nomology, Ontology, and Phenomenology of Law and Technology*, 8 MINN. J. L. SCI. & TECH. 449 (2007).

[2] For instance, on the important issue of cybersecurity, for instance, the United States Congress has been considering legislation proposed by President Obama. *See generally* Jennifer Steinhauer, *House Passes Cybersecurity Bill After Companies Fall Victim to Data Breaches*, N.Y. TIMES ONLINE (April 22. 2015) *available at* http://www.nytimes.com/2015/04/23/us/politics/computer-attacks-spur-congress-to-act-on-cybersecurity-bill-years-in-making.html?_r=0 (last accessed May 4, 2015).

only on the rules that would apply during times of peace, but also on those applicable to the use of force by States and other actors (*jus ad bellum*), as well as those specialized rules that would apply during periods of armed conflict (*jus in bello*).[3]

Approaches have generally, although not exclusively, fallen into two categories. Either the underlying issues presented by cyber operations can be integrated into the existing system of rules and underlying structures of international law, or cyber operations present something so fundamentally new that a whole new set of rules and structures is required. A third approach, resisting the pull of either extreme, views existing structures as generally sufficient, but reasons that in order to be effective, the discipline must reject those parts of itself that are incompatible with this new subject matter—in effect, going *against* existing doctrine in an effort to address the new demands of cyber operations.

The viability of these three approaches remains contested. Each offers a way forward that may attract adherence from particular stakeholders and resistance from others. States with a relative comparative advantage with respect to cyber operations may prefer to maintain the *status quo*. For them, existing rules and structures, which leave many specific cyber-related questions unanswered, are preferred because they are free to act unencumbered by specific prohibitions. To the extent that the rules are unclear, the existing structure of international law also grants States a near monopoly of legislative power, and they will not see any incentive to abdicate this influential role. Proponents of a fundamentally different approach, which, among other things, may allow for a greater law-making role for private actors and individuals, will advocate for rules and structures that more realistically represent the power dynamics that actually exist. Such rules and structures would constitute a special legal regime, outside that which already exists, which would galvanize the private actors, such as large technology corporations, that exercise a unique role with respect to cyber operations, while also putting forth a specialized set of legal rules to regulate cyber operations. The third approach, which utilizes existing structures of international law to reject its own incompatible elements, would integrate both new rules and new actors, essentially reforming the rules and structures of international law from the inside out. Its supporters are those who would value the stability of existing international law, but who also recognize the necessity of epochal change.

---

[3] For an overview of the subject *see generally* Vida Antolin-Jenkins, *Defining the Parameters of Cyberwar Operations: Looking for Law in All the Wrong Places?* 51 Naval L. Rev. 132 (2005); Jason Barkham, *Information Warfare and International Law on the Use of Force*, 34 N.Y.U. J. Int'l L. & Pol'y 57 (2001–2002); Susan Brenner, *"At Light Speed": Attribution and Response to Cybercrime/Terrorism/Warfare*, 97 J. L. & Criminology 379 (2007); Yoram Dinstein, *Cyber War and International Law: Concluding Remarks at the 2012 Naval War College International Law Conference*, 89 Int'l L. Stud. 276 (2013); Herbert Lin, *Offensive Cyber Operations and the Use of Force*, 4 J. Nat. Sec. L. & Pol'y 63 (2010); Paul Rosenzweig, *International Law and Private Actors Cyber Defensive Measures*, 50 Stan. J. Int'l L. 103 (2014); Matthew Waxman, *Cyber Attacks and the Use of Force: Back to the Future of Article 2(4)*, 36 Yale J. Int'l L. 421 (2011); Michael Schmitt, *The Law of Cyberwarfare:* Quo Vadis? 25 Stan. L. & Pol'y Rev. 269 (2014); Oona Hathaway et al. *The Law of Cyber Attack*, 100 Cal. L. Rev. 817 (2012); Steven G. Bradbury, *The Developing Legal Framework of Defensive and Offensive Cyber Operations*, 2 Harv. Nat. Sec. J. (2011).

Which of these three approaches—"in," "out" or "against"[4] the box—is the right one? A critical examination of existing law, policy and practice in this area yields interesting, if not entirely satisfying, answers.

6.2 Discussion

6.2.1 Inside the Box

The view that existing international rules and structures are sufficient to account for the challenges posed by cyber operations emphasizes the general applicability of the relevant rules, irrespective of the means or methods used, as well as technological advances from the past, such as nuclear and chemical weapons, lasers, outer space exploration and similarly disruptive inventions, that, despite their unique characteristics, have nonetheless been to a large extent integrated into the existing rules.

The precise content of the rules and their relationship with new technologies has depended on the substantive context in which they are applied. For present purposes, it is instructive to focus on the *jus ad bellum* and the *jus in bello*.

The *jus ad bellum* regulates the recourse to the use of force by States. As reflected in the Charter of the United Nations (the "Charter"), it provides that States "shall refrain in their international relations from the threat or use of force against the territorial integrity or political independence of any state, or in any other manner inconsistent with the Purposes of the United Nations".[5] This general prohibition includes two main exceptions.[6] The first, reflected in Article 2, paragraph 7 and Chapter VII of the Charter, is that the United Nations Security Council may, upon its determination of the existence of "any threat to the peace, breach of the peace, or act of aggression…decide what measures shall be taken…to maintain international peace and security", including the use of force. The second, reflected in Article 51 of the Charter, is that all States enjoy an "inherent right of individual or collective self-defence if an armed attack occurs…".

The main challenges posed by cyber operations to the *jus ad bellum* concern the particular terminology that is used. For instance, Article 2, paragraph 4 of the Charter, which sets out the general prohibition on the threat or use of force by States in their international relations, refers to the "territorial integrity or political independence" of any State. Cyber operations, which often take place in virtual space and

---

[4] The consideration of "against" the box approaches are influenced considerably by the writing of David Kennedy. *See generally* David Kennedy, *When Renewal Repeats: Thinking against the Box*, 32. N.Y.U. J. Int'l Law & Pol. 335 (1999–2000). This is addressed in greater detail in Section II.C below.

[5] Charter of the United Nations, Article 2, paragraph 4.

[6] A third exception, related to humanitarian interventions, also enjoys significant support. To date, however, it remains *de lege ferenda*.

may not have the physical impact of traditional, kinetic attacks, are difficult to reconcile with "territorial integrity" and "political independence" as those terms have traditionally been understood.

With respect to the exercise of self-defense by States, Article 51 includes two notable qualifiers. The first is that States shall enjoy an "inherent" right of self-defense and the second is the right shall apply where "an armed attack occurs". The use of the term "inherent" has been interpreted to broaden the right to include situations where an attack is imminent, such that the defensive action itself can be exercised pre-emptively.[7] Such an interpretation is seemingly at odds with the plain language of Article 51, which requires the actual occurrence of an armed attack. It has nonetheless gained widespread, if not universal, acceptance.

In cases where an intervention or other type of attack does in fact occur, the requirement for the lawful exercise of self-defense in accordance with Article 51 is that it rises to the level of an "armed attack". The term has no settled definition; however, it has been interpreted to require certain levels of intensity, scale and effects[8] characteristic of traditional uses of military force. Where cyber operations are directed at the destruction or disruption of data, or otherwise take aim at virtual targets, it remains unclear whether they could constitute armed attacks. This is despite the fact that such operations may have an impact, broadly defined, that exceeds the destruction caused by traditional, kinetic attacks, including the use of rockets or other weapons.

The response to these challenges from the advocates of the first technique has been, to the extent possible, to reinterpret existing terms to account for cyber operations. Various "effects" tests for cyber operations have been proposed. The influential Tallinn Manual on the International Law Applicable to Cyber Warfare (the "Tallinn Manual"), developed by the North American Treaty Organization (NATO) Cooperative Cyber Defence Centre of Excellence (CCDCOE), adopted the following straightforward rule on the exercise of self-defense against an armed attack:

> A State that is the target of a cyber operation that rises to the level of an armed attack may exercise the inherent right to self-defence. Whether a cyber operation rises to the level of an armed attack depends on its scale and effects.

The Tallinn Manual then addresses the factors that would be considered in determining whether the particular "scale and effects" of a cyber operation would rise to

---

[7] These arguments are premised largely on the formulation of self-defense put forth by Daniel Webster with respect to the *Caroline* incident in 1837, where Webster justified the pre-emptive use of force in self-defense not only in response to attacks but also in cases where the threat of attack was "…instant and overwhelming, leaving no choice of means, and no moment for deliberation".

[8] For the relevant pronouncements of the International Court of Justice *see* Military and Paramilitary Activities in und against Nicaragua (Nicaragua v. United States of America). Merits, Judgment. I.C.J. Reports 1986, p. 14; Oil Platforms (Islamic Republic of Iran v. United States of America), Judgment, I. C. J. Reports 2003, p. 161; Armed Activities on the Territory of the Congo (Democratic Republic of the Congo v. Uganda), Judgment, I.C.J. Reports 2005, p. 168. *See also* Eritrea-Ethiopia Claims Commission, Partial Award, *Jus ad bellum*, Ethiopia's Claims 1–8 (December 19, 2005).

the level of an armed attack. While noting that opinions on the matter remain divided, it states, as a general matter, that:

> [T]he parameters of the scale and effects criteria remain unsettled beyond the indication that they need to be grave. That said, some cases are clear…any use of force that injures or kills persons or damages or destroys property would satisfy the scale and effects requirement…acts of cyber intelligence gathering and cyber theft, as well as cyber operations that involve brief or periodic interruption of non-essential cyber services, do not qualify as armed attacks.

Between these two ends of the spectrum lies a great unresolved middle, where a large percentage of the most pernicious cyber operations may occur. The lack of clarity benefits States with the strongest cyber capacities because it creates room for doubt, deliberation and argument. In essence, the determination of whether a particular cyber operation rises to the level of an armed attack rests with the targeted State. Any exercise of self-defense would also require attribution. To date, many victims of have been hesitant to declare the occurrence of an attack out of apprehension of revealing vulnerabilities. Attacking States and their proxies have also remained largely silent on their capabilities and practice. The result is widespread speculation and supposition. The Security Council represents the only international deliberative body with the authority to pronounce on the occurrence of an armed attack (or a threat thereof)—but it too is subject to power politics. The United States, China and Russia, three States with the most advanced cyber capabilities and the most extensive repositories of practice, exercise veto power on the Council, and they may see little benefit to airing sensitive or strategic information before the international community as a whole.

Below the threshold of an armed attack, lesser uses of force, assuming they are wrongful, may also give rise to the right of victim States to exercise certain countermeasures. Such countermeasures are subject to the criteria of proportionality. They must also be limited to procuring cessation of the wrongful act and achieving reparation for the injury suffered. The mainstream view further holds that any countermeasures exercised by victim States must be non-forcible, temporary and reversible.[9]

The use of countermeasures in response to cyber operations presents similar interpretive problems to the use of force in self-defense to an armed attack. While the definition of the use of force, generally, is more expansive than armed attack, the test used for determining its existence would still rely largely on the physical effects caused. Forcible interventions against a State's cyber infrastructure or data networks may constitute uses of force, or they may not. Once again, the line between an attack and an act of espionage or mere interference is either muddled or disappears completely. Proponents of so-called "active defense" in response to cyber attacks, which may be triggered automatically by forcible intrusions, have justified such

---

[9] *See* International Law Commission, Draft Articles on Responsibility of States for Internationally Wrongful Acts (2001) (A/RES/56/83, annex).

actions under the rubric of countermeasures,[10] but in order to be effective, active defense may itself require the use of force. This situation likely constitutes a form of self-defense *lite*[11] rather than an exercise of countermeasures, the legality of which is subject to dispute.

The *jus in bello*, applicable during periods of armed conflict, is more amenable to the integration of cyber operations than the *jus ad bellum*; however, significant challenges persist. This area of law relies largely on the Hague Conventions of 1907 and the Geneva Conventions of 1949, as well as their additional protocols. In addition, a significant corpus of customary international law supports, reinforces and strengthens the codified rules.[12]

While the *jus in bello* includes many specific instructions on the conduct of hostilities, its fundamental precepts are that parties to a conflict must distinguish between civilians and combatants at all times and that unnecessary damage or suffering must be avoided. The principle of distinction, whereby attacks may only be directed against combatants, and may not be directed against civilians, is operationalized through the obligations of (i) proportionality; (i) precaution and (iii) military necessity.

Proportionality requires that any damage caused to civilians, including loss of life, injury or damage to civilian objects, must not be excessive in relation to the concrete and direct military advantage anticipated. Precaution requires that all feasible measures must be taken to avoid and/or minimize harm to civilians. Finally, military necessity limits permissible measures to those that are necessary to accomplish a legitimate military purpose. Similar to the principle of proportionality, it is premised on the rationale that the purpose of attacks must be limited to weakening the military capacity of the other parties to the conflict.

The *jus in bello* also prohibits certain means and methods of warfare. The general rule is that weapons that may cause superfluous injury or unnecessary suffering are prohibited. On this basis, poisonous gas, chemical, biological and nuclear weapons,[13] exploding and expanding bullets and lasers are generally outlawed. Specific arms control agreements have also been concluded to regulate the stockpile and use of such weapons.

To the extent that cyber operations are used by a party to a conflict to achieve a military objective, the *jus in bello* would apply. While conceptually clear, the appli-

---

[10] For an excellent overview of the arguments both for and against justifying active defense as countermeasures *see generally* Oona Hathaway, *The Drawbacks and Dangers of Active Defense*, 6th International Conference on Cyber Conflict, NATO CCDOE (2014) *available at* https://ccdcoe.org/cycon/2014/proceedings/d2r1s5_hathaway.pdf.

[11] *See* Oil Platforms (Islamic Republic of Iran v. United States of America), I. C. J. Reports 2003, p. 324 (Separate Op. J. Simma).

[12] *See generally* JEAN-MARIE HENCKAERTS & LOUISE DOSWALD-BECK, CUSTOMARY INTERNATIONAL HUMANITARIAN LAW, VOLUME I: RULES (2005).

[13] While not illegal *per se*, the International Court of Justice has found that "the threat or use of nuclear weapons would generally be contrary to the rules of international law applicable in armed conflict, and in particular the principles and rules of humanitarian law". Legality of the Threat or Use of Nuclear Weapons, Advisory Opinion, I.C.J. Reports 1996, p. 226.

cation of the *jus in bello* to the actual use of cyber weapons as well as cyber targets presents considerable challenges.

For instance, the interconnected nature of cyber infrastructure, and the sharing of networks among public and private actors, raises the difficulty in distinguishing between civilian and military targets. The effects of cyber attacks may also be uniquely difficult to control. Accordingly, it may not be possible to satisfy the principle of proportionality with any reasonable degree of certainty. Determining participation in hostilities also becomes difficult. If a civilian launches a cyber attack far from the territory where the armed conflict occurs, he or she may arguably be subject to lawful targeting by the affected party, a result which would, among other things, undermine the usual time, space and targeting confines of the *jus in bello*.

These challenges suggest that the use of cyber weapons may not be permitted under many, if not most, circumstances where the *jus in bello* applies; however, that result also seems incongruent with the principles upon which the body of rules is founded. If a military target could be neutralized without the firing of a single munition or the destruction of any physical property, would this not represent, on the whole, a more favourable outcome, even if the effects of such an operation could not be conclusively controlled prior to the act? The existing rules and structures of the *jus in bello*, as applied to the current state of the art in cyber weapons, struggle to yield satisfactory answers to such questions.

6.2.2 Outside the Box

The proponents of the second technique see many of the same challenges as the proponents of the first. In their prescriptions, however, they manifest a much different approach. Rather than adapting their arguments to existing rules and structures, they seek to break new ground by establishing an entirely new system.

As noted in the introduction, the second technique presumes that cyber operations present something so fundamentally new that an entirely new set of international legal rules and structures is required. This presumption recognizes the seeming incompatibility of the existing *jus ad bellum* and the *jus in bello*, which were established mainly if not exclusively by States, with the actual power dynamics of cyberspace, as well as the governance and regulatory arrangements that apply to peaceful uses of the technology.

States are not the only, nor in most cases the most powerful, stakeholders in cyberspace. In fact, cyberspace is effectively "ungoverned" in the formal sense; it exists outside the regulatory control of any State or group of States. The space is ruled by a complex interaction of public and private actors. Efforts to bring cyberspace under the auspices of international control have been met with fierce opposition. For the time being, at least, it appears that changes to the current governance arrangement are unlikely to emerge as viable options.

The ungoverned, or, more appropriately, alternatively governed, nature of cyberspace, presents distinct challenges and opportunities. National governments and powerful software corporations have leveraged their respective technological advantages to exercise a certain amount of control over the space. Because of this unique mix, a great deal of informal governance and order already exists. Private actors lack the formal authority of the State but they nonetheless serve as checks on illicit activities. The informal form of government is reminiscent of the *lex mercatoria*, or private economic law, used by European merchants during the medieval period; it is driven less by formal categories than by factual exercises of power by whoever participates in the space.[14]

In accordance with an "outside the box" approach, the elaboration of a new international convention or novel set of rules specifically addressing cyber operations achieves a measure of the desired regulatory specificity. Various proposals along these lines have been put forward.[15] Such an approach may result in rules of sufficient content; however, a comprehensive solution would also require structural changes to the international legal system itself, specifically on the matter of international legal subjectivity. Actors currently outside the system, such as corporations and other private actors that exercise *de facto* authority in cyberspace, must be brought under its auspices. This would require the affirmative support of both States and the private actors themselves. Aligning the interests of the various actors and incentivizing their participation in the system represents a significant obstacle to the success of the approach.

Any separate arrangement must also reconcile itself with the existing rules. Even if such rules represent a point of departure, the fact remains that the subjects of a cyber-specific arrangement will, to a large extent, also have certain obligations under the existing regime. New rules on the use of force or the conduct of armed hostilities, for instance, must interact with the existing *jus ad bellum* and *jus in bello*. Conflicts may arise, particularly where obligations overlap or intersect. Expanding the range of actors that may legitimately exercise force beyond State may also create substantial apprehension.

The issue of subjects' obligations *inter se* under such an arrangement presents further challenges. As presently constituted, the international legal system is horizontally integrated in the sense that States (and other formal subject of international law, such as international and regional organizations) owe obligations to one

---

[14] *See generally* Gunther Teubner, *Societal Constitutionalism: Alternatives to State-centered Constitutional Theory*, in Christian Joerges, Inger-Johanne Sand & Gunther Teubner (eds.), Constitutionalism and Transnational Governance (2004) 3; *see also* Reidenberg, *supra* note 2, at 554.

[15] For a recent example *see e.g.* Letter dated 9 January 2015 from the Permanent Representatives of China, Kazakhstan, Kyrgyzstan, the Russian Federation, Tajikistan and Uzbekistan to the United Nations addressed to the Secretary-General of the United Nations (A/69/723) (proposing an international code of conduct for information security for adoption by United Nations General Assembly). *See also* Davis Brown, *Proposal for an International Convention to Regulate the Use of Information Systems in Armed Conflict*, 47 Harv. J. Int'l L. 179 (2006).

another. It is vertically integrated in the sense that States owe certain obligations to citizens and other private actors under their jurisdiction or otherwise subject to their control. The principle of sovereignty separates international from domestic obligations. An arrangement open to private actors, which infuses them with international legal personality, resets the current vertical and horizontal dynamics, casting sovereignty in an entirely different light.

6.2.3 Against the Box

Going against the box with respect to the international regulation of cyber operations requires a revision of existing rules and structures. In effect, the international legal system must reject those parts of itself that are incompatible with the challenges presented. The approach utilizes existing structures to integrate both new rules and new actors, essentially reforming international law from the inside out.

With respect to the *jus ad bellum*, this would necessitate an amendment to the applicable regime, namely the Charter. The mechanisms for amending the Charter involve massive political mobilization. Article 108 of the Charter requires amendments to be adopted by a vote of two thirds of the General Assembly and ratified domestically by two thirds of the Members of the United Nations, including all of the permanent members of the Security Council. Accordingly, the United Kingdom, the United States, France, Russia and China exercise a veto over the amendment process. On rare occasions, the Charter has been amended, including with respect to the composition of the Council; however, any amendment to the fundamental provisions contained in Article 2, paragraphs 4 and 7, or Chapter VII of the Charter appears, in the present circumstances, to be extremely unlikely.

Assuming such an amendment were possible, the rules on the use of force by States could be revised to address the gaps mentioned above, such as the application of the right of self-defense in response to a range of attacks. Rather than relying on challenging—not to mention anachronistic—interpretations of the scale and effects test with reference to the armed attack requirement, new measures for determining the impact of particular operations could be developed. In essence, the existing prohibitions on the use of force contained in Article 2, paragraph 4 of the Charter, as well as the preconditions for the exercise of self-defense in Article 51, would be redrafted in their entirely. The prevalence of cyber operations in military arsenals, and the willingness of powerful States to use them, would, under the third approach, lead to a wholesale revision of the existing *jus ad bellum*.

In the context of the *jus in bello* a similar epochal shift would be required. Supposing that States would look for ways to include cyber weapons in their arsenals and regulate, rather than prohibit, their use, a shift along these lines would likely entail the elaboration of a new additional protocol to the Geneva Conventions. Such a protocol would specify the application of the principles of distinction, proportionality, precaution, and military necessity to cyber operations, determining, for instance, what level of certainty would be required to affirmatively meet the proportionality test.

As Kennedy has written, new thinking that truly goes against the box must not merely reinforce existing biases and blindspots.[16] Otherwise what presents itself as new or innovative actually repeats the limitations that served to restrict such thinking in the first place. The challenge is to innovate without falling back into the established disciplinary discourse or going outside the field.[17]

Accordingly, an important element of the third approach is the development of a new vocabulary—in essence, a new way of describing the international regulation of cyber operations—that avoids the kinds of logical stalemates of the existing regime. One way to accomplish such an objective is to expand the range of actors invited to participate in the elaboration of the rules. The drivers of development in the cyber field, such as programmers, military leaders and other technical experts, could collaborate with international lawyers and policy-makers to derive accurate and enforceable rules.

Similar to the second approach, thinking against the box in the cyber context also necessitates new rules on international legal subjectivity. Once again, actors currently outside the system, such as corporations and other private actors, must be brought under its auspices. In order to avoid the renewal-as-repetition trap, the third approach must also establish new methods for revision and amendment. The stability of the system must be balanced against institutional mechanisms that facilitate, and encourage, change.

6.3 Analysis

The approach of integrating cyber operations into existing international legal rules and structures has clear intuitive appeal. The rules on the *jus ad bellum* and *jus in bello* have developed over the course of human experience to entrench the fundamental prohibitions on (i) the use of force in international relations, save for certain emergency situations; and (ii) the targeting of civilians or the causation of unnecessary harm during the conduct of such hostilities. The rules have been interpreted, to varying levels of success, to account for prior evolutions in weapons and capabilities. Accordingly, only a major development—something that may, for instance, alter the way in which humans lives their lives—could justify an elemental departure from the existing regime.

Cyber operations may constitute such a development. The truth is that the outer reaches of the technology—where it might evolve in the next 20, 30 or 100 years—remain the subject of conjecture and, at best, informed speculation. At any rate, it is abundantly clear that the existing rules do not fully address the possibilities. The insistence on integrating cyber operations into the settled *jus ad bellum* and *jus in bello* leaves significant gaps that no creative reasoning or interpretation can fill completely. The result is a zone of doubt that can only be addressed through practice and contestation. A result that may lead to significant suffering and damage before any single interpretation gains consensus adherence.

---

[16] *See generally* Kennedy, *supra* note 5.
[17] *Id.*

The approach of creating a stand-alone regime applicable to cyber operations also has some practical appeal. For instance, it could be drafted to specifically address the aspects of cyber operations, such as the test for triggering the right to self-defense or the requirement of proportionality, that are unaddressed by the present *jus ad bellum* and *jus in bello*. This alone would bring the applicable international rules closer to reality.

The challenge becomes reconciling a new regime with the body of rules and underlying structures that already exist. Because the second approach would leave the present rules intact, it would, in essence, fragment the applicable regimes. Moreover, the requirement of reformatting the structure of international legal subjectivity to introduce new, informal, actors represents a significant operational hurdle.

Going against the box would also provide a number of benefits. Rather than requiring the establishment of a new regime, it would use the existing structures to revise existing rules, thus constituting a revitalized and cohesive whole. It would also avoid the limitations of the past by incorporating new law-making structures and methods for revision and amendment. Without the requisite political will, however, the approach stands minimal chances of success. The third approach must not only displace the established rules but also integrate new actors, all while developing a new vocabulary to describe cyber operations within the international legal discourse. Collectively, these challenges are daunting, to say the least.

6.4 Conclusion

The foregoing present various approaches for the international legal regulation of cyber operations. The absence of a clean and comprehensive solution highlights the challenges raised by the issue. While it can be counterproductive to stare at an obstacle and marvel at its sheer immovability, cyber operations are, in fact, profound in their subversion of the international legal system. Until the capabilities of cyber operations are better known, the effort to regulate them through international law will remain a significant work-in-progress. Rather than abdicating their role, however, international lawyers would be well-served to remain vigilant and attuned to new developments. The time will come when the issue ripens for international regulation and at that moment the accumulated experience of testing the various approaches will prove invaluable.

Matthew Hoisington BA (Dartmouth College), JD (Boston College), LLM (the Fletcher School of Law and Diplomacy at Tufts University), is currently an associate legal officer in the United Nations Office of Legal Affairs, Office of the Legal Counsel, where he focuses on issues of peacekeeping, counterterrorism, sanctions, privileges and immunities and international accountability. From 2011 to 2012, he was a contributor to *Explorations in Cyber International Relations* as part of Project Minerva at MIT/Harvard. He has published numerous articles on the international regulation of the use of force in cyberspace, including 'Cyberwarfare and the Use of Force Giving Rise to the Right of Self-Defense' and 'International Law and Ungoverned Space'.

Chapter 7
Military Objectives in Cyber Warfare

Marco Roscini

Abstract This Chapter discusses the possible problems arising from the application of the principle of distinction under the law of armed conflict to cyber attacks. It first identifies when cyber attacks qualify as 'attacks' under the law of armed conflict and then examines the two elements of the definition of 'military objective' contained in Article 52(2) of the 1977 Protocol I additional to the 1949 Geneva Conventions on the Protection of Victims of War. The Chapter concludes that this definition is flexible enough to apply in the cyber context without significant problems and that none of the challenges that characterize cyber attacks hinders the application of the principle of distinction.

Keywords Cyber attacks • Cyber operations • International law • Law of armed conflict • Military objective • Principle of distinction

7.1 The Principle of Distinction in the Law of Armed Conflict

Article 48 of the 1977 Protocol I additional to the 1949 Geneva Conventions on the Protection of Victims of War provides for the 'basic' obligation of the belligerents to 'at all times distinguish between the civilian population and combatants and between civilian objects and military objectives and accordingly [to] direct their operations only against military objectives'. This obligation of distinction applies both in international and non-international armed conflicts (Henckaerts and Doswald-Beck 2005 (vol. I): 3) and its customary status has been firmly upheld by national and international courts. In particular, according to the 1996 International Court of Justice (ICJ)'s Advisory Opinion on the *Legality of the threat or use of nuclear weapons*, the obligation to protect the civilian population and civilian objects is a 'cardinal' principle of international humanitarian law (*Legality of the*

This Chapter is largely based, with some amendments and updates, on the author's book, Roscini 2014a: 176–192.

M. Roscini (✉)
International Law, Westminster Law School, London, UK
e-mail: M.Roscini@westminster.ac.uk

Treat or Use of Nuclear Weapons, Advisory Opinion, 1996, para 78). The International Criminal Tribunal for the former Yugoslavia (ICTY) confirmed that 'it is now a universally recognised principle...that deliberate attacks on civilians or civilian objects are absolutely prohibited by international humanitarian law' (*Prosecutor v Kupreškić* 2000, para 521).

While both Article 48 and Article 51(1) of Additional Protocol I refer to the notion of 'military operations', which includes 'all movements and acts related to hostilities that are undertaken by armed forces' (Sandoz et al. 1987: para 1875),[1] the Protocol's Commentary makes clear that the application of the principle of distinction is limited 'to military operations during which violence is used', i.e. 'attacks' (Sandoz et al. 1987: para 1875). This is confirmed by the language of Articles 51(2) and 52(1) of Additional Protocol I, according to which civilians, the civilian population and civilian objects 'shall not be the object of *attack*' (emphasis added). With regard to other military operations, only a more general obligation of 'constant care ... to spare the civilian population, civilians and civilian objects' applies (Additional Protocol I, 1977, Article 57(1)). Rules 1 and 7 of the International Committee of the Red Cross (ICRC)'s Study on Customary International Humanitarian Law, which incorporate the principle of distinction, also refer to 'attacks', and not to 'military operations' (Henckaerts and Doswald-Beck 2005 (vol. I): 3 and 25).

Whenever cyber operations conducted during an armed conflict and having a nexus with it amount to 'attacks', then, they are subject to the principle of distinction and may only be directed against military objectives. Before discussing what a military objective is in the cyber context, however, we need to examine when cyber operations qualify as 'attacks' under the law of armed conflict.

7.2 Cyber Operations Qualifying as 'Attacks' Under the Law of Armed Conflict

'Attacks' are defined in Article 49(1) of Additional Protocol I as 'acts of violence against the adversary, whether in offence or in defence'. In other words, attacks under the law of armed conflict are only those acts of hostilities characterized by 'violence': unlike other acts of hostilities, like military espionage, non-violent military harm is not sufficient. It is not the author, the target or the intention that define an 'act of violence'. Rather, a cyber operation amounts to an 'attack' in the sense of Article 49(1) of Additional Protocol I when it employs cyber means or methods of warfare that result or are reasonably likely to result in violent effects (Roscini 2014a: 179). If a cyber operation causes or is likely to cause loss of life or injury to

---

[1] *The Commentary* subsequently rephrases the definition as 'any movements, manœvres and other activities whatsoever carried out by the armed forces with a view to combat' (Sandoz et al. 1987: para 2191). Inconsistently, the Commentary of Art 13 of Additional Protocol II defines 'military operations' more narrowly as 'movements of attack or defence by the armed forces in action' (ibid., para 4769).

persons or more than minimal material damage to property, then, it is an 'attack' and the law of targeting fully applies, including the principle of distinction.[2] Had it been conducted in the context of an armed conflict between Iran and those states allegedly responsible for the cyber operation, for instance, Stuxnet would have been an example of such an 'attack' because of the damage it caused to the centrifuges of Iran's Natanz uranium enrichment facility. Similarly, the cyber attack that allegedly damaged a steel mill in Germany (Zetter 2015) would have qualified as an 'attack' under Article 49(1) if it had had a belligerent nexus with an armed conflict. The relevant violent effects of a cyber attack include 'any reasonably foreseeable consequential damage, destruction, injury, or death', whether or not the computer system is damaged or data corrupted (*Tallinn Manual* 2013: 107). If the attack is intercepted and the reasonably expected violent effects do not occur, or occur to a lesser degree, the operation would still qualify as an 'attack' for the purposes of Article 49(1) (*Tallinn Manual* 2013: 109–110).

There is disagreement, however, on whether cyber operations that merely disrupt the functionality of infrastructures without causing material damage also amount to 'attacks' in the sense of Article 49(1) of Additional Protocol I. Rule 30 of the *Tallinn Manual on the International Law Applicable to Cyber Warfare* appears to rule this out. The majority of the experts that drafted the Manual maintained that disruptive cyber operations may be 'attacks' only 'if restoration of functionality requires replacement of physical components' (*Tallinn Manual* 2013: 108). The problem with this view, which still relies on the occurrence of some physical damage, is that the attacker may not be able to know in advance whether the restoration of functionality will require replacement of physical components or mere reinstallation of the operating system: the attacker could claim, therefore, that it was not aware that it was conducting an 'attack' and thus avoid the application of the law of targeting.

The limits of the doctrine of kinetic equivalence, which requires the occurrence of physical consequences for a cyber operation to be an 'attack', become evident if one considers that, under the Tallinn Manual's approach, a cyber attack that shuts down the national grid or erases the data of the entire banking system of a state would not be an 'attack', while the physical destruction of one server would. Some commentators have therefore tried to extend the notion of 'attack' to include at least some disruptive cyber operations. Dörmann, for instance, recalls that the definition of 'military objective' in Article 52(2) of Additional Protocol I mentions not only destruction but also 'neutralization' of the object and concludes that, when the object (person or property) is civilian, '[i]t is irrelevant whether [it] is disabled through destruction or in any other way.' (Dörmann 2004). Therefore, the incapacitation of an object, like a civilian power station, without destroying it would still qualify as an 'attack'. Melzer adopts a different approach to reach the same conclusion and argues that the principles of distinction, proportionality and

---

[2] Rule 30 of the *Tallinn Manual* for instance, defines a cyber attack as 'a cyber operation, whether offensive or defensive, that is reasonably expected to cause injury or death to persons or damage or destruction to objects' (*Tallinn Manual* 2013: 106). *The Manual* includes 'serious illness and severe mental suffering' in the notion of 'injury' (*Tallinn Manual* 2013: 108).

precautions apply not to 'attacks', but rather to the broader notion of 'hostilities': therefore, 'the applicability of the restraints imposed by IHL [international humanitarian law] on the conduct of hostilities to cyber operations depends not on whether the operations in question qualify as "attacks" (that is, the predominant form of conducting hostilities), but on whether they constitute part of the "hostilities" within the meaning of IHL' (Melzer 2011: 11). According to this view, cyber operations disrupting the enemy radar system would not amount to 'attack' because of the lack of violent consequences, but, as an act of hostilities, they would still be subject to the restrictions on the choice and use of methods and means of warfare (Melzer 2011). This position, however, is inconsistent with the above mentioned prevailing view according to which the rules contained in Part IV, Section I of Additional Protocol I essentially apply to 'attacks' and not to 'hostilities' or 'military operations'.

It is submitted that a better way of including at least certain disruptive cyber operations in the definition of 'attack' under Article 49(1) of Additional Protocol I is to interpret the provision in an evolutionary way taking into account the recent technological developments and to expand the concept of 'violence' to include not only material damage to objects, but also severe incapacitation of physical infrastructures without destruction.[3] This is suggested by Panama in its views on cyber security submitted to the UN Secretary-General, where it qualifies cyber operations as a 'new form of violence' (UN Doc A/57/166/add.1, 2002: 5). Indeed, the dependency of modern societies on computers, computer systems and networks has made it possible to cause significant harm through non-destructive means: cyber technologies can produce results comparable to those of kinetic weapons without the need for physical damage. After all, if the use of graphite bombs, which spread a cloud of extremely fine carbon filaments over electrical components, thus causing a short-circuit and a disruption of the electrical supply, would undoubtedly be considered an 'attack' even though it does not cause more than nominal physical damage to the infrastructure, one cannot see why the same conclusion should not apply to the use of viruses and other malware that achieve the same effect. It is, however, only those cyber operations that go beyond transient effects and mere inconvenience and cause significant functional harm to infrastructures that can qualify as 'attacks' in the sense of Article 49(1) of Additional Protocol I During the crisis between Ukraine and Russia over Crimea, a limited disruption of Ukrainian mobile communications through Distributed Denial of Service (DDoS) attacks and the defacement of certain state-run news websites and social media (the content of which was replaced with pro-Russian propaganda) were reported: because of their limited disruptive effects, such operations would not be 'attacks' under Article 49(1).[4] Only the less stringent

---

[3] On evolutionary interpretation in the cyber context, see Roscini 2014a: 20–24, 280–281.

[4] Roscini 2014b. Denial of service (DoS) attacks, of which 'flood attacks' are an example, aim to inundate the targeted system with excessive calls, messages, enquiries or requests in order to overload it and force its shut down. Permanent DoS attacks are particularly serious attacks that damage the system and cause its replacement or reinstallation of hardware. When the DoS attack is carried out by a large number of computers organized in botnets, it is referred to as a 'distributed denial of service' (DDoS) attack.

obligation of 'constant care … to spare the civilian population, civilians and civilian objects' (Additional Protocol I, 1977, Article 57(1)) would apply to such cyber operations short of attack, but not the prohibition to conduct them against civilians and civilian objects.

7.3 The Definition of 'Military Objective'

The principle of distinction requires that cyber operations conducted during an armed conflict by the belligerents against each other and amounting to 'attacks' be directed only against military objectives.[5] The first definition of 'military objective' to appear in a legal text can be found in the 1923 Hague Rules on Air Warfare: 'an object of which the destruction or injury would constitute a distinct military advantage to the belligerent' (Article 24(1)).[6] To clarify the definition, the Rules provided an illustrative list of military objectives (Article 24(2)).[7] The Rules, however, have never been adopted in treaty form. No definition appears in the 1949 Geneva Conventions, although the term is employed.[8] On the other hand, 'military objectives' are expressly defined in Article 52(2) of the 1977 Additional Protocol I as

> those objects which by their nature, location, purpose or use make an effective contribution to military action and whose total and partial destruction, capture or neutralization, in the circumstances ruling at the time, offers a definite military advantage.

This definition has been incorporated into several military manuals and, in spite of the unclear position of certain states like the United States (which will be discussed below), it is largely thought to reflect customary international law (Henckaerts and Doswald-Beck 2005 (vol. I): 30). The definition is also applicable in non-international armed conflicts (Henckaerts and Doswald-Beck 2005 (vol. I): 29).

---

[5] For a discussion of the application of the principle of distinction to the targeting of individuals (as opposed to objects), see Roscini 2014a: 192–215.

[6] Without referring to the notion of military objective, Art 2 of the 1907 Hague Convention IX Concerning Bombardment by Naval Forces in Time of War contains a list of objects that can be destroyed.

[7] The list includes military forces; military works; military establishments or depots; factories constituting important and well-known centres engaged in the manufacture of arms, ammunition or distinctively military supplies; lines of communication or transportation used for military purposes. It is doubtful whether the list is exhaustive (Rogers 2004: 60).

[8] See Arts 4, 19(2) of Geneva Convention I and Arts 4, 18(5) of Geneva Convention IV. The 1956 New Delhi Draft Rules for the Limitation of the Dangers Incurred by the Civilian Population in Time of War, drafted by the ICRC, proposed a list of military objectives, to be reviewed at intervals of no more than 10 years by a group of experts; however, even if an object had belonged to one of the listed categories, it would not have been a military objective if its total or partial destruction, in the circumstances ruling at the time, had offered no military advantage (Art 7). Another attempt to define the concept of 'military objective' was made by the Institute of International Law in 1969 (Annuaire de l'Institut de droit international (1969–II).359).

If one applies the above definition of 'military objective' to targeting in the cyber context, however, some interpretative problems arise. First of all, what 'objects' are relevant? Cyber operations can be directed at cyber targets, i.e. data, software, or networks, and/or hard targets, i.e. information hardware (e.g. computers, servers, routers, fibre-optic cables, and satellites), physical infrastructures, or persons.[9] When the cyber operation aims to cause material damage to physical property or persons or incapacitation of infrastructures, or such effects are foreseeable, the attacked 'object' is not only, and not mainly, the information itself, but rather the persons, property or infrastructure attacked *through* cyberspace (*Tallinn Manual* 2013: 108). In the case of Stuxnet, for instance, the relevant 'object' was not the Siemens software that operated the centrifuges at the Natanz uranium enrichment facility in Iran, but the centrifuges themselves. Similarly, in the previously mentioned cyber attack against a steel mill in Germany, the object of the attack was the mill, not its operating system. Commentators have debated whether data are per se an 'object' for the purpose of Article 52(2) of Additional Protocol I (Schmitt 2011 and Lubell 2013). The Experts that drafted the Tallinn Manual did not manage to achieve consensus on this point so no solution was incorporated in the black-letter rules (*Tallinn Manual* 2013: 127). The problem should not be overestimated. As already said, if the cyber operation deletes, corrupts or alters data in order to cause damage to or disrupt the functioning of an infrastructure, it is such infrastructure that is the intended 'object' of the attack. Similarly, if the cyber operation deletes or alters medical records, so that patients receive the wrong treatment, it is those individuals that are (also) targeted. If, on the other hand, the cyber operation only results in the corruption, deletion, or alteration of data without destructive or disruptive consequences on physical infrastructures, it will not be an 'attack' in the sense discussed above, and the law of targeting and the notion of 'military objective' will therefore not apply, whether or not the data are an 'object'. Cyber attacks not qualifying as 'attacks' under the law of armed conflict will only be subject to the general duty of 'constant care' when conducting military operations provided in Article 57(1) of Additional Protocol I and to the rules providing for special protection if applicable, such as those on cultural property for certain digital art and on the protection of diplomatic archives and correspondence.

7.3.1 'Effective Contribution to Military Action'

According to Article 52(2) of Additional Protocol I, two cumulative elements must be present for an 'object' to be a military objective and therefore targetable: it must effectively contribute to military action *and* its total or partial destruction, capture or neutralization, in the circumstances ruling at the time, must offer a

---

[9] Hard targets can be attacked both by kinetic or cyber means, while software and data can be attacked only by cyber means (Rauscher and Korotkov 2011: 19).

definite military advantage. Article 52(2) indicates the criteria to evaluate whether the object effectively contributes to military action, i.e. nature, location, purpose, or use.[10] Effective contribution to military action by nature characterizes those objects which are inherently military and cannot be employed but for military purposes, for instance computers designed specifically to be used as components of weapon systems or to facilitate logistic operations (Dinstein 2013).[11] Other examples include military command, communication, and control networks used for the transmission of orders or tactical data and military air defence networks.[12] The premises from where the military cyber operations are conducted (such as USCYBERCOM headquarters at Fort Mead or the 12-storey building in the Pudong New Area of Shanghai which is allegedly the home of the People's Liberation Army's Unit 61398) (Mandiant 2013) are also military objectives by nature (Turns 2012). An example of effective contribution by use would be a server normally used for civilian purposes which is taken over by the military, even if it is used for non-combat purposes (Dinstein 2013). If the server is about to be used by the military but this has not occurred yet, it may be a military objective by purpose.[13] As to military objectives by location, the Commentary to Rule 38 of the Tallinn Manual makes the example of a cyber attack on a water reservoir's Supervisory Control and Data Acquisition (SCADA) system to cause the release of water and thus prevent the use of a certain area by the enemy (*Tallinn Manual* 2013: 128).

The use of an object by the military is then sufficient to make it a military objective (providing that its destruction or neutralization also offer a definite military advantage in the circumstances ruling at the time). Most cyber infrastructures, however, are dual-use, i.e. at the same time used by civilians and the military. It is well known, for instance, that about 98% of US government communications travel through civilian-owned or civilian-operated networks (Geiß and Lahmannn 2012). Servers, fibre-optic cables, satellites, and other physical components of cyberspace are also almost entirely dual-use, as well as most technology and software used in this field: everyday applications such as web

---

[10] Confusingly, the Commentary to Rule 38 of the *Tallinn Manual* makes the example of a cyber operation against a website that inspires 'patriotic sentiments' among the population as a case of non-effective contribution to military action (*Tallinn Manual* 2013:13); however, such an operation would not be an 'attack' in the sense of either Art 49(1) of Additional Protocol I or Rule 30 of the *Manual* itself.

[11] It is however normally the software rather than the hardware that turns a computer into a military objective (Dinstein 2012: 263).

[12] The three US Department of Defense's internal networks, for instance, would be examples of networks that are military objectives by nature. In particular, the Secret Internet Protocol Router Network (SIPRNet), which is not connected to the internet, is used for classified information and to transmit military orders, while the Joint Worldwide Intelligence Communications System (JWICS) is used to communicate intelligence information to the military. On the three DoD networks, see Clarke and Knake 2010: 171–3.

[13] According to the ICRC *Commentary*, purpose is 'the intended future use of an object, while that of *use* is concerned with its present function' (Sandoz et al. 1987: para 2022, emphasis in the original).

browser, e-mail client and even command line (cmd.exe) can be used as an instrument for cyber attacks. The advent of cloud computing, where military and civilian data are stored side by side, is nothing but the latest manifestation of the dual-use character of information technology.[14] The fact that an object is *also* used for civilian purposes does not affect its qualification under the principle of distinction: if the two requirements provided in Article 52(2) of Additional Protocol I are present, the object is a military objective but the neutralization of its civilian component needs to be taken into account when assessing the incidental damage on civilians and civilian property under the principle of proportionality.[15] What is prohibited is to attack the dual-use cyber infrastructure *because* of its civilian function or to attack a dual use facility where the anticipated concrete and direct military advantage of the attack is outweighed by the expected civilian damage and/or injury. It should be recalled that, under Article 52(3) of Additional Protocol I, '[i]n case of doubt whether an object which is normally dedicated to civilian purposes…is being used to make an effective contribution to military action, it shall be presumed not to be so used'.[16] Unlike its counterpart with regard to persons (Article 50(1) of Additional Protocol I), the customary status of this provision is, however, dubious: it is, for instance, not included in the ICRC Study on Customary International Humanitarian Law. It has also been observed that satellites, cables, routers, and servers are not 'normally dedicated to civilian purposes', as they are also widely used by the military (Geiß and Lahmannn 2012: 386).

The effective contribution must be to 'military action'. 'Military action' has a broad meaning that corresponds to the 'general prosecution of the war' (Rogers 2004: 67). The United States' definition of 'military objective' employs language different from that contained in Additional Protocol I and it includes all objects which 'effectively contribute to the enemy's war-fighting or war sustaining capability' (US Navy, U.S Marine Corps, U.S Coast Guard 2007: para 8.2). If 'war fighting' can be considered equivalent to 'military action', 'war sustaining' is much broader and includes activities not directly connected to the hostilities: it would therefore allow attacks aimed to incapacitate political and economic targets in order to 'persuade' the enemy to stop fighting.[17]

The US definition of 'military objective' is also reflected in the cyber context. According to the US Air Force's Cornerstones of Information Warfare, the United

---

[14] Jensen has for instance claimed that 'Microsoft Corporation Headquarters in Washington State is a valid dual-use target, based on the support it provides to the U.S. war effort by facilitating U.S. military operations' (Jensen 2002–2003: 1160). However, he eventually denies that it is a lawful military objective because of doubts with regard to the military advantage that can be gained from its destruction or neutralization (1167–6).

[15] See Rule 39, *Tallinn Manual*. As has been observed, an 'object becomes a military objective even if its military use is only marginal compared to its civilian use' (Droege 2012: 563).

[16] The provision only applies when doubt concerns the use of the object, not its nature, location or purpose (Boothby 2012: 71).

[17] For critical comments of the US position and the documents in which it appears, see Bartolini 2006: 235–6.

States 'may target any of the adversary's information functions that have a bearing on his will or capability to fight' (Department of Air Force 1997: footnote 5).[18] This view would, for instance, legitimize attacks like the 2012 cyber operations against Saudi Aramco, the world's largest oil producer, which destroyed the data of about 30,000 company computers and, according to Saudi Arabia, targeted the country's economy with the purpose of preventing the pumping of oil into domestic and international markets.[19] The US CYBERCOM former Head also declared that power grids, banks, and other financial institutions and networks, transportation-related networks, and national telecommunication networks are 'all potential targets of military attack, both kinetic and cyber, under the right circumstances', although only when 'used solely to support enemy military operations'.[20]

This expanded notion of military objective is at odds with the definition contained in Article 52(2) of Additional Protocol I, which is largely considered to reflect customary international law. It should be noted that the 1976 US Air Force Pamphlet incorporated a definition of military objective analogous to that contained in the Protocol (U.S. Air Force 1976: para 5–3(b)(1)). The subsequent 1998 USAF Intelligence Targeting Guide also incorporates the Protocol's definition to the letter (U.S. Air Force 1998: para 1.7.1) and the 1997 edition of the Report on US practice in international law notes that '[t]he *opinio juris* of the U.S. government recognizes the definition of military objectives in Article 52 of Additional Protocol I as customary law', although it adds that 'United States practice gives a broad reading to this definition, and would include areas of land, objects screening other military objectives, and war-supporting economic facilities as military objectives' (Henckaerts and Doswald-Beck 2005 (vol. II): 188).

7.3.2 'Definite Military Advantage'

Even objects that, because of their nature, use, purpose, or location, effectively contribute to military action are not, as such, military objectives unless their total or partial destruction, capture or neutralization, in the circumstances ruling at the time, are militarily necessary, i.e. offer a 'definite military advantage' (Boothby 2012: 103).[21] Article 52(2) envisages not only the total or partial destruction of the attacked

---

[18] It also appears that China sees cyber operations against financial systems, power generation, transmission facilities, and other NCIs as part of a conflict with another state (Owens et al. 2009: 333).

[19] Al Arabiya News: 2012. Oil production, however, remained uninterrupted.

[20] Responses to advance questions, Nomination of Lt Gen Keith Alexander for Commander, US Cyber Command, US Senate Committee on Armed Services, 15 April 2010, 13, http://armed-services.senate.gov/statemnt/2010/04%20April/Alexander%2004-15-10.pdf.

[21] DeSaussure refers to the examples of the 1972 Christmas bombing of Hanoi or the never implemented bombing of a depot in the heart of Argentina during the Falklands war, which would have not helped the British reoccupy the islands (DeSaussure 1987: 513).

object, but also its capture or neutralization, which includes 'an attack for the purpose of denying the use of an object to the enemy without necessarily destroying it' (Bothe et al. 1982: 325). These words fit cyber operations that incapacitate but do not destroy infrastructures like a glove. As the ICRC has observed, 'the fact that a cyber operation does not lead to the destruction of an attacked object is…irrelevant': the definition of military objective, which refers to neutralization, 'implies that it is immaterial whether an object is disabled through destruction or in any other way' (ICRC 2011: 37).

The advantage must be of a military nature and 'definite', i.e. not speculative or indirect: the Commentary explains that 'it is not legitimate to launch an attack which only offers potential or indeterminate advantages' (Sandoz et al. 1987: para 2024). Shutting down the computer system operating the adversary's air defences would, for instance, provide an evident 'definite' military advantage. By contrast, a cyber attack aimed at demoralizing the civilian population would be unlawful. The problem with establishing the definite military advantage requirement in the cyber context is that measurement of effects can often be difficult: it is still not confirmed, for instance, whether Stuxnet did destroy any centrifuges at Natanz and, if so, with what consequences on the Iranian nuclear programme. Indeed, while Iran denied that the incident caused significant damage, the IAEA reported that Iran stopped feeding uranium into thousands of centrifuges (Broad 2010): it is however unclear whether this was due to Stuxnet or to technical malfunctions inherent to the equipment used (Ziolkowski 2012). The fact that a definite military advantage is eventually not gained from the operation, however, does not necessarily deprive the object of its qualification as a military objective, as long as the attacker had a reasonable expectation that the intended results would occur.

The destruction, capture or neutralization of the object must offer a definite military advantage 'in the circumstances ruling at the time'. This excludes any potential future advantage but also implies that an object which could not normally be considered a military objective may become one if it is used in direct support of the hostilities. The time between the identification of the target, the planning of the attack and the execution of the attack must therefore be reasonably short, as the circumstances could rapidly change and an object that qualified as a military objective at a certain time may subsequently turn into a civilian one (Vierucci 2006).

7.4 Is the Internet a Military Objective?

The internet can be disrupted by attacking its hardware or software components. The former type of attack targets servers, fibre-optic cables and other physical internet infrastructure used to ensure connectivity, while the latter affects systems like the Domain Name System (DNS), which translates domain names into IP addresses: if the DNS is compromised, the web browser would not know where to direct the visit. The China Internet Network Information Centre (CNNIC), for instance, reported that the national domain name resolution registry came under a series of a sustained DDoS attack on 25 August 2013, which interrupted or slowed down visits (Vincent 2013). A cyber operation against either the hardware or software components of the internet would qualify as an 'attack' in the sense of Article 49(1) of Additional Protocol I if material damage or significant loss of functionality of infrastructure ensue.

The internet is a computer network, which is a type of communication network: the question whether or not the internet is a military objective, then, can first be approached reasoning by analogy with more traditional means of communication (Doswald-Beck 2002; Dörmann 2001). The 1956 ICRC list of military objectives includes 'the lines and means of communication, installations of broadcasting and television stations, telephone and telegraph exchanges of fundamental military importance'.[22] Military manuals also include 'communication installations used for military purposes' as an example of military objective (UK Ministry of Defence 2004: 57). Targeting communication nodes has been a high priority in all recent armed conflicts. Media and broadcasting systems were included in the target list both in Operation Desert Storm and in Operation Allied Force (Fenrick 1996–1997). In the former case, the attacks were justified by the United States not only on the ground that the facilities were part of the military communications network, but also because they were used for Iraqi propaganda (Bartolini 2006). On 23 April 1999, NATO aircraft bombed the headquarters of the Radio Television of Serbia (RTS) in Belgrade (Aldrich 1999). According to the Organization, it was a lawful target, since the station was used for military purposes as part of the control mechanism and of the propaganda machinery of the Milošević government (Amnesty International 2000). The ICTY Final Report concluded that the attack was lawful because it was aimed mainly at disabling the Serbian military command and control system and at destroying the apparatus that kept Milošević in power (ICTY *Final Report* 2000). On 12 November 2001, the Kabul office of Al-Jazeera news television was hit by a guided bomb during Operation Enduring Freedom (Herold 2002) and other radio/television stations were attacked because they were used as means of propaganda by the Taliban (Cryer 2002). In Operation Iraqi Freedom, the United States bombed the Ministry of Information, the Baghdad Television Studio and Broadcast Facility and the Abu Ghraib Television Antennae Broadcast Facility (Human Rights Watch 2003). Unlike in the 1991 operation, however, it seems that in 2003 the emphasis in the legal justification of the attacks was more on the facilities being part of the military communications network than on their use to spread propaganda (Bartolini 2006). The antennas of Libya's state broadcaster were also attacked by NATO aircraft during the 2011 Operation Unified Protector (Bartolini 2013). The attack had the purpose 'of degrading Qadhafi's use of satellite television as a means to intimidate the Libyan people and incite acts of violence against them' and was motivated on the fact that 'TV was being used as an integral component of the regime apparatus designed to systematically oppress and threaten civilians and

---

[22] The list is reprinted in ICTY *Final Report,* 2000, para 39.

to incite attacks against them' (NATO Operation *Unified Protector,* Statement by Colonel Lavoie, 2011).

Although, as a means of communication, the internet could potentially qualify as a military objective, it would still have to meet the two requirements of Article 52(2) of Additional Protocol I. If internet disruption had the sole purpose of stopping propaganda, undermining civilian morale or psychologically harass the population, its neutralization would not offer a 'definite' military advantage (even if it weakened the political support for the enemy government) (ICTY *Final Report* 2000: para 55). On the other hand, if the internet had become part of the adversary's military communication system, it would effectively contribute to military action, but, if connection can be easily and promptly restored, it can be doubted that its neutralization or destruction would provide a definite military advantage. As has been observed, 'an attacker nowadays must probably destroy a network of telecommunication *in toto* (or at least its central connection points) in order to paralyse the command and control structures of the enemy armed forces, which in themselves clearly constitute a legitimate military objective' (Oeter 2013: 174).[23] This may be particularly difficult to achieve in the case of the internet, which is notoriously characterized by a high level of resilience: if certain channels become unusable as a consequence of a cyber or kinetic attack on servers, the data flow will simply find another path to reach its destination and it might well be that the destruction or neutralization of certain internet infrastructure has no practical effect at all (Geiß and Lahmannn 2012). Connection would probably slow down, but it would still continue through mirror servers, mobile phones, or satellites.[24]

The ICTY Final Report on the NATO bombing campaign against Yugoslavia suggests that a broadcasting station may also constitute a military objective when it is employed to incite the population to commit war crimes or crimes against humanity as in the case of Radio Mille Collines in Rwanda in 1994 (ICTY *Final Report*, 2000). As has been seen, this argument was also used by NATO to justify the attack on Libya's state television in 2011, but the justification was offered more for the purposes of the Operation's protective mandate than from an international humanitarian law perspective (Bartolini 2013). Although it is doubtful that these views are consistent with the *lex lata*, as propaganda or incitement to commit crimes do not amount to 'effective contribution to military action', it cannot be excluded that, in parallel with the developments in international criminal law, a customary international law rule is emerging that allows attacks against these heinous uses of means of communication. If this is the case, and should the internet be used for such

---

[23] The ICTY *Final Report* on the NATO bombing campaign against Yugoslavia emphasized that, even if the RTS building in Belgrade was considered a military objective, broadcasting was interrupted only for a brief period and in any case Yugoslavia's command and control network, of which the RTS building was allegedly a part, could not be disabled with a single strike (ICTY *Final Report*, 2000, para 78).

[24] Resilience does not, however, mean invulnerability. For instance, 88 per cent of Egyptian internet access was shut down as a consequence of the withdrawal of 3500 Border Gateway Protocol (BGP) routes by Egyptian ISPs (Williams 2011).

purposes, connectivity could be disrupted through a kinetic or cyber operation against its components.

The internet, however, is not only a means of communication, but also an important economic resource. As the German *Law of Armed Conflict Manual* recalls, only economic objectives that make an effective contribution to military action can be considered lawful targets (Federal Ministry of Defence, *Law of Armed Conflict Manual*, 2013: para. 407) and the same view is contained in the 1998 USAF Intelligence Targeting Guide (*Intelligence Targeting Guide*, para A4.2.2.1). According to the 2002 US Joint Doctrine for Targeting, however, lawful targets also include economic facilities that '*indirectly but effectively* support and sustain the enemy's warfighting capability' (*US Joint Doctrine for Targeting:* A-3; emphasis added). The Eritrea-Ethiopia Claims Commission (EECC) also affirmed that '[t]he infliction of economic losses from attacks against military objectives is a lawful means of achieving a definite military advantage' and that 'there can be few military advantages more evident than effective pressure to end an armed conflict' (EECC, Partial Award, 2005 para 121). While it is accepted that certain economic targets may be military objectives when they meet the two requirements provided in Article 52(2) of Additional Protocol I, the EECC seems to justify attacks against *any* economic target (Vierucci 2006). This view goes too far and is not consistent with the definition of 'military objective' contained in Additional Protocol I, which reflects customary international law.

It is worth noting that, even if the internet qualified in certain situations as a military objective, the attacker would still have to take into account the disruption caused to its civilian function and to neutrals under the principle of proportionality (Article 51(5)(b) of Additional Protocol I). The possibility of shutting down specific segments, websites, or networks, therefore, should always be explored first (*Tallinn Manual*, 2013: 136).

7.5 Conclusions

As the United States has observed, the application in the cyber context of the law of armed conflict, which was conceived with kinetic weaponry in mind, presents 'new and unique challenges that will require consultation and cooperation among nations' (UN Doc A/66/152, 2011: 19). This essay has discussed some of these challenges, in particular those related to the application of the principle of distinction and of the definition of 'military objective' to cyber operations conducted during an armed conflict and having a nexus with it. This essay has concluded that none of these challenges significantly affects the application of the principle of distinction. Indeed, the definition of 'military objective' contained in Article 52(2) of Additional Protocol I is flexible enough to apply in the cyber context, in spite of the peculiar characteristics of this new domain of warfare. The dual-use character of most cyber infrastructures, i.e. the fact that they are at the same time used both by civilians and the military, is also an overestimated problem which is not unique to cyberspace and which does not render the existing rules obsolete.

Bibliography

Books and Articles

Aldrich, George H. 1999. Yugoslavia's television studios as military objectives. *International Law Forum du droit International* 1: 149–150.

Bartolini, Giulio. 2006. Air operations against Iraq (1991 and 2003). In *The law of air warfare-contemporary issues*, ed. Natalino Ronzitti and Gabriella Venturini, 227–272. Utrecht: Eleven Publishing.

Bartolini, Giulio. 2013. Air targeting in operation *unified protector* in Libya. *Jus ad bellum* and IHL issues: An external perspective. In *Legal interoperability and ensuring observance of the law applicable in multinational deployments*, ed. Stanislas Horvat and Marco Benatar, 242–279. Brussels: International Society for Military Law and the Law of War.

Boothby, William H. 2012. *The law of targeting*. Oxford: Oxford University Press.

Bothe, Michael, Karl J. Partsch, and Waldemar A. Solf. 1982. *New rules for victims of armed conflicts: Commentary on the two 1977 protocols additional to the Geneva conventions of 1949*. The Hague: Nijhoff.

Clarke, Richard A., and Robert K. Knake. 2010. *Cyber war. The next threat to national security and what to do about it?* New York: Harper Collins.

Cryer, Robert. 2002. The fine art of friendship: *Jus in Bello* in Afghanistan. *Journal of Conflict and Security Law* 7(1): 37–83.

DeSaussure, Hamilton. 1987. Remarks at the Six Annual American Red Cross-Washington College of Law Conference on International Humanitarian Law: A workshop on customary international law and the 1977 protocols additional to the 1949 Geneva conventions. *American University Journal of International Law and Policy* 2: 415–538.

Dinstein, Yoram. 2012. The principle of distinction in cyber war in international armed conflicts. *Journal of Conflict and Security Law* 17: 261–278.

Dinstein, Yoram. 2013. Cyber war and international law: Concluding remarks at the 2012 Naval War College international conference. *International Law Studies* 89: 276–287.

Doswald-Beck, Louise. 2002. Some thoughts on computer network attack and the international law of armed conflict. *International Law Studies* 76: 163–186.

Droege, Cordula. 2012. Get off my cloud: Cyber warfare, international humanitarian law, and the protection of civilians. *International Review of the Red Cross* 94: 533–578.

Fenrick, William J. 1996–1997. Attacking the enemy civilian as a punishable offence. *Duke Journal of Comparative and International Law* 7:539–570.

Geiß, Robin, and Henning Lahmannn. 2012. Cyber warfare: Applying the principle of distinction in an interconnected space. *Israel Law Review* 45: 381–400.

Henckaerts, Jean Marie, and Louise Doswald-Beck. (ed.). 2005. *Customary International Humanitarian Law*. Cambridge: Cambridge University Press (two volumes).

Jensen, Eric Talbot. 2002–2003. Unexpected consequences from knock-on effects: A different standard for computer network operations? *American University International Law Review* 18: 1145–1188

Lubell, Noam. 2013. Lawful targets in cyber operations: Does the principle of distinction Apply? *International Law Studies* 89: 252–275.

Oeter, Stefan. 2013. Methods and means of combat. In *The handbook of international humanitarian law*, ed. Dieter Fleck, 115–230. Oxford: Oxford University Press.

Owens, William, A. Dam, W. Kenneth, and Herbert S. Lin. 2009. *Technology, policy, law, and ethics regarding U.S. acquisition and use of cyberattack capabilities*. Washington, DC: The National Academies Press.

Rauscher, Karl F., and Andrey Korotkov. 2011. *Working towards rules for governing cyber conflict. Rendering the Geneva and Hague conventions in cyberspace*. New York: East West Institute.

Rogers, A.P.V. 2004. *Law on the battlefield*. Manchester: Manchester University Press.
Roscini, Marco. 2014a. *Cyber operations and the use of force in international law*. Oxford: Oxford University Press.
Roscini, Marco. 2014b, March 31. *Is there a 'Cyber War' between Ukraine and Russia*. OUPBlog. http://blog.oup.com/2014/03/is-there-a-cyber-war-between-ukraine-and-russia-pil/.
Sandoz, Yves, Christophe Swinarski, and Bruno Zimmermann (eds.). 1987. *Commentary on the additional protocols of 8 June 1977 to the Geneva conventions of 12 August 1949*. Dordrecht: Nijhoff.
Schmitt, Michael N. 2011. Cyber operations and the *Jus in Bello*: Key issues. *International Law Studies* 87: 89–112.
Schmitt, Michael N. (ed.). 2013. *Tallinn manual on the international law applicable to cyber warfare*. Cambridge: Cambridge University Press.
Turns, Davis. 2012. Cyber warfare and the notion of direct participation in hostilities. *Journal of Conflict and Security Law* 17: 297–300.
Vierucci, Luisa. 2006. Sulla nozione di obiettivo militare nella guerra aerea: recenti sviluppi della giurisprudenza internazionale. *Rivista di diritto Internazionale* 89: 693–735.

National Military Manuals and Operational Handbooks

Department of Air Force. 1997. *Cornerstones of information warfare*. http://www.google.co.uk/url?url=http://www.dtic.mil/cgi-bin/GetTRDoc?Ad.
Federal Ministry of Defence. 2013. *Law of armed conflict manual*. Joint Service Regulation 15/2. http://www.bmvg.de/resource/resource/MzEzNTM4MmUzMzMyMmUzMTM1MzMyZTM2MzIzMDMwMzAzMDMwMzAzMDY5MzIzNDZmN2E3NjZmNjgyMDIwMjAyMDIw/Law%20of%20Armed%20Conflict_Manual_2013.pdf.
UK Ministry of Defence. 2004. *The manual of the law of armed conflict*. Oxford: Oxford University Press.
US Air Force. 1998, February 1. *Intelligence targeting guide*. Air force pamphlet 14-210 intelligence. http://www.fas.org/irp/dodir/usaf/afpam14-210/part17.htm.
U.S Air Force Pamphlet 110-31. 1976. *International law – The conduct of armed conflict and air operations*. https://www.cna.org/sites/default/files/research/5500045700.pdf
US Navy, U.S Marine Corps, U.S Coast Guard. 2007. *The commander's handbook on the law of naval operations*. http://www.fichl.org/uploads/media/US_Navy_Commander_s_Handbook_1995.pdf.

Reports

Amnesty International. 2000, June. 'Collateral Damage' or 'Unlawful Killings?' Violations of the laws of war by NATO during operation *Allied Force*. AI-Index EUR70/18/00 https://www.amnesty.org/en/documents/document/?indexNumber=EUR70%2F018%2F2000&language=en.
Final Report to the Prosecutor by the Committee Established to Review the NATO Bombing Campaign Against the Federal Republic of Yugoslavia. 8 June 2000. http://www.icty.org/x/file/Press/nato061300.pdf.
Human Rights Watch. 2003. *Off target. The conduct of the war and civilian casualties in Iraq*. http://www.hrw.org/reports/2003/usa1203/usa1203.pdf.
ICRC. 2011, October. *International humanitarian law and the challenges of contemporary armed conflicts*. https://www.icrc.org/eng/assets/files/red-cross-crescent-movement/31st-international-conference/31-int-conference-ihl-challenges-report-11-5-1-2-en.pdf.

Mandiant. 2013. APT1. *Exposing one of China's cyber espionage units.* http://intelreport.mandiant.com/Mandiant_APT1_Report.pdf.
Melzer, Nils. 2011. *Cyberwarfare and international law.* UNIDIR, 27. http://www.isn.ethz.ch/Digital-Library/Publications/Detail/?lang=en&id=134218.
Ziolkowski, Katharina. 2012. Stuxnet – Legal Considerations. *NATO CCD COE Publications,* 5. https://ccdcoe.org/search.html

Cases

Legality of the Threat or Use of Nuclear Weapons, Advisory Opinion, 8 July 1996, ICJ Reports 1996.
Partial Award, *Western Front, Aerial Bombardment and Related Claims, Eritrea's Claims 1, 3, 5, 9–13, 14, 21, 25 and 26*, 19 December 2005, Reports of International Arbitral Awards. Vol. XXVI, Part VIII
Prosecutor v Kupreškić, Case No IT-96-16-T, Trial Chamber Judgment, 14 January 2000.

On-Line Publications and Other Documents

Al Arabiya News. 2012, December 9. *Saudi Aramco says cyber attack targeted kingdom's economy.* http://www.alarabiya.net/articles/2012/12/09/254162.html.
Broad, William J. 2010, November 23. Report suggests problems with Iran's nuclear Efforts. *The New York Times.* http://www.nytimes.com/2012/11/24/world/middleeast/24nuke.html.
Dörmann, Knut. 2001, May 19. *Computer network attack and international humanitarian law.* http://www.icrc.org/eng/resources/documents/misc/5p2alj.htm.
Dörmann, Knut. 2004, November 9. *Applicability of the additional protocols to computer network attacks.* https://www.icrc.org/eng/assets/files/other/applicabilityofihltocna.pdf.
Herold, Marc W. 2002, March. *A Dossier on civilian victims of United States' aerial bombing of Afghanistan: A comprehensive accounting.* http://www.cursor.org/stories/civilian_deaths.htm.
Statement by the Spokesperson for NATO Operation Unified Protector, Colonel Roland Lavoie, Regarding Air Strike in Tripoli. 30 July 2011. NATO Strikes Libyan State TV Satellite Facility. http://www.nato.int/cps/en/natolive/news_76776.htm.
Vincent, James. 2013, August 27. Chinese domains downed by 'Largest Ever' cyber attack. *The Independent.* http://www.independent.co.uk/life-style/gadgets-and-tech/news/chinese-domains-downed-by-largest-ever-cyberattack-8786091.html.
Williams, Christopher. 2011, January 28. How Egypt shut down the internet. *The Telegraph.* http://www.telegraph.co.uk/news/worldnews/africaandindianocean/egypt/8288163/How-Egypt-shut-down-the-internet.html.
Zetter, Kim. 2015, August 1. A cyber attack has caused confirmed physical damage for the second time ever. *Wired.* https://www.wired.com/2015/01/german-steel-mill-hack-destruction/.

Marco Roscini has a PhD from the University of Rome 'La Sapienza' and is currently professor of international law at the University of Westminster, London. Professor Roscini has written extensively on international security law, including nuclear disarmament, the law of armed conflict and cyberwarfare. His most recent book, *Cyber Operations and the Use of Force in International Law*, was published by Oxford University Press in 2014.

Chapter 8
Defining Cybersecurity Due Diligence Under International Law: Lessons from the Private Sector

Scott J. Shackelford, Scott Russell, and Andreas Kuehn

Abstract Although there has been a relative abundance of work done on exploring the contours of the law of cyber war, far less attention has been paid to defining a law of cyber peace applicable below the armed attack threshold. Among the most important unanswered questions is what exactly nations' due diligence obligations are to their respective private sectors and to one another. The International Court of Justice ("ICJ") has not explicitly considered the legality of cyber weapons to this point, though it has ruled in the *Corfu Channel* case that one country's territory should not be "used for acts that unlawfully harm other States." But what steps exactly do nations and companies under their jurisdiction have to take under international law to secure their networks, and what of the rights and responsibilities of transit states? This chapter reviews the arguments surrounding the creation of a cybersecurity due diligence norm and argues for a proactive regime that takes into account the common but differentiated responsibilities of public- and private-sector actors in cyberspace. The analogy is drawn to cybersecurity due diligence in the

---

An updated and expanded version of this article featuring both private and public-sector experiences with building out due diligence was published as Scott J. Shackelford, Scott Russell, & Andreas Kuehn, *Unpacking the International Law on Cybersecurity Due Diligence: Lessons from the Public and Private Sectors*, 17 CHICAGO JOURNAL OF INTERNATIONAL LAW 1(2016).

S.J. Shackelford (✉)
Business Law and Ethics, Indiana University, Bloomington, IN, USA

Center for Applied Cybersecurity Research, Indiana University, Bloomington, IN, USA

Hoover Institution, Stanford University, Stanford, CA, USA
e-mail: sjshacke@indiana.edu

S. Russell
Center for Applied Cybersecurity Research, Indiana University, Bloomington, IN, USA

A. Kuehn
Zukerman Cybersecurity Predoctoral Fellow, Center for International Security and Cooperation, Stanford University, Stanford, CA, USA

School of Information Studies, Syracuse University, Syracuse, NY, USA

private sector and the experience of the 2014 National Institute of Standards and Technology ("NIST") Framework to help guide and broaden the discussion.

Keywords Cybersecurity • Cyber attack • Due diligence • International law

8.1 Introduction

Rarely does it seem that some facet of cybersecurity is not front-page news. From compromises of the SWIFT banking code system to cyber attacks on Ukraine's power grid, cybersecurity is increasingly taking center stage in diverse arenas of geopolitics, international economics, security, and law. Yet the field of international cybersecurity law and policy remains relatively immature despite some important steps forward in recent years such as the *Tallinn Manual* and *Tallinn 2.0* projects. For example, although there has been a relative abundance of work done on exploring the contours of the law of cyber war, far less attention has been paid to defining a law of cyber peace applicable below the armed attack threshold (Schmitt 2013). This is surprising since the vast majority of cyber attacks do not cross the armed attack threshold. Among the most important unanswered questions is what exactly are nations' due diligence obligations to secure their networks and prosecute or extradite cyber attackers. The International Court of Justice ("ICJ") has some guiding jurisprudence on this point, such as the *Corfu Channel* case that one country's territory should not be "used for acts that unlawfully harm other States" (Corfu Channel Case 1949, para. 22). But analogizing is required, and these cases are not by themselves dispositive. A wealth of information is available in the arena of cybersecurity due diligence from both the public and private sectors that has to date been largely untapped to help answer the question of what steps nations and companies under their jurisdiction have to take to secure their networks, along with clarifying the rights and responsibilities of transit states.

This chapter reviews the arguments surrounding the creation of a cybersecurity due diligence norm and argues for a proactive regime that takes into account the common but differentiated responsibilities of various stakeholders in cyberspace. The analogy is drawn to cybersecurity due diligence in the private sector and the experience of the 2014 National Institute of Standards and Technology Cybersecurity Framework ("NIST Framework") to help guide and enrich the discussion (Ensign 2014). Ultimately we argue that international jurisprudence has an invaluable role to play, but the experience of national regulators and the private sector is also informative in this space especially given the robust and necessary public-private cross-pollination occurring with regards to clarifying and spreading cybersecurity best practices.

This chapter is structured as follows. We begin by reviewing the applicable ICJ jurisprudence and literature on cybersecurity due diligence under international law. We then turn to national case studies to help flesh out a potential cybersecurity due diligence norm focusing on the United States, Germany, and China. Finally, we review lessons from the private-sector cybersecurity due diligence context focusing on mergers and acquisitions to better understand where the rubber meets the road and conclude with some implications for managers and policymakers.

8.2 Unpacking Due Diligence Under International Law

International law may be defined as "the body of legal rules," norms, and standards that applies "between sovereign states" and non-state actors, including international organizations and multinational companies, enjoying legal personality (International Labor Organization 2013). The primary sources of international law are treaties, general principles of law, and custom, the third of which requires evidence of state practice that nations follow out of a sense of legal obligation (Statute of the International Court of Justice 1945, art. 38). The subsidiary sources of international law include judicial decisions and scholarly writing. Given the recent nature and rapid development of cyber-capabilities, there are comparatively few treaties that specifically address the rights and obligations of States vis-a-vis these cyber-capabilities with the notable exception of the Council of European Convention on Cybercrime (also known as the Budapest Convention) discussed below. Absent a robust treaty regime and given the geopolitical difficulties of negotiating new agreements in this area, it is vital to clarify the role of customary international law as it relates to due diligence.

8.2.1 An Introduction to Customary International Cybersecurity Law

A vital component of customary international law was articulated by the ICJ case of *Nicaragua v. United States*, which involved a dispute over the United States' involvement with the Contra rebellion in Nicaragua (Nicaragua Case 1986). In *Nicaragua*, the ICJ held that customary international obligations would arise from the consistent, widespread practice of States to engage in specific acts or omissions, performed out of a sense of obligation that such acts or omissions were required by international law (*opinio juris*). The combination of State practice and *opinio juris*, performed by a significant number of States and without the express disavowal by a significant number of states, would give rise to international obligations under customary international law. The underlying rationale is that this combination reflects a consensus in the international community that the actions taken represent an unspoken international obligation. Depending on the type of norm involved, that state practice needs to be more or less widespread. For new norms, such as regarding cybersecurity, the standard generally is "virtually uniform" State practice (N. Sea Continental Shelf 1969). That is a high bar, needless to say, as we discuss further below.

Despite *Nicaragua's* clear articulation of the rule, in practice the development of customary international law presents a temporal dilemma, since for a State to engage in actions out of a sense of legal duty, this presupposes the existence of such a duty, and therefore the prior existence of the customary international law (Bradley

2013). To help resolve this dilemma, Professor Frederic Kirgis, in response to *Nicaragua*, argued for what he called a "sliding scale approach" (Kirgis 1987). Professor Kirgis argues that State practice and *opinio juris* need to be understood on a spectrum, wherein the requirement for *opinio juris* increases as the evidence of State practice decreases. Rather than impose strict requirements for both State practice and *opinio juris*, the sliding scale approach argues that a strong history of State practice can give rise to international obligations absent *opinio juris*, and that likewise compelling *opinio juris* could give rise to international obligations with little evidence of State practice conforming thereto. This sliding scale approach may prove particularly important in the realm of cyber activities, as these novel technologies have arisen too rapidly for evidence of widespread State practice to emerge, yet compelling *opinio juris* may still exist as the basis for international obligations.

Proving *opinio juris*, however, is a difficult task, especially in the cyber realm. The temporal dilemma makes pointing to existing rules complicated, so the preferred method is to identify broad principles. The ICJ suggests that these broad principles can be found by looking to treaties, as multilateral treaties evidence a widespread agreement among States, and indeed most courts rely on treaties to identify *opinio juris*, often exclusively so (Gulati 2013). Yet in the cyber realm, treaties thus far have largely focused on implementing domestic cybercrime laws, and have done relatively little to address cybersecurity standards leaving such decisions to the private sector and standards bodies as embodied in the NIST Framework discussed below. The Budapest Convention, the African Union Convention on Cybersecurity and Data Protection, and the various ASEAN working groups on cybercrime all could serve as helpful sources of State practices on enacting and enforcing cybercrime laws within their territories and cooperating to prosecute and extradite cybercriminals. However, these agreements often lack binding language. Similarly, the Organization of American States has also encouraged Member States to join the Budapest Convention and to ratchet up regional cooperation to mitigate cybercrime, whereas a nonbinding UN General Assembly Resolution calls on states to "eliminate safe havens" for cybercriminals (G.A. Res. 55/63, 2001). While it is unlikely that a non-signatory State would be bound to the specific terms of a treaty to which it did not sign through customary principles—particularly in the short term—that treaty may still serve to identify broad principles that inform *opinio juris*, and thereby can form the foundation for emerging international obligations.

The search for *opinio juris* is further complicated by the widespread use of State-sponsored cyber-activity, from cyber espionage to State-sponsored cyber crime. While the classification of State cyber-activities is a well-known problem, the mere fact that these activities are so widespread suggests a lack of *opinio juris* against aggressive State cyber-activity below the armed-attack threshold. This is reinforced by the *Tallinn Manual's* discussion of the international law relating to espionage, which is ostensibly legal as a matter of international law (Schmitt 2013). Similarly, domestic cybersecurity practices are highly variable and can involve the surreptitious installation of malware, as alleged of Chinese telecommunications providers and the U.S. National Security Agency (NSA) alike (Gruener 2012). Given the relative lack of multilateral progress, claiming a widespread consensus for an underlying cybersecurity principle would be challenging in this area.

8.2.2 ICJ Jurisprudence as It Relates to Cybersecurity Due Diligence

Although the ICJ has never directly addressed cybersecurity due diligence requirements, the cases discussing due diligence generally can serve as broad guideposts for States from which we may infer cyber-specific applications. It is worth noting that these cases all arose prior to the rise of cyber attacks, but some of the principles that underlay them may still have some applicability including *Corfu Channel*, *Trail Smelter*, and *Nicaragua*.[1] Before briefly reviewing these cases it is first important, though, to attempt a definition of "cybersecurity due diligence." In the transactional context, this term has been defined as "the review of the governance, processes and controls that are used to secure information assets" (Ryan and Navarro 2015). The concept as it is used here builds from this definition and may be understood as the customary national and international obligations of both state and non-state actors to help identify and instill cybersecurity best practices and governance mechanisms so as to promote cyber peace through enhancing the security of computers, networks, and ICT infrastructure. Cybersecurity due diligence obligations may exist between States, between non-state actors (e.g., private corporations, end-users), and between State and non-state actors. Applicable instruments include technical standards, legal requirements born from treaty or custom, as well as national policies and private-sector industry norms discussed below.

8.2.2.1 Corfu Channel

One of the earliest ICJ cases on the issue of international due diligence standards was the 1947 resolution of the Corfu Channel dispute (Corfu Channel Case 1949). In this dispute, two British warships struck mines and were sunk in the Corfu Channel, an international strait located in Albanian territorial waters. The British brought the case before the ICJ, which focused primarily on the right of innocent passage and on the duty of the Albanian government to warn the British of the mines' existence. Although the Court ruled that there was insufficient evidence to conclude that the Albanian government had placed the mines itself, it did conclude that the Albanian government should have known of the mines' existence, and therefore had a duty to warn the British warships. The ICJ based its decision on "certain general and well-recognized principles," specifically "every State's obligation not to allow knowingly its territory to be used for acts contrary to the rights of other States" (Corfu Channel Case 1949, 22).

---

[1] However, it should be noted that other jurisprudence is also on point and is not discussed here due to space constraints, including: *Legality of the Threat or Use of Nuclear Weapons, Advisory Opinion* – General Assembly, ICJ Reports, 8 July 1996, at 22, para. 29; *Gabcikovo-Nagymaros Project* (Hungary v. Slovakia), Judgment, 25 September 1997, ICJ Reports (1997), at 7, para. 53; *Case concerning pulp mills on the river Uruguay* (Argentina v. Uruguay), Judgment, 20 April 2010, para. 193.

This obligation, although articulated in the context of domestic waterways, has carryover into the cybersecurity realm. The most direct cyber-parallel would be a duty to warn other States operating within the subject State's domestic networks of vulnerabilities known to exist on those networks, but this might extend more generally to a duty to warn other States of vulnerabilities detected in that other State's networks (Tikk 2011). While this principle is unlikely to require the warning State to identify vulnerabilities with particularity, it could require that State to warn other States of the existence of the equivalent of 'cyber mines' (such as logic bombs). The underlying principle of these duties, drawn from Corfu, is that States have a duty to warn other States of known or foreseeable harms, particularly when those harms arise from within the warning State's sovereign territory. However, whether such duties could effectively coexist with the current international standards regarding espionage, discussed above, and the exceptions for national security, discussed below, is not yet apparent. Nor is it immediately clear how this reasoning jives with the increasing use of cloud-based computing by companies and governments and the related jurisdictional issues raised, or the G20's 2015 list of cybersecurity norms that includes a duty to assist victim nations.

Of particular note in *Corfu Channel* is that the ICJ articulated different standards of proof for direct State actions and omissions. The standard required to prove a State action was not specifically stated, although the ICJ noted that it required "a degree of certainty not shown here," whereas to prove an omission required "no room for reasonable doubt" (Corfu Channel Case 1949, 17–18). Some commentators have suggested that this reflects a higher burden of proof for omissions than direct actions (Mar, 2012). Nonetheless, omissions are likely to be easier to prove in practice, as the ICJ is more willing to accept circumstantial evidence in these instances, particularly when the opposing party controls the direct evidence. Consequently in *Corfu*, although the British government failed to meet the standard of proof that the Albanian government had placed the mines, it nonetheless was able to satisfy the evidentiary burden to prove that the Albanian government would have known of the mines' existence. This issue is relevant to cyber attacks since even though a given exploit may be launched from within a State's territorial boundaries, attributing it back to that State is no easy feat (Mudrinich 2012).

The attribution problem may become less burdensome, however, when attempting to prove the State's knowledge of attackers within its territory, as Corfu's allowance for "more liberal recourse to inferences of fact and circumstantial evidence" when the evidence is controlled by the opposing State may make proving knowledge easier. Although the mere fact that the activity occurred in the State's territory is not evidence of knowledge, activities such as the use of the State's non-commercial critical infrastructure may serve as a rebuttable presumption that the State had knowledge of the attack (Heinegg 2013). Some commentators go even further, and assert that States can be held accountable without actual or presumed knowledge if that State failed to enact or enforce appropriate cyber-legislation, citing a failure to satisfy a State's duty to prevent cyber attacks within its own territory (Sklerov 2009). Regardless of the viability of such an expansive view of State responsibility, the principle of *Corfu* is that the ICJ will not absolve States of liability for actions occurring within their territory solely due to a lack of direct attribution to the State.

8.2.2.2 Trail Smelter

The ICJ also dealt with the issue of due diligence in the *Trail Smelter* dispute, which involved the emission of environmentally hazardous materials across the U.S.-Canadian border, raising the question of what obligations States owe neighboring States. This case thus placed the principle of territorial sovereignty at loggerheads with newer conceptions surrounding effects jurisdiction. Ultimately, the ICJ held that "no State has the right to use or permit the use of its territory ... to cause injury by fumes ... to the territory of another ... when the case is of serious consequence and the injury is established by clear and convincing evidence" (Trail Smelter Case 1938, 1965). Although directed towards the emission of "fumes," the Trail Smelter case has come to represent the broader "no harm" principle, which requires of States "that activities within their jurisdiction or control respect the environment of other states ..." (Bodle 2012, 457).

This no harm principle, although directed towards environmental harms, enjoys parallels with cybersecurity, and may serve as the foundation for a broader State obligation not to permit domestic activity that results in "serious consequences" internationally. Specifically, the analogy could be drawn such that if noxious activity from one State causes serious repercussions in another, then the host state has a duty to mitigate the threat. Indeed, as with environmental pollution, overuse can occur in cyberspace, such as when spam messages consume limited bandwidth, which have been called a form of "information pollution," and distributed denial of service attacks, which can cause targeted websites to crash through too many requests for site access (Ophardt 2010; Hurwitz 2009). However, though recognized by the ICJ, this precedent does not enjoy significant State practice since recognizing it widely would likely require the end of much transboundary pollution, a laudable goal to be sure but an impracticable one for the foreseeable future. Yet Trail Smelter's reference to cases of "serious consequence" ultimately suggests that State practice may exist in maintaining noxious domestic activity below a certain threshold of permissibility, albeit a high one, and therefore could support a broader no harm principle in customary international law applicable to cyber attacks.

8.2.2.3 Nicaragua

Perhaps the least clear, yet potentially most far-reaching international cybersecurity due diligence obligation from the ICJ is the one articulated in *Nicaragua*: that of State sovereignty. In deciding against the United States in that case introduced above, the ICJ articulated the obligation of States not to intervene in the domestic affairs of other States if that intervention related to "the choice of a political, economic social and cultural system, and the formulation of foreign policy" (Nicaragua

1986, 106–08). This principle of State sovereignty may be read as being in contradiction to the effects jurisdiction basis of the Court's decision in *Trail Smelter*. However, it is an important debate in the cybersecurity context with some nations asserting varying degrees of national sovereignty over their domestic intranets even as others espouse the virtues of a "global networked commons" (Clinton 2010). Indeed, several dozen nations now routinely filtering traffic, threatening the dawn of a new age of Internet sovereignty (Lewis 2011a). How multi-stakeholder Internet governance will jive with classic conceptions of State sovereignty is unclear, but the potential for domestic cyber policies to have international ramifications has never been greater; a fact that may entail obligations on the cyber powers in particular, some of which are discussed below.

8.2.2.4 Cybersecurity Due Diligence Obligations of Transit States

Cyber attacks are frequently routed through several transit states before reaching their ultimate targets, both to obfuscate the attack's origin and to stir international tensions as well as because of the distributed nature of the Internet's architecture (Mudrinich 2012). As with attacks launched from within a State, the obligations of States that merely transmit malicious Internet traffic originating elsewhere will likely depend upon that State's knowledge of the attack. The obligations of a State that knowingly allows a cyber attack to be transmitted through its domestic networks will likely be greater than those that do so without knowledge. Among those States that transmit the attack unwittingly, different standards could be applied to those that comply with cybersecurity best practices and those that fail to do so (Heinegg 2013). Furthermore, repeated or continuous cyber-activity through a State's domestic networks may give rise to a presumption of knowledge, and direct use of State controlled critical infrastructure could serve as evidence that the transit State knew or should have known of a cyber attack in progress (Heinegg 2013).

Yet State knowledge must be understood in context, as the individual packets transmitted through the State's network may, taken alone, be innocuous (Heinegg 2013). Cyber attacks are complex, often made up of myriad components, and so knowledge of an individual exploit does not necessarily equate to knowledge of the overarching campaign. Attacks may be broken apart into bits of seemingly innocent or unintelligible code, only to be recognizable as a cyber threat when reconstructed at a particular target. Stuxnet, for example, was designed in such a way that it would only be activated on specific hardware and specialized systems (Zetter 2014).

As for the specific duties that may be required of transit States, these would likely reflect the role that a given nation's infrastructure played in the attack. The highest level of due diligence that could reasonably be required would be an affirmative obligation to monitor a nation's networks for cyber attacks and to mitigate any such threat such as through a duty to assist. This would be akin to requirements of neutral States in time of war, which are told to disallow and resist any belligerent force from transporting troops or munitions through a neutral territory. Less potentially onerous, yet far more likely requirements would be a duty to warn target

States of attacks detected on their networks (without a hard requirement to monitor and eliminate) and a duty to cooperate with cyber-forensics conducted by the target State to identify the cyber-attack's source (Heinegg 2013). The transit State may still be under a general obligation to enact and enforce domestic cybercrime legislation, as discussed above, although this is unlikely to be relevant for mere transmission. Most broadly, the State may be subject to a generalized duty to maintain a minimum standard of cybersecurity care, as discussed above for the States in which the attack originated.

The role of transit States ultimately will reflect the degree to which their actions and omissions contributed to the attack. While these obligations are certainly less demanding than those of the State where the attack originated, transit States nonetheless may have some obligations, and must consider the international implications of their domestic cybersecurity strategies. However, it should be noted that as command and control servers move to target nations, due diligence standards may shift (Mcafee 2013). And regardless, there is a need to clarify the international law of neutrality more broadly to define whether or not victim nations can or should hold neutral nations through which cyber attacks transited accountable for not being diligent in repelling attackers (Schmitt 2013).

8.2.2.5 Caveats

Although all of these cases address the concept of international due diligence, it is unclear to what extent these opinions should shape international cybersecurity law and policy. Both *Corfu Channel* and *Trail Smelter* are arguably distinguishable on the grounds of physical proximity. *Corfu Channel* involved a State's obligations in their bordering sovereign waters and addressed issues raised by ships of other nations physically occupying those waters, while *Trail Smelter* involved environmental discharge across a neighbor's borders. Both cases recognize that actions undertaken by a State within its own territory can have consequences beyond that territory, but are nonetheless constrained to geographically proximate territories. But this geographical constraint is not reflected in the realm of cybersecurity, wherein actions taken within one's borders can impact anywhere accessible via global networks. This substantial expansion of the territory on which harmful activity may occur may be the slippery slope that derails this aspect of cybersecurity due diligence requirements for States. After all, if it were otherwise many nations would be in breach of the environmental obligations to one another through the emission of greenhouse gases responsible for global climate change.

Yet perhaps this aspect of international due diligence should be an arena of *lex feranda* that could lead to a change in attitudes within the international community. International environmental obligations, although originally geographically constrained, have increased in their scope of impact, with major environmental catastrophes such as the Fukushima Nuclear Reactor and the Deepwater Horizon oil spill showing that a single stakeholder's environmental actions and omissions can lead to global environmental challenges. As the world shrinks through environmental and technological changes, geographic isolation, perhaps, should no longer be a viable excuse for neglecting common "no harm" obligations. Indeed, some commentators have already argued "that states have an obligation of due diligence to prevent significant transboundary cyberharm to another state's intellectual property" (Messerschmidt 2013, 279).

Another caveat to the above discussion that should be addressed is the exemption for national security under international law. Customary international law recognizes four national security exceptions: change of circumstances, the law of reprisal, self-defense, and the doctrine of necessity (Rose-Ackerman 2008). Each of these exceptions recognizes instances in which a State's international obligations can be stayed due to the actions or threat of action of another State. While narrow in scope, these exceptions insert more uncertainty into an already uncertain arena, as none have been clarified in the realm of cyber-activities, which often implicate issues of national security (Schmitt 2014). For instance, the World Trade Organization ("WTO"), incorporating the General Agreement on Tariffs and Trade ("GATT"), employs a broad exception for "essential security interests," which effectively serves as an un-appealable, self-determined "get out of jail free card" (GATT 1994). Despite the GATT's restriction on unilateral economic sanctions, the United States has on multiple occasions used the national security exception to impose unilateral economic sanctions, most recently against Russia. This exception for national security is a frequently bemoaned aspect of international law, but nevertheless suggests a fundamental valuation in the international community that State sovereignty is to be given preference on issues implicating essential security interests. Therefore, any cybersecurity due diligence standards must be understood to likely contain a national security exception, and the ever-increasing importance of cybersecurity for national security interests may lead such an exception to ultimately swallow the rule.

Ultimately, the existence of these caveats and exceptions makes any definitive statement regarding the status of international due diligence standards that much more difficult, leading to the necessity of examining public- and private-sector approaches to help clarify some of the missing elements to a cybersecurity due diligence norm.

8.3 National and Private-Sector Approaches to Cybersecurity Due Diligence

As was discussed in the previous section international law, while informative, does not spell out in detail how nations should go about enhancing their cybersecurity to account for emerging due diligence obligations. As a result, it is helpful to consider both public- and private sector approaches for defining due diligence. Such national strategies could, in time, crystallize into customary international law as state practice clarifies (Henckaerts et al. 2005). Similarly, given the extensive public-private cross-pollination of cybersecurity best practices, private-sector efforts aimed at enhancing cybersecurity are similarly informative to consider given the extent to which they are informing national policymaking with the NIST Framework being a case in point. Thus, this final section begins by discussing several national case studies of cybersecurity due diligence including the United States, Germany, and China as a first step to uncovering a due diligence governance spectrum.[2] We then move on to discuss the extent to which cybersecurity is entering the due diligence process of mergers and acquisitions in the U.S. private sector context. Finally, we conclude with several observations about how industry cybersecurity norms are being translated into national policy, and what that means for managers, policymakers, and the field of cybersecurity due diligence generally.

8.3.1 National Approaches to Regulating Cybersecurity Due Diligence

This sub-section briefly reviews the national approaches of the United States, Germany, and China with regards to cybersecurity due diligence regulation. These case studies were chosen to provide common and civil law, as well as developed and emerging market perspectives on this issue. This analysis is not meant to be dispositive of the topic under consideration, but rather to provide a snapshot for how this influential subset of nations is approaching the topic of cybersecurity due diligence. Further research is required to flesh out whether the noted trends are playing out globally.[3]

8.3.1.1 United States

The topic of cybersecurity due diligence per se has not received an inordinate amount of attention by the Obama Administration, though it has referenced the topic in its 2011 International Strategy for Cyberspace. In it, the Administration states of cybersecurity due diligence that: "States should recognize and act on their responsibility to protect information infrastructures and secure national systems from damage or misuse" (Obama 2011, 10). This represents an effort to help crystallize a cybersecurity due diligence norm in international law essential to broader efforts to promote cyber peace (Shackelford 2014). The argument goes that due to the practical and political difficulties surrounding multilateral treaty development in the cybersecurity arena, norm creation provides an opportunity to enhance global cybersecurity without waiting for a comprehensive global agreement, which could

---

[2] For further information on how cybersecurity governance is playing out in the arena of critical infrastructure protection around the world, see generally Shackelford and Craig 2014.

[3] A more comprehensive comparative case study of these cyber powers—including a cybersecurity due diligence matrix—is included in Shackelford, Russell, & Kuehn, *supra* note 1.

come too late if at all. Yet despite general agreement as to the value of cybersecurity norms including due diligence, still "even simple norms face serious opposition. Conflicting political agendas, covert military actions, espionage[,] and competition for global influence" have created a difficult context for cyber norm development and diffusion (Lewis 2011b, 58); a situation that NSA revelations arguably exacerbated. As a result, to be successful in such a difficult climate, norms must be "clear, useful, and do-able ...." (Finnemore 2011, 90). What would a cybersecurity due diligence norm look like, then? It is helpful to briefly review U.S. approaches to this topic in order to provide a build out the framework for discussion discussed above.

The United States in many ways pioneered cybersecurity at the national level, beginning with the creation of the first Cyber Emergency Response Team at Carnegie Mellon University in 1988 in response to a growing number of network intrusions. Today, though, the field is crowded with an alphabet soup of agencies and organizations responsible for various aspects of the nation's cybersecurity. The Department of Defense alonem reportedly operates more than 15,000 networks in 4000 installations spread across some 88 countries (Lord and Sharp 2011). Yet the majority of U.S. efforts in this space have been focused on securing vulnerable critical infrastructure ("CI"). Although Congress have been active in this regard, successive administrations—including those of Presidents Clinton, Bush, and Obama—have pushed the ball forward on securing vulnerable CI.

Most recently, President Obama declared the U.S. CI to be a "strategic national asset" in 2009 though a fully integrated U.S. cybersecurity policy has yet to be established (Obama 2009). In the face of Congressional inaction, President Obama issued an executive order that, among other things, expanded public-private information sharing and established the NIST Framework comprised partly of private-sector best practices that companies could adopt to better secure CI (Obama 2013). This Framework is important since, even though its critics argue that it helps to solidify a reactive stance to the nation's cybersecurity challenges (Armerding 2014), it is arguably spurring the development of a standard of cybersecurity care in the United States that plays into discussions of due diligence. In particular, the NIST Framework harmonizes consensus standards and industry best practices to provide, its proponents argue, a flexible and cost-effective approach to enhancing cybersecurity that assists owners and operators of critical infrastructure in assessing and managing cyber risk. Although the NIST Framework has only been out for a relatively short time, already some private-sector clients are receiving the advice that if their "cybersecurity practices were ever questioned during litigation or a regulatory investigation, the 'standard' for 'due diligence' was now the NIST Cybersecurity Framework" (Verry 2014). Over time, the NIST Framework not only has the potential to shape a standard of care for domestic critical infrastructure organizations but also could help to harmonize global cybersecurity best practices for the private sector writ large given active NIST collaborations with a number of nations including the United Kingdom, Japan, Korea, Estonia, Israel, and Germany.

8.3.1.2 Germany

Germany's cybersecurity due diligence efforts rely on close collaboration between the public and private sectors, nationally and globally (German Federal Ministry of the Interior 2011). Long known for its strong national data protection law with fines up to EUR 300,000, Germany is moving now to: (a) mandate strict cybersecurity standards for CI, and (b) assign the responsibility to protect users and secure CI to service providers and operators of CI, respectively (Bundesministerium des Innern 2008). In particular, the federal government approved the German Cybersecurity Strategy ("Cyber-Sicherheitsstrategie für Deutschland") in February 2011. The Strategy recognizes cyberspace as an essential domain for the German state, economy, and society, and emphasizes the protection of CI as a core cybersecurity policy priority. Moreover, the Strategy addresses cybersecurity due diligence by recognizing that "incidents in other countries' information infrastructures may also indirectly affect Germany" (German Federal Ministry of the Interior 2011, 4). It also calls for a code of conduct, international legal harmonization and cooperation, and states that service providers may need to assume greater responsibility for the security of their digital products and users (German Federal Ministry of the Interior 2011, 4–7).

Germany has also been active in identifying and spreading cybersecurity best practices in a similar vein as the NIST Framework. The Federal Office for Information Security ("Bundesamt für Sicherheit in der Informationstechnik", BSI) first released its IT Baseline Protection ("IT-Grundschutz") in 1994. This set of BSI standards contains recommendations for cybersecurity and has been adopted by German corporations and international stakeholders; some of the standards are now available in English, Swedish, and Estonian. These standards are best practice recommendations that have become "de-facto standards for [the German] IT security" (OWASP Review), but are not legally enforceable save for data protection fines mentioned earlier.

Efforts are also underway in Germany's private sector to widen the discussion and dissemination of cybersecurity best practices. For example, established in 2012, the Alliance for Cybersecurity ("Allianz für Cybersicherheit") is an initiative under the aegis of the Federal Office for Information Security. It brings together more than a thousand public and private participating entities to share best practices and further the cause of German cybersecurity due diligence. The Alliance encourages voluntary reporting of cyber incidents and attacks to collect information about current cyber threats against German organizations. These private efforts help to shape industry norms and contribute towards responsible cyber behavior.

Germany's Minister of the Interior Dr. Thomas de Maizière recently addressed the topic of cybersecurity due diligence in particular during the 2014 Global Cyberspace Cooperation Summit in Berlin. Referring to the need to carefully consider the principle of responsibility in cyberspace, de Maizière, pointed to a basic tenet in law: he who creates a risk for others is responsible for it. The greater the risk, the larger the responsibility ("[…] wer ein Risiko für andere schafft, trägt dafür Verantwortung. Je größer das Risiko ist, umso höher die Verantwortung") (de

Maizière 2014).[4] Partly in response to this sentiment (and the 2013 NSA revelations), the German government drafted the IT Security Act ("Entwurf eines Gesetzes zur Erhöhung der Sicherheit informationstechnischer Systeme (IT-Sicherheitsgesetz)"), which is pending as of this writing. If enacted, the new law would require companies to employ state of the art security standards to secure their websites – or be held liable in the event of a breach. More stringent security requirements and responsibilities would apply for CI operators. The designated CI sectors are responsible for developing appropriate security standards (similar to the NIST Framework's approach), pending the Ministry of the Interior's approval. CI operators would also be obligated to inform the authorities of cyber attacks. These cybersecurity policy efforts are estimated to create a need for between 200 and 425 new jobs across the federal government and cost for personnel and resources of up to EUR 38 million annually (Greis 2014).

8.3.1.3 China

China applies tight controls over its domestic Internet in order to advance the Communist party's economic, political, and military interests and to secure its rule (Wong 2014). On the international stage, it continuous to seek cooperation "to promote the building of a peaceful, secure, open, and cooperative cyberspace" and attempts to shape international norms, particularly with regard to the sovereign state's control over the domestic Internet and censorship under the disguise of information security (Sceats 2015).[5] At the same time, there are increasing tensions between the U.S. and China about mutually alleged cyber exploitations. In 2014, the U.S. indicted five hackers of the People's Liberation Army for economic cyber espionage; China protested sharply (Weihua 2014). The U.S. government has billed China as the "world's most active and persistent perpetrators of economic espionage" (DNI 2011), while in June 2013, President Obama warned that the continuation of U.S. intellectual property theft is a serious matter that will hinder the further development of economic trade relations with China. The U.S reaction can be conceived as an approach to shape norms on cybersecurity due diligence, by calling out China to take responsibilities for alleged cyber exploitations. Indeed, there has been progress on this score with the G2 cybersecurity code of conduct being announced in 2015, which among other things commits the U.S. and China to collaboratively stop bilateral economic espionage. Such norms have a strong political dimension, as the China case study shows.

---

[4] This sentiment may also be considered another manifestation of the sliding scale approach discussed in Section 16.2.1.

[5] China is pursuing cyber diplomacy on an array of fronts. Among other actions, China is furthering the multilateral cybersecurity initiative with the Shanghai Cooperation Organization, is negotiating a bilateral cybersecurity treaty with Russia, is involved in a U.S.-China working group to diffuse tensions around mutually alleged cyber exploitations, and has been drafting cybersecurity-relevant proposals and declarations to garner support from like-minded states at the 2014 World Internet Conference in China and at various UN meetings.

As with the U.S., China's cybersecurity strategy is fragmented, but its development and implementation has recently garnered political support of high-ranking senior government officials. In early 2014, Chinese President Xi Jinping stressed that a uniform and comprehensive approach to "network security" is necessary to turn China into a "cyber power" (Jinping 2014). The speech coincided with the establishment of the "Central Cyber Security and Informatization Leading Group," which under the leadership of President Xi Jinping will guide China's cybersecurity policy efforts.

In many ways, China's cybersecurity strategy is broader in scope than either its U.S. or German counterparts. In addition to addressing the security of networks and computers, it includes censorship of content and information control to a far greater extent than is the case in these Western nations. It is the Chinese government's official position that "properly guiding Internet opinion is a major measure for protecting Internet information security" (Buckley and Hornby 2010). China's take on cybersecurity is reflected in the idea of Internet sovereignty and its use of the Internet as a means to build up a domestic information economy and secure network infrastructure that benefits domestic development and political stability.

China's first cybersecurity strategy dates back to 2003. It is referred to as "Document 27: Opinions for Strengthening Information Security Assurance Work and covers – inter alia – CI protection (Segal 2012). The current 2012 cybersecurity strategy continues some of the earlier cybersecurity considerations (including CI protection) while also addressing China's dependency on foreign technology as a security issue, the promotion of Chinese cryptography standards, and the build-up of broadband infrastructure, next-generation mobile technology, and e-government services (Gierow 2014). Observers have criticized the document as an inconsistent "grab bag of vague policy proposals" (Segal 2012).

Some of these measures are in line with cybersecurity due diligence efforts; others are broader in scope and have raised concerns, particularly from U.S. and European counterparts. For example, in 2007, China established a set of security standards, the "Regulations on Classified Protection of Information Security" (which are also referred to as the Multi-Level Protection Scheme, "MLPS") with the objective of safeguarding information and protecting national security (Ahrens 2012). Western firms and organizations repeatedly expressed their disapproval since these technical standards are incompatible with international IT security standards. Rather than protecting national security, these standards have been perceived as protectionist measures that shield Chinese domestic IT firms from global competition. Some argue that such efforts have actually resulted in *less* secure Chinese standards and technology (Gierow 2014). Leading cybersecurity companies such as Kaspersky and Symantec have been barred from competing in China's corporate market for financial institutions and power utilities, for instance. Such developments may help open the door for cyber attacks on China's CI; a detriment to the cause of cybersecurity due diligence.

Similar to MLPS, and as part of its economic policy, China has attempted to establish its own wireless network standard ("WAPI"). In reaction to NSA revelations, it announced work on an independent, Chinese operating systems for desktop computers as well as mobile devices (Xinhunet 2014). Other recent or pending Chinese legislation portend still more protection, such as requiring technology companies that sell to China's banks to submit their source code for government inspection and to even divulge encryption keys and install backdoors to give Chinese authorities access to secured data and communication (Mozur 2015). Such policies impact Western tech firms, and could even bar them from China's growing market (Gierow 2014).

In summary, China expresses the need for the control of information and exclusion of foreign owned-security technologies in order to protect its societal stability. As a result, its strategy focuses on national security and economic advancement. Elements of cyber due diligence consequently look quite different when compared to the U.S. or German cases, demonstrating the difficulty of crafting a global norm in this space. However, one could potentially construe a Chinese version of cybersecurity due diligence that is at the other end of a possible spectrum and that includes domestic economic rationales and protectionist measures as opposed to a narrower focus on securing CI. In fact, many of the policy objectives are similar across the three case studies; what differs are the means.

Custom requires widespread state practice that is undertaken out of a sense of legal obligation. Depending on the type of norm involved, that state practice needs to be more or less widespread. For new norms, such as cybersecurity, the standard generally is "virtually uniform" state practice.[6] This threshold has not yet arguably been reached in the cybersecurity due diligence context, as may be seen by the three approaches taken by these nations with the U.S. being more voluntary, Germany taking a relatively more regulatory approach, and China's broader economic and national security efforts. Yet aside from national case studies, there are also valuable lessons from the private sector that could inform the eventual shape of a cybersecurity due diligence norm, which we turn to next.

8.3.2 Lessons from the Private Sector

Among the criticisms of the NIST Framework is that, although it does a good job at promoting general "cyber hygiene" for those organizations that implement it, it is less well suited to protecting firms from sophisticated and targeted cyber attacks sometimes called Advanced Persistent Threats ("APTs"). Indeed, there is a cybersecurity due diligence industry emerging in which the NIST Framework, and for that matter the German BSI Standards, play a role but are only one aspect of a larger decision-making process that companies contemplating all sorts of business decisions from mergers and acquisitions to supply chain management must consider. This section investigates some hallmarks of this trend primarily in the U.S. mergers and acquisitions context but with other related asides as space permits.

---

[6] N. Sea Continental Shelf (F.R.G./Den. v. Neth.) 1969 I.C.J. 41, 72 (Feb. 20).

U.S. law includes a host of relevant legal questions faced by the private sector as part of an overarching cybersecurity due diligence process (Barnett 2014). It is critical for companies, for example, to have detailed cybersecurity strategies in place on what employee and customer data has been retained and used, and how that data is secured. If unsatisfactorily undertaken, potential resulting causes of action include negligence, breach of contract, breach of fiduciary duty, and invasion of privacy, to name a few. This can lead to the ousting of managers up to and including the C-Suite as seen in the aftermath of the Target and Sony cyber attacks, but still many organizations have not taken the necessary steps to internalize cybersecurity due diligence (PwC 2012). One arena in which some progress is being made, though, is mergers and acquisitions.

Jason Weinstein, former deputy assistant attorney general at the U.S. Department of Justice, summarized the issue of cybersecurity due diligence succinctly when he said: "When you buy a company, you're buying their data, and you could be buying their data-security problems" (Ensign 2014). In other words, "Cyber risk should be considered right along with financial and legal due diligence considerations" (Ayres 2014). Already a majority of respondents in one 2014 survey reported that cybersecurity challenges are altering the M&A landscape, while 82 % said that cyber risk would become more predominant over the following 18 months (Ayres 2014). A majority of surveyed firms also said that a cyber attack during the M&A negotiation process could scuttle the deal, which is a concern given the range of serious cyber attacks coming to light on a regular basis in an era of increasing mergers (Ayres 2014). Managers now considering what form cybersecurity due diligence should take have a wealth of resources (as well as a growing array of compliance obligations) to consider. These include, in the U.S. context, the NIST Framework, as well as guidance from the Securities and Exchange Commission, International Standards Organization, National Association of Corporate Directors, and the Payment Card Industry (PCI) Security Standards Council (Ayres 2014). Together, these frameworks, and others, provide the beginnings of a cybersecurity due diligence standard guiding judges as they work through causes of action such as breach of fiduciary duty and negligence resulting from data breaches.[7] The same goes for partnerships with vendors. The Target breach, for example, which wound up exposing some 40 million credit card numbers, was the result of lax security from a HVAC (heating, ventilation, and air conditioning) vendor that for some reason had access to myriad Target systems well beyond HVAC networks.

The end result of all this is that there is a push among IT professionals to go beyond mere due diligence and move toward the use of real-time analytics and other cybersecurity best practices to monitor vendors' systems (Norton 2014). The lesson here is constant vigilance, e.g., letting an initial process of cybersecurity due diligence be the first, and not the last, word in an ongoing proactive and comprehensive

---

[7] *Cf.* Willingham v. Global Payment, 2013 WL 440,702 at 19 (N.D. Ga 2013) (unreported) (reflecting an alternative view in which courts are reluctant rely on data security standards as a means of determine whether a duty was owed, let alone whether they should be used to determine a reasonable standards of care).

cybersecurity policy that promotes cyber hygiene along with the best practices essential for battling APTs. Such a policy should be widely disseminated and regularly vetted as part of an overarching enterprise risk management process, along with having an incident response plan in place that includes private and public information sharing mechanisms.[8]

8.3.3 A Polycentric Approach to Cybersecurity Due Diligence

These private sector best practices should inform national and indeed international debates playing out in the field of cybersecurity due diligence. This multi-level, multi-purpose, multi-functional, and multi-sectoral model (McGinnis 2011), championed by scholars including Nobel Laureate Elinor Ostrom and Professor Vincent Ostrom, challenges orthodoxy by demonstrating the benefits of self-organization, networking regulations "at multiple scales," and examining the extent to which national and private control can in some cases coexist with communal management (Ostrom 2008). It also posits that, due to the existence of free riders in a multipolar world, "a single governmental unit" is often incapable of managing "global collective action problems" such as cyber attacks (Ostrom 2009, 35). Instead, a polycentric approach recognizes that diverse organizations working at multiple levels can create different types of policies that can increase levels of cooperation and compliance, enhancing "flexibility across issues and adaptability over time" (Keohane and Victor 2011, 9). This conceptual framework, in other words, recognizes both the common but differentiated responsibilities of public- and private-sector stakeholders as well as the potential for best practices to be identified and spread organically generating positive network effects that could, in time, result in the emergence of a cascade toward a cybersecurity due diligence norm.[9] Such a norm should not only focus on the cyber hygiene referenced in the NIST Framework but should also encourage the uptake of proactive cybersecurity best practices referenced above so as to secure our networks along with clarifying the rights and responsibilities of transit states.

8.4 Conclusion

The field of international cybersecurity due diligence remains a complex, demanding, and difficult arena, but one that requires sustained academic, private, and public engagement if progress is to be made. There is an array of paths forward. For

---
[8] For more on this topic, see Amanda N. Craig, Scott J. Shackelford, & Janine Hiller, *Proactive Cybersecurity: A Comparative Industry and Regulatory Analysis*, 18 Am. Bus. L. J. 721 (2015).
[9] *See* Martha Finnemore & Kathryn Sikkink, *International Norm Dynamics and Political Change*, 52 Int'l Org. 887, 895–98 (1998).

example, States could exercise due diligence through passive means, promote resiliency in domestic and partner nation's networks (Edwards 2007). Warning systems for various types of cyber attacks facilitated by cyber emergency response teams, active (and two-way) private-sector information sharing and collaboration on identifying and spreading cybersecurity best practices, and a robust cyber hygiene campaign may be considered other essential elements of cybersecurity due diligence. Other best practices include partitioning access to code and systems, audits and regular penetration testing, and promoting redundancy and parallel network construction to build further resiliency, as well as harnessing cybersecurity expertise beyond one's own organizational boundaries through bug bounty and vulnerability reward programs (Westervelt 2013; Kuehn and Mueller 2014). The NIST Framework, and the related standards it references, provides a conceptual toolbox to identify gaps in an organization's cybersecurity readiness that both public and private sector actors should be aware, along with the German BSI Standards and Chinese equivalents. There is plenty of low-hanging fruit. After all, the Australian government hasreportedly been succcessful in preventing 85 % of cyber attacks through following three common sense techniques: application whitelisting (only permitting pre-approved programs to operate on networks), regularly patching applications and operating systems, and "minimizing the number of people on a network who have 'administrator' privileges" (Lewis 2013).

Over time, as legal harmonization progresses, there will be increasing opportunities to build out cybersecurity norms, including those surrounding the question of due diligence. Already, a number of national governments referenced above, and even some companies such as Microsoft, have released lists of draft norms for stakeholder consideration (McKay et al. 2014). Given both the rich cross-pollination of cybersecurity best practices and the cyber threat posed by a huge range of attackers to the public and private sectors, conceptions of cybersecurity due diligence should be gleaned from existing customary international law but built out through a review of industry norms that are in turn informing national policies. Achieving some measure of cyber peace requires the active involvement of public and private stakeholders. It may be time for more international lawyers to reach out to CISOs, and vice versa.

References

Ahrens, Nathaniel. 2012. National security and China's information security standards: Of Shoes, Buttons, and Routers. Center for Strategic and International Studies, November 8. http://csis.org/publication/national-security-and-chinas-information-security-standards. Accessed 26 Mar 2015.

Armerding, Taylor. 2014. NIST's finalized cybersecurity framework receives mixed reviews. *CSO*, January 31. http://www.csoonline.com/article/2134338/security-leadership/nist-s-finalized-cybersecurity-framework-receives-mixed-reviews.html. Accessed 26 Mar 2015.

Ayres, Erin. 2014. Cybersecurity easing its way into M&A due diligence. *Cyber Risk Network*, Aug. 22. http://www.cyberrisknetwork.com/2014/08/22/cybersecurity-easing-way-ma-process/. Accessed 26 Mar 2015.

Barnett et al. 2014. *Cybersecurity issues in dealmaking: What you need to know*. ACG. http://www.acg.org/UserFiles/file/Cybersecurity%20Webinar%20-Final.pdf. Accessed 26 Mar 2015.

Bodle, Ralph. 2012. Climate Law and geoengineering. In *Climate change and the law, Ius Gentium: Comparative perspectives on law and justice*, ed. Erkki Hollo, Kati Kulovesi, and Michael Mehling, 447–470. Dordrecht: Springer.

Botnet Control Servers Span the Globe. Mcafee. https://blogs.mcafee.com/mcafee-labs/botnet-control-servers-span-the-globe. Accessed 23 Jan 2013.

Bradley, Curtis A. 2013. The chronological paradox, state preferences, and Opinio Juris. Duke law. http://law.duke.edu/cicl/pdf/opiniojuris/panel_1-bradley-the_chronological_paradox,_state_preferences,_and_opinio_juris.pdf. Accessed 26 Mar 2015.

Buckley, Chris and Lucy Hornby. 2010. China defends censorship after Google threat. *Reuters*, January 14. http://www.reuters.com/article/2010/01/14/us-china-usa-google-idUSTRE60C1TR20100114. Accessed 26 Mar 2015.

Bundesministerium des Innern. 2008. *Schutz Kritischer Infrastrukturen – Risiko- und Krisenmanagement: Leitfaden für Unternehmen und Behörden*. http://www.bmi.bund.de/SharedDocs/Downloads/DE/Broschueren/2008/Leitfaden_Schutz_kritischer_Infrastrukturen.pdf?__blob=publicationFile. Accessed 26 Mar 2015.

Case Concerning the Military and Paramilitary Activities In and Against Nicaragua (Nicaragua v. U.S.), 1986. I.C.J.14, 183 (June 27).

Chinese OS expected to debut in October. Xinhunet, August 24, 2014. http://news.xinhuanet.com/english/china/2014-08/24/c_133580158.htm. Accessed 26 Mar 2015.

Clinton, Hillary Rodham. 2010. Remarks on internet freedom. U.S. Department of State. http://www.state.gov/secretary/20092013clinton/rm/2010/01/135519.htm. Accessed 26 Mar 2015.

Corfu Channel Case (United Kingdom v. Albania), 1949. I.C.J. 244 (Dec.15).

de Maizière, Thomas. 2014. Sichere Informationsinfrastrukturen in einem Cyber-Raum der Chancen und der Freiheit. http://www.bmi.bund.de/SharedDocs/Reden/DE/2014/12/east-west-cyber-summit.html?nn=3314802. Accessed 26 Mar 2015.

Definition of International Law. Int'l Labor Org. http://www.actrav.itcilo.org/actrav-english/telearn/global/ilo/law/lablaw.htm. Accessed 25 Mar 2015.

Del Mar, Katherine. 2012. The international court of justice and standards of proof. In *The ICJ and the evolution of international law: The enduring impact of the Corfu channel case*, ed. Karine Bannelier, 98–123. London: Routledge.

DNI, Office of the National Counterintelligence Executive, Foreign Spies Stealing U.S. Economic Secrets in Cyberspace, Report to Congress on Foreign Economic Collection and Industrial Espionage: 2009–2011, October 2011.

Edwards, Dennis et al. 2007. Prevention, detection and recovery from cyber-attacks using a multilevel agent architecture. *System of systems engineering* 1, 1 (2007). doi:10.1109/SYSOSE.2007.4304228.

Ensign, Rachel Louise. 2014. Cybersecurity due diligence key in M&A deals. *W all Street Journal*, April 24. http://blogs.wsj.com/riskandcompliance/2014/04/24/cybersecurity-due-diligence-key-in-ma-deals.

Eye of the Storm: Key Findings from the 2012 Global State of Information Security Survey. PwC. http://www.pwc.co.nz/global-state-of-information-survey.aspx. Accessed 26 Mar 2015.

Finnemore, Martha. 2011. Cultivating international cyber norms. In *America's cyber future: Security and prosperity in the information age*, eds. Kristin M. Lord and Travis Sharp, 87–102. Washington, DC: CNAS.

GATT 1994: General Agreement on Tariffs and Trade 1994, Apr. 15, 1994, Marrakesh Agreement Establishing the World Trade Organization, Annex 1A, THE LEGAL TEXTS: THE RESULTS OF THE URUGUAY ROUND OF MULTILATERAL TRADE NEGOTIATIONS 17 (1999), 1867 U.N.T.S. 187, 33 I.L.M. 1153 (1994).

General Assembly resolution 55/63, Combatting the criminal use of information technologies, A/RES/55/63 (22 Jan 2001). http://www.itu.int/ITU-D/cyb/cybersecurity/docs/UN_resolution_55_63.pdf. Accessed 26 Mar 2015.

German Federal Ministry of the Interior. 2011. Cyber-Sicherheitsstrategie für Deutschland. http://www.bmi.bund.de/DE/Themen/IT-Netzpolitik/IT-Cybersicherheit/Cybersicherheitsstrategie/cybersicherheitsstrategie_node.html. Accessed 26 Mar 2015.

Gierow, Hauke Johannes. 2014. *Cyber security in China: New political leadership focuses on boosting national security*. Mercator Institute for China Studies. http://www.merics.org/fileadmin/templates/download/china-monitor/China_Monitor_No_20_eng.pdf. Accessed 26 Mar 2015.

Greis, Friendhelm. 2014. Kabinett beschließt Meldepflicht für Cyberangriffe. *Golem.de*, December 17. http://www.golem.de/news/it-sicherheitsgesetz-regierung-beschliesst-meldepflicht-fuer-cyberangriffe-1412-111234.html. Accessed 26 Mar 2015.

Gruener, Wolfgang. 2012. Many new PCs in China come with malware preinstalled. *Tom's Hardware*, September 24. http://www.tomshardware.com/news/microsoft-pc-windows-security-china,17758.html. Accessed 26 Mar 2015.

Gulati, Mitu. 2013. How do courts find international custom? *Duke Law*. http://law.duke.edu/cicl/pdf/opiniojuris/panel_6-gulati-how_do_courts_find_international_custom.pdf. Accessed 26 Mar 2015.

Henckaerts, Jean-Marie, and Doswald-Beck, Louise. 2005. *Assessment of customary international law*. ICRC. http://www.icrc.org/customary-ihl/eng/docs/v1_rul_in_asofcuin. Accessed 26 Mar 2015.

Hurwitz, Roger. 2009. The prospects for regulating cyberspace: A schematic analysis on the basis of Elinor Ostrom. MIT. http://web.mit.edu/ecir/pdf/hurwitz-ostrom.pdf. Accessed 26 Mar 2015.

International Labor Organization. 2013. Definition of international law. *International Labour Organization*. http://www.actrav.itcilo.org/actrav-english/telearn/global/ilo/law/lablaw.htm. Accessed 25 Mar 2015.

Jinping, Xi: China must evolve from a large internet nation to a powerful internet nation. *Xinhuanet.com*, February 27, 2014. http://news.xinhuanet.com/politics/2014-02/27/c_119538788.htm. Accessed 26 Mar 2015.

Keohane, Robert O., and David G. Victor. 2011. The regime complex for climate change. *Perspectives on Policy* 9: 7–23.

Kirgis, Frederic L. 1987. Custom on a sliding scale. *The American Journal of International Law* 81(1): 146–151.

Kuehn, A., and M. Mueller. 2014. Analyzing bug bounty programs: An institutional perspective on the economics of software vulnerabilities. *Proceedings of the 42nd research conference on communication, information, and internet policy*. 12–14 September, 2014, Arlington, VA. http://papers.ssrn.com/sol3/papers.cfm?abstract_id=2418812.

Lewis, James A. 2011a. Why privacy and cyber security clash. In *America's cyber future: security and prosperity in the information age*, ed. Kristin M. Lord and Travis Sharp, 123–142. Washington, DC: CNAS.

Lewis, James A. 2011b. Confidence-Building and international agreement in cybersecurity. In *Disarmament forum: Confronting cyberconflict*, 51–59. United Nations Institute for Disarmament Research. http://www.unidir.org/files/publications/pdfs/confronting-cyberconflict-en-317.pdf. Accessed 26 Mar 2015.

Lewis, James A. 2013. Raising the bar for cybersecurity. CSIS. http://csis.org/files/publication/130212_Lewis_RaisingBarCybersecurity.pdf. Accessed 26 Mar 2015.

Lord, Kristin M., and Travis Sharp. 2011. Executive summary. In *America's cyber future: Security and prosperity in the information age*. Washington, DC: CNAS.

McGinnis, Michael D. 2011. An introduction to IAD and the language of the Ostrom workshop: A simple guide to a complex framework. *Policy Studies Journal* 39(1): 169–183.

McKay et al. 2014. *International cybersecurity norms: Reducing conflict in an Internet-dependent world*. Microsoft. http://tinyurl.com/ogv9qzq. Accessed 26 Mar 2015.

Messerschmidt, Jan E. 2013. Hackback: Permitting retaliatory hacking by non-state actors as proportionate countermeasures to transboundary cyberharm. *Columbia Journal of Transnational Law* 52: 275–323.

Mozur, Paul. 2015. New rules in China upset Western Tech Companies. *New York Times*, January 28. http://www.nytimes.com/2015/01/29/technology/in-china-new-cybersecurity-rules--perturb-western-tech-companies.html. Accessed 26 Mar 2015.

Mudrinich, Erik M. 2012. Cyber 3.0: The department of defense strategy for operating in cyberspace and the attribution problem. *The Air Force Law Review* 68: 167–206.

N. Sea Continental Shelf (F.R.G./Den. v. Neth.), 1969. I.C.J. 41, 72 (Feb. 20).

Norton, Steven. 2014. Going beyond due diligence to monitor vendor cybersecurity. *Wall Street Journal*, March 21. http://blogs.wsj.com/cio/2014/03/21/going-beyond-due-diligence-to--monitor-vendor-cybersecurity/. Accessed 26 Mar 2015.

Obama, Barack. 2009. Remarks by the president on securing our nation's cyber infrastructure. White House, Office of the Press Secretary. http://www.whitehouse.gov/the-press-office/remarks-president-securing-our-nations-cyber-infrastructure. Accessed 26 Mar 2015.

Obama, Barack. 2011. *International strategy for cyberspace: Prosperity, security, and openness in a networked world*. White House. https://www.whitehouse.gov/sites/default/files/rss_viewer/international_strategy_for_cyberspace.pdf. Accessed 26 Mar 2015.

Obama, Barack. 2013. *Executive order on improving critical infrastructure cybersecurity*. White House, Office of the Press Secretary. http://www.whitehouse.gov/the-press-office/2013/02/12/executive-order-improving-critical-infrastructure-cybersecurity. Accessed 26 Mar 2015.

Ophardt, Jonathan A. 2010. Cyber warfare and the crime of aggression: The need for individual accountability on tomorrow's battlefield. *Duke Law and Technology Review* 3: 1–76.

Ostrom, Elinor. 2008. Polycentric systems as one approach for solving collective-action problems. Indiana University. http://dlc.dlib.indiana.edu/dlc/bitstream/handle/10535/4417/W08-6_Ostrom_DLC.pdf?sequence=1. Accessed 26 Mar 2015.

Ostrom, Elinor. 2009. A polycentric approach for coping with climate change. The World Bank. http://www.iadb.org/intal/intalcdi/pe/2009/04268.pdf. Accessed 26 Mar 2015.

OWASP Review BSI IT-Grundschutz Baustein Webanwendungen. https://www.owasp.org/index.php/OWASP_Review_BSI_IT-Grundschutz_Baustein_Webanwendungen. Accessed 26 Mar 2015.

PwC. 2012. Eye of the storm: Key findings from the 2012 global state of information security survey. PwC. http://www.pwc.co.nz/global-state-of-information-survey.aspx. Accessed 26 Mar 2015.

Rose-Ackerman, Susan, and Benjamin Billa. 2008. Treaties and national security. *New York University Journal of International Law and Politics* 40: 437–495.

Ryan, Tim, and Leonard Navarro. 2015. Cyber due diligence: Pre-transaction assessments can uncover costly risks. Kroll, January 28. http://blog.kroll.com/2015/cyber-due-diligence-pre-transaction-assessments-can-uncover-costly-risks/.

Sceats, Sonya. 2015. China's cyber diplomacy: A taste of law to come? *The Diplomat*, January 14. http://thediplomat.com/2015/01/chinas-cyber-diplomacy-a-taste-of-law-to-come/.

Schmitt, Michael N. 2013. *Tallinn manual on the international law applicable to cyber warfare*. Cambridge: Cambridge University Press.

Schmitt, Michael N. 2014. "Below the threshold" cyber operations: The countermeasures response option and international Law. *Virginia Journal of International Law* 54: 697–732.

Segal, Adam. 2012. China moves forward on cybersecurity policy. *Council on Foreign Relations*, July 24. http://blogs.cfr.org/asia/2012/07/24/china-moves-forward-on-cybersecurity-policy/. Accessed 26 Mar 2015.

Shackelford, Scott J. 2014. *Managing cyber attacks in international law, business, and relations: In search of cyber peace*. Cambridge: Cambridge University Press.

Shackelford, Scott J., and Amanda N. Craig. 2014. Beyond the new 'digital divide': Analyzing the evolving role of governments in internet governance and enhancing cybersecurity. *Stanford Journal of International Law* 50: 119–184.

Sklerov, Matthew J. 2009. Solving the dilemma of state responses to cyberattacks: A justification for the use of active defenses against states who neglect their duty to prevent. *Military Law Review* 201: 1–84.
Statute of the International Court of Justice, art. 38, June 26, 1945, 59 Stat. 1055. http://www.icj-cij.org/documents/index.php?p1=4&p2=2&p3=0.
Tikk, Eneken. 2011. Ten rules of behavior for cyber security. *Survival* 53(3): 119–132.
Trail Smelter Case. 1938, 1965. Trail Smelter Arbitration (U.S. v. Can.), 3 Rep. Int'l Arb Awards (R.I.A.A.) 1905 (1941).
Verry, John. 2014. *Why the NIST cybersecurity framework isn't really voluntary*. Information Security Blogs. http://www.pivotpointsecurity.com/risky-business/nist-cybersecurity-framework. Accessed 26 Mar 2015.
von Heinegg, Wolff Heintschel. 2013. Territorial sovereignty and neutrality in cyberspace. *International Law Studies* 89: 123–156.
Weihua, Chen. 2014. China protests against US indictment. *China Daily*, May 20. http://usa.chinadaily.com.cn/world/2014-05/20/content_17519650.htm. Accessed 26 Mar 2015.
Westervelt, Robert. 2013. Kaspersky: Redundancy, offline backup critical for cyberdefense. *CRN*, February 8. http://www.crn.com/news/security/240148219/kaspersky-redundancy-offline-backup-critical-for-cyberdefense.htm. Accessed 26 Mar 2015.
Wong, Edward. 2014. For China, cybersecurity is part of strategy for protecting the communist party. *New York Times*, December 3. http://sinosphere.blogs.nytimes.com/2014/12/03/for-china-cybersecurity-is-part-of-strategy-for-protecting-the-communist-party/. Accessed 26 Mar 2015.
Zetter, Kim. 2014. *Countdown to zero day*. New York: Random House.

Scott J. Shackelford is a professor at the faculty of Indiana University along with being a research fellow at the Harvard Kennedy School's Belfer Center for Science and International Affairs, a senior fellow at the Center for Applied Cybersecurity Research, a visiting scholar at Stanford Law School and a term member at the Council on Foreign Relations. Professor Shackelford has written more than 100 books, articles and essays for diverse outlets ranging from the *University of Illinois Law Review* and the *American Business Law Journal* to the *Christian Science Monitor*, the *Huffington Post* and the *San Francisco Chronicle*. He is also the author of *Managing Cyber Attacks in International Law, Business, and Relations: In Search of Cyber Peace* (Cambridge University Press, 2014). Both Prof. Shackelford's academic work and teaching have been recognized with numerous awards, including a Hoover Institution National Fellowship, a Notre Dame Institute for Advanced Study Distinguished Fellowship, the 2014 Indiana University Outstanding Junior Faculty Award and the 2015 Elinor Ostrom Award.

Scott Russell is an attorney and postdoctoral fellow at Indiana University with the Center for Applied Cybersecurity Research. Scott specializes in information privacy, cybersecurity and digital surveillance law. His scholarship has included defining private-sector cybersecurity best practices, data aggregation and the Fourth Amendment and international data governance. In addition, Scott manages and frequently contributes to the CACR blog, the CACR Supplement, which provides in-depth discussion of current privacy and security issues. Scott received his bachelors from the University of Virginia and his JD from the Indiana University Maurer School of Law.

Andreas Kuehn is a predoctoral cybersecurity fellow at the Center for International Security and Cooperation at Stanford University and a PhD candidate in information science and technology and a Fulbright Scholar at the School of Information Studies at Syracuse University. His current research on cybersecurity focuses on the formation of software vulnerability and exploit markets, the Wassenaar Arrangement as it relates to the control of cyber technology as well as changes in information security norms and practices and the organizational and institutional aspects of vulnerability disclosure. Andreas has worked and consulted with various government agencies and regulators in Austria and Switzerland, private firms and think tanks on broad areas including cybersecurity, Internet governance, technology policy and digital government. Andreas has an MSc in information systems from the University of Zurich, Switzerland.

Chapter 9
Cyber Warfare and Organised Crime. A Regulatory Model and Meta-Model for Open Source Intelligence (OSINT)

Pompeu Casanovas

Abstract OSINT stands for *Open Source Intelligence*, (O)SI for *(Open) Social Intelligence*, and PbD for *Privacy by Design*. The CAPER EU project has built an OSINT solution oriented to the prevention of organized crime. How to balance freedom and security? This chapter describes a way to embed the legal and ethical issues raised by the General Data Reform Package (GDRP) in Europe into security and surveillance platforms. It focuses on the indirect strategy to flesh out ethical principles through Semantic Web Regulatory Models (SWRM), and discusses the possibility to extend them to Cyber Warfare. Institutional design and the possibility to build up a Meta-rule of law are also discussed.

Keywords OSINT • Social Intelligence • Privacy • Security • Semantic Web • Regulatory Models

9.1 Preliminaries: The Legal and Political Problem

There are many ways to conceptualise intelligence —individual, collective, swarm, etc. — to describe and research how it works or how to use it in courses of action. Since 2001, this practical side has received a strong boost. The explosion of Internet, the wide use of HTML protocols, and the speeding of the Semantic Web through W3C standards, is related to it (Casanovas et al. 2016a). But inaugurating the century with the *global terrorist threat* after September 11th was key to fund new research programs for military use. Some of these programs are focused on Open Source Intelligence (OSINT).

P. Casanovas (✉)
Institute of Law and Technology, Autonomous University of Barcelona, Barcelona, Spain

Faculty of Business and Law, Data to Decisions Cooperative Research Centre, Deakin University, Geelong, VIC, Australia
e-mail: pompeu.casanovas@uab.com; p.casanovasromeu@deakin.edu.au

The word is somehow misleading, mostly due to the venerable history of Open Source (OS) in computing. There is also an ongoing discussion in legal theory on the role that OS plays in intellectual property, licenses, publishing and patents. Yet, when applied to intelligence, OSINT does not simply refer to the origin of the digital outcome, but to the legal and political sphere of the community where the "intelligent" outcome is encapsulated, distributed, reused and transformed.

What does it mean for a document, an image, a video to be qualified as OSINT? We could state that it basically means to place it in a public domain or, better, in *no man's land* domain, free to be grabbed and manipulated *for public reasons* —by LEAs (Law Enforcement Agencies), Intelligence Services, State Agencies… [1] But, as I will contend later on, many restrictions apply. There is no clear-cut line separating the *private* and *public* domains. Rather, there is a grey continuum zone bridging the two areas.

This chapter deals with the relationships and differences between (Open) Social Intelligence (OSI) and Open Source Intelligence (OSINT) or, in other words, with how to combine freedom and surveillance. This is currently one of the hot topics in European legislation and it is worthwhile to face it from a regulatory point of view.

The General Data Protection Reform package (GDPR) is at stake. European data protection law has been under review for a long time, eventually resulting in the recently approved General Data Protection Regulation (April 14th 2016).[2] The new rules intend to put citizens back in control of their data, notably through: (i) the right to be forgotten (when you no longer want your data to be processed and there are no legitimate grounds for retaining it, the data will be deleted); (ii) easier access to your own data (a right to data portability to make easier to transfer personal data between service providers); (iii) putting citizens in control (requirement of explicit consent to process personal data), (iv) Privacy by Design (PbD) and Privacy by Default (PbD) —as they are becoming essential principles in EU Data Protection Rules (EU Commission 2014).

The Opinion 28 released by the European Group on Ethics of 20 May 2014 described Ethics of Security and Surveillance Technologies (EGE 2014a). The Opinion advanced a set of sixteen concrete recommendations for the attention of the EU, member states, and a range of public and private stakeholders. It *"challenges the notion that 'security' and 'freedom' can be traded against one another"*, and *"calls for a more nuanced approach, in which the proportionality and effectiveness of security and surveillance technologies are subject to rigorous assessment, and in which rights are prioritized rather than traded"*. Certain core principles, such as human dignity, cannot be traded (EGE 2014b).

---

[1] This chapter is partially based on my work at the EU Network on Social Intelligence (SINTELNET) http://www.sintelnet.eu/. I revised some of my previous positions on OSINT (Casanovas 2014).

[2] Cfr. *Proposal for a Regulation of the European Parliament and of the Council on the protection of individuals with regard to the processing of personal data and on the free movement of such data* (General Data Protection Regulation) COM/2012/011 final – 2012/0011 (COD). After 4 years, the final draft of April 6th was finally approved by the EU Parliement on April 14th 2016. See http://data.consilium.europa.eu/doc/document/ST-5419-2016-INIT/en/pdf. For a useful and short summary of its content, see Albrecht (December, Albrecht 2015) and de Hert and Papakonstantinou (2016).

In the same vein, the EU Data Protection authorities and the Article 29 Working Party, on its plenary meeting of 25 November 2014, adopted a Declaration on European Values with 16 points (A29, 2014a).[3] Point one and two state that:

1. *The protection of personal data is a fundamental right. Personal data (which includes metadata) may not be treated solely as an object of trade, an economic asset or a common good.*
2. *Data protection rights must be balanced with other fundamental rights, including non-discrimination and freedom of expression, which are of equal value in a democratic society. They must also be balanced with the need for security.*

What does it exactly mean? Replacing the mechanism of security as a *general exception* to rules by another approach in which other principles apply adds some complexity to the balance between freedom and security.[4] The Common Law tradition (both British and American) is not considering protection of privacy as a fundamental right so far (Donohue 2005-06; Moshirnia 2013). Along with the upcoming Directive on personal data processing in criminal matters,[5] GDPR shapes a new general framework for the protection and exercise of rights. The final Regulation consists of 99 articles and 179 Recitals.[6]

I will contend that the epistemic approach to strike such a balance requires an additional level of analysis to figure out a general structure to compare the outcomes

---

[3] This Opinion must be completed with the WP29 Opinion on the legal grounds of surveillance of electronic communications for intelligence and national security purposes that was adopted on April 10th 2014. The origins of the statement are clearly expressed: "*The focus of this Opinion lies with the follow up that is needed after the Snowden revelations.*" A major part of the Working Document discusses the applicability of the transfer regime of Directive 95/46/EC.

[4] Quoting Marju Lauristin (Raporteur) at the recent Debate on the protection of individuals with regard to the processing of personal data for the purposes of crime prevention (Strasbourg, Wednesday, 13 April 2016): "*In this framework, the very important thing is that the general principles of proportionality, legitimacy and purpose-limitation are included in police work. That means that no form of mass surveillance is possible. The collection of data is not possible. Retention for an unlimited or unclear period is not possible. Another important point is that we foresee the inclusion of data protection professionals in the police institutional setting: specifically, in police work.*" http://www.europarl.europa.eu/sides/getDoc.do?pubRef=-//EP//TEXT+CRE+20160413+ITEM-015+DOC+XML+V0//EN&amp;amp;language=en&amp;amp;query=INTERV&amp;detail=3-515-000

[5] *Directive on the protection of individuals with regard to the processing of personal data by competent authorities for the purposes of prevention, investigation, detection or prosecution of criminal offences or the execution of criminal penalties, and the free movement of such data and repealing Council Framework Decision 2008/977/JHA (05418/1/2016 – C8-0139/2016 – 2012/0010(COD))*{SEC(2012) 72 final}. See the text of the draft adopted on March 14th 2014 at the first reading at http://www.europarl.europa.eu/sides/getDoc.do?type=TA&language=EN&reference=P8-TA-2016-0126, and at http://www.europarl.europa.eu/sides/getDoc.do?pubRef=-//EP//NONSGML+TA+P7-TA-2014-0219+0+DOC+PDF+V0//EN. It is now at the second reading now.

[6] Recital 19 states that GDPR does not apply to "*the processing of personal data by competent authorities for the purposes of the prevention, investigation, detection or prosecution of criminal offences or the execution of criminal penalties, including the safeguarding against and the prevention of threats to public security and the free movement of such data*".

that I will call *Meta-rule of law*. This is also needed when addressing the issue of using OSINT not only to fight organised crime, but cyber-attacks and terrorism.

This chapter is divided into four sections. The first one draws a distinction between Open Source and Social Intelligence. The second one is centred on Privacy by Design polices. In Section 3 I introduce the CAPER regulatory model to facilitate the regulation of intelligence-driven policing platforms to fight organised crime.[7] Finally, I will defend, with some limitations, the extension of such a model to cyber warfare, advancing ten preliminary observations.

As a synthesis of my position: (i) even in surveillance toolkits and serious security issues some feasible ways to bridge PbD principles and citizens' rights are possible; (ii) PbD can be broadly understood as a form of institutional design; (iii) to balance security and freedom, to comply with existing regulations, and to foster trust, *intermediate regulatory models* based on hard law, policies, soft law (standards) and ethics are required; (iv) these models could also be used to solve some of the regulatory and legal puzzles raised by the *Tallinn Manual on the International Law applicable to Cyber Warfare* (2013).

9.2 Open Source Intelligence (OSINT) and Social Intelligence (OSI)

9.2.1 OSINT

OSINT has a military origin. Gathering knowledge from open source information lies on very practical reasons, not only on the development of data mining, big data and cloud computing. Mark Pythian (2009) observes that collection techniques that worked well in a Cold War context are not that useful in the end of the twentieth century. Recruiting agents within Al-Qaeda, or attempting to infiltrate Islamic radical groups is a nearly impossible task for Western agencies.[8]

Our technological approach should acknowledge that the process to gather publically available information is not new either. Web 2.0 and 3.0 are just *enhancing* the heuristic behaviour of producing knowledge through all possible sources. For instance, libraries have always been a source of information to turn it into usable knowledge after intensive queries. Now, they have emerged again on the international scene as a critical source of soft power (McCary 2013).

---

[7] *Collaborative Information, Acquisition, Processing and Reporting for the Prevention of Organized Crime* (CAPER) http://www.fp7-caper.eu/
[8] Phythian (2009: 68–69) graphically quotes a former CIA operative about this, down to earth:

> The CIA probably doesn't have a single truly qualified Arabic-speaking officer of Middle Eastern background who can play a believable Muslim fundamentalist who would volunteer to spend years of his life with shitty food and no women in the mountains of Afghanistan. For Christ's sake, most case officers live in the suburbs of Virginia. We don't do that kind of thing.

There is no homogeneous definition of the term OSINT. It depends on the field, purposes and actions in which it is used. Within the intelligence community, OSINT is usually defined as unclassified information obtained from any publicly available source in print, electronic, or verbal form (radio, television, newspapers, journals, internet, commercial databases, and video). The process to gather intelligence in this way begins with raw information from primary sources assembled through filtering and editing processes. OSINT is then "constructed". Only after the process has been completed, OSINT is created (Burke 2007).

Intelligence Services refer to OSINT according to military uses as "unclassified information that has been deliberately discovered (…) to a select audience" (Steele 2007).[9] Not all collections and accesses to information sources fall equally under this definition. Several requirements should be satisfied as preliminary conditions (Jardines 2015). The information must be: (i) publicly available, (ii) lawful, (iii) properly vetted, (iv) acquired second hand, (v) and be produced to satisfy an intelligence requirement. In this sense, it has been already incorporated into the context of US Military as a new useful analytical dimension in addition to human intelligence (HUMINT), signals intelligence (SIGINT), imagery intelligence (IMINT) and measurement and signatures intelligence (MASINT).

The NATO early published in 2001 and 2002 three OSINT Handbooks, now accessible online: the *NATO Open Source Intelligence Reader*, the *NATO OSINT Reader*, and the *Intelligence Exploitation of the Internet*.

Reports for the US Congress are quite clear about its wide adoption: "A consensus now exists that OSINT must be systematically collected and should constitute an essential component of analytical products" (Best and Cummings 2008).

9.2.2 *(Open) Social Intelligence (OSI)*

OSINT is considered also for other non-military purposes as a cluster of tools to browse the web, aggregate information, and getting reliable profiles from websites, blogs, social networks, and other public digital spaces. From this broader point of view, it may be defined synthetically as "the retrieval, extraction and analysis of information from publicly available sources" (Best 2008), without further requirements. This approach is taken by many to get, structure and manage information in a broad array of social domains —e.g. media (Bradbury 2011), education (Kim et al. 2013), business (Fleisher 2008), disaster management (Backfried et al. 2012), and fire services (Robson 2009). It entails a definition of the concept referring to

---

[9] This is the official definition (*US Army FM 2-0 Intelligence March 2010*), based on National Defence Autorization Act for FY 2006, & 931: 1. *Open-source intelligence is the discipline that pertains to intelligence produced from publicly available information that is collected, exploited, and disseminated in a timely manner to an appropriate audience for the purpose of addressing a specific intelligence requirement. Open-source intelligence (OSINT) is derived from the systematic collection, processing, and analysis of publicly available, relevant information in response to intelligence requirements.*

functions being performed by a computer system —retrieval, extraction and analysis of information. Thus, it is ostensive in nature, offering a descriptive meaning.

It is worth to notice that OSINT consists of reusable and reused (second hand) information, embedded into broader courses of action, with a display of possible frameworks and changing scenarios. The difference between military and non-military uses lies on the type of frameworks, stakeholders, players, and organisations involved in sharing and reusing data and metadata, rather than on the information content. Two more features matter: the type of technology used, and the direction of the workflow in the communication framework (top down or bottom up).

Mobile technologies are usually linked to crowdsourced usages of content (Poblet 2011). Crowdsourced platforms, micro-tasking, crisis mapping, and cooperative organisations face OSINT from a *cooperative* and a *collective* point of view, empowering people to obtain a common end (Poblet et al. 2014). Large amounts of data can be gathered, analysed and conveyed online and in near real time.

This is not incompatible with police or military functions, as long as they monitor crowd participation. E.g. in the London and Vancouver riots of 2011, Legal Enforcement Agents (LEAs) obtained such voluntary cooperation (Keane and Bell 2013).

Let's follow this thread. *Collective* or *Social Intelligence* are scientific terms, developed by research communities in Artificial Intelligence, Cognitive Science and Social and Political Sciences.[10] OSINT is a rather functional, pragmatic term used when collecting open source information for some specific purposes. Could OSINT be related to (Open) Social Intelligence (OSI)?

Both concepts —OSI and OSINT— have an operational side and denote the circulation and transformation of non-structured information into structured information on the Web. In their history of the concept, Glassman and Klang (2012) offer a communicative and cultural approach —"the Web as an extension of mind", OSINT as the interface between *fluid intelligence* and *crystallized intelligence* (Backfried et al. 2012).

If this is so, the field, methodology and theory of Social Intelligence could comprehend what is referred by OSINT, as the social mind is faced as a set of social affordances that can be represented, described and reproduced computationally as *inner* mechanisms, as social *artefacts* performing a collective work (Castelfranchi 2014). Social Intelligence focuses on the human/machine coordination of *artificial socio-cognitive technical systems*, assuming that they interact in a shared web-mediated space with aims, purposes, intentions, etc. and are amenable to models *and* meta-models from a theoretical point of view (Noriega et al. 2014).

If the concept of OSINT is used to describe the operational functionalities of a computational system, this use could be embedded into a conceptually broader set of notions to be effective. In artificial socio-cognitive systems "rationality is based on the model that agents have of the other agents in the system" (Noriega and d'Iverno 2014). This epistemic assumption is not necessary for OSINT systems,

---

[10] Cfr. http://www.sintelnet.eu/

more pragmatically oriented and centred on visual analytics and on the interface between intra and inter-organizational teams.

I will only address the political side of the problem, as governance and the issues related to international and humanitarian law are crucial. Which meta-model would be needed to build *at the same time* military and police uses, and the institutional design of privacy and civil rights protections (striking a balance between freedom and security)? This is certainly a challenge, for which I will make use of a sociocognitive perspective based on open social intelligence (OSI).

9.3 The CAPER strategy

9.3.1 Intelligence-Led Policing and the Law

Organised crime is a difficult subject of study. Authors have found that priority setting and strategic planning in the field of organised crime is inherently characterized by uncertainty (Verfaille and Beken 2008). The big numbers of illegal activities and cybercrime in particular, are hard to assess accurately. In this new field, "the global nature of the cybercrime industry inherently downplays the role of localized law-enforcement agencies" (Kshetri 2010, 247).

To fight it, from some time now, European LEAs have adopted intelligence-led policing as a method: *"the application of criminal intelligence analysis in order to facilitate crime reduction and prevention in a criminal environment through effective policing strategies and external partnership projects"* (Ratcliffe 2003, 2008). O'Connnor (2006) adds to this definition the use of intelligence products for decision-making both at the tactical and strategic level. Strategic intelligence refers to pattern and trend analysis of crime indicators, as opposed to tactical intelligence, which is anything evidential or helpful in making a case. There are four primary types of analytic outcomes provided by an intelligence-led police department on individual or group behaviour: (i) profiles, (ii) briefs (fact patterns related to investigative hypothesis), (iii) assessments, (iv) estimates (forecast or predictive statements) (O'Connor 2006).

OSINT is not limited to police departments. Specialized companies, attentive to geopolitical indicators, leverage OSINT and combine it with big data analytics. Zeeshan-Ul-Hassan Usmani (2014) underlines the point of terrorist predictions:

We need to focus more on the OSINT, the 'open-source intelligence' databases, the things that we can gather from online blogs, online magazines-for example, if we can see the way they recruit. We can get quite a few hints here and there from the online blogs and chat rooms. So we need to account for that. Second, technically you need to account for the geopolitical indicators, defined as GPIs. [...] Here is another bizarre example: Pakistan and India are natural rivals when it comes to cricket matches. Both teams had played 18 one-day matches since 2007. India won ten and nothing happened, Pakistan won eight and the probability of getting a terrorist attack within 24 h of winning against India is 100 %. Correlation doesn't mean causation, so there might be other factors in play. For example, when Pakistan wins we have thousands of people on street dancing, which makes them an easy target. Perhaps separatist groups do not like the nation to be happy about anything or some other reason, but we know we have this probability and we can use it for better protection and warnings.

Social media data renders social life more visible to police and other investigators (Trottier 2014). When it comes to security and surveillance issues, national and transnational differences are notable. In Europe, principles such as consent, subject access, and accountability are core to current legislation and to the General Data Protection Reform package (GDRP). The individual is deemed to keep control over the personal data being collected. Informational rights are usually known as the ARCO rights (access, rectification, cancellation, and objection). LEAs' behaviour must be compliant with regional, national, and EU laws.

The problem of dealing with law is that *law* is not a well-defined field. Rules, norms, principles and values are expressed in natural language, and there is a real problem to address them analytically, for the same statute, article, principle or concept might be *interpreted* in different ways whenever instantiated in a decision or a ruling (Casanovas 2014).

Technically, this can be faced as an "interoperability" problem, except for the fact that law always resists complete modelling, as shown by the integration of domain and core ontologies into upper-level ones (Casellas et al. 2005), and the interactive dimension of ontology-building (Casanovas et al. 2007). Consequently, since there are objective limitations to formalise legal statements, the analyst is forced to complete this missing part by settling some general framework by her own.

Practical decisions and implementation of norms are usually ground in some theory. From an epistemic point of view, the analyst is simultaneously working with an operational language and the structuring meta-system for such a language. Models and meta-models come together, and one of the most interesting tasks is to reveal the inner structure of the framework (the meta-model) of legal interpretations.

CAPER is an OSINT platform to fight organised crime and to facilitate transnational cooperation (Aliprandi et al. 2014). The meta-model for CAPER has been detailed in eight Deliverables and some articles and papers (Casanovas et al. 2014a, b; González-Conejero et al. 2014; Casanovas 2014, 2015a, b). In the CAPER workflow, privacy by design policies and ethical guidelines to protect citizens' rights have been worked out simultaneously to build a CAPER platform compliant with legal and ethical specifications.

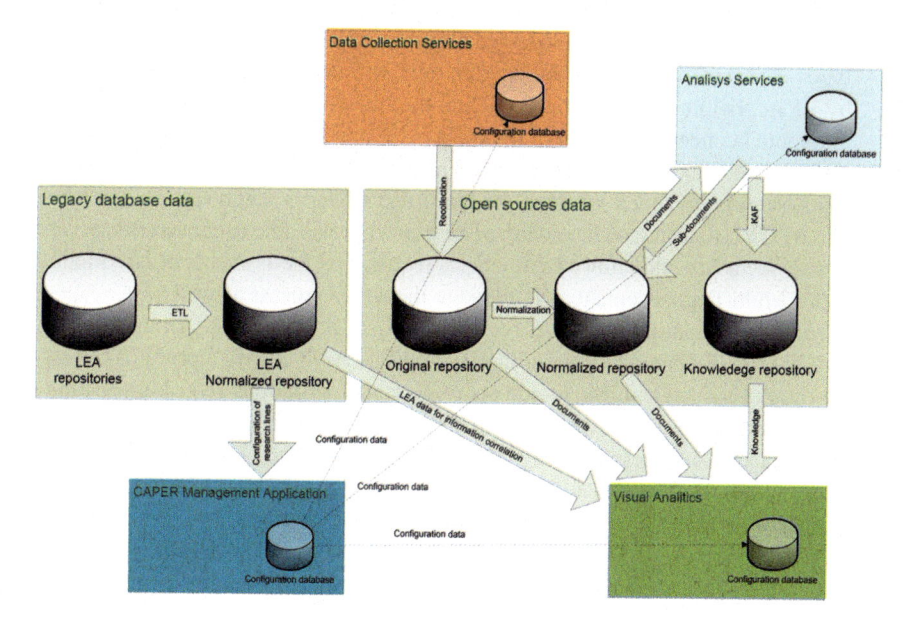

Fig. 9.1 CAPER databases overview. Source: http://www.fp7-caper.eu/results.html

9.3.2 The CAPER Workflow

The functionalities of the CAPER platform are manifold[11]: (i) implementing a framework to perform the task of connecting multiple data sources with multiple visualization techniques via a standardized data interface, including support for data-mining components; (ii) enabling a quick import of data types from disparate data sources in order to improve the ability of different LEAs to work collaboratively; (iii) supporting pattern discovery, documentation and reuse, thus increasing progressively detection capabilities. The architecture design has four components: (i) data harvesting (knowledge acquisition: data gathering), (ii) analysis (content processing), (iii) semantic storage and retrieval, (iv) and advanced visualization and visual analytics of data. Figure 9.1 below shows the interaction and workflow between databases.

Besides a specific Privacy Impact Assessment (PIA), we explored several related strategies that constitute an *indirect* approach to PbD principles (Casanovas et al. 2014a, b). This is the final result:

1. The CAPER workflow is addressed to four different LEA's analysts: (i) The *Generic Analyst* (GA) (ii) the *Advanced Analyst* (LEA-AA) (iii) the *System Administrator*, (iv) LEA's External User (LEU).

---

[11] Most Deliverables were confidential. I am offering here a standard synthetic description, as in Casanovas et al. (2014b).

2. Well-defined scenarios are extracted from the experience of managing specific types of crimes. CAPER tools operate only within the investigation conducted by LEAs, helping them to better define the lines of research, but avoiding any automated conceptualisation of them.
3. Well-defined module interdependencies are drawn in advance. The CAPER crawling system is sustained by three modules: (i) crawler by keyboard, (ii) by URL, (iii) by URL focusing on keyboards. The crawler is also able to convert images and videos metadata into allowed mimetypes required by the *Visual Analytics* module (VA). Multi-lingual semantics is added to the whole process as well.
4. Two different ontologies have been built: (i) a Multi-lingual Crime Ontology (MCO) for 13 languages, including Hebrew and Arabic, with a proof of concept on drugs (346 nodes). MCO adjusts to country legislations (e.g. possession of drugs is a crime in UK, but not in Spain); (ii) a legal ontology focusing on European LEAs Interoperability (ELIO). ELIO has been built with the aim to improve the acquisition and sharing of information between European LEAs (González-Conejero et al. 2014).

9.3.3 PbD and Security

The concepts of Privacy by Design (PbD) —and Data Protection by Design and by Default— are well known in the computer science and legal research communities (Cavoukian 2010). This conceptual body aims at developing the Principles of Fair Information Practices (FIPs)[12] that follow from the Alan Westin tradition in private law,[13] and the technological idea of a meta-layer to manage and secure the identity of users on the web set by the Microsoft architect Kim Cameron (2005).

Ann Cavoukian (2012) asserts that *"it is not true that privacy and security are mutually opposing"*, and that big and smart data *"proactively builds privacy and security in"*. It might be true, but it is not evident in the fight against organised crime. Both in the military and humanitarian fields, to be effective OSINT tools have been designed just for what they should be controlled: spotting as much as possible and getting personal information about individuals and organizations.

Bert-Jaap Koops, Jaap-Henk Hoepman and Ronald Leenes (2013, 2014) experienced this void on the sidelines of law and technology when they faced the problem of modelling the protections of General Data Reform Package into OSINT platforms.[14] As might be expected, they found that privacy regulations cannot be hardcoded —*"'privacy by design' should not be interpreted as trying to achieve rule*

---

[12] 1. Openess and Transparency, 2. Individual Participation. 3. Collection Limitation, 4. Data Quality, 5. Use Limitation, 6. Reasonable Security. 7. Accountability.

[13] These historical origins must still be retraced and reconstructed carefully. I am grateful to Graham Greenleaf for this observation.

[14] VIRTUOSO (*Versatile InfoRmation Toolkit for end-Users oriented Open-Sources explOitations*), http://www.virtuoso.eu/.

compliance by techno-regulation. Instead, fostering the right mindset of those responsible for developing and running data processing systems may prove to be more productive. Therefore, in terms of the regulatory tool-box, privacy by design should be approached less from a 'code' perspective, but rather from the perspective of 'communication' strategies [emphasis added]". Quite recently, the concept of "tactics" has been introduced by Colesky et al. (2016) to bridge the gap between legal data protection requirements and system development practice.

I subscribe the authors' guidelines, but perhaps another conclusion could be drawn from these limitations. There are other possibilities to embed PbD into surveillance platforms, albeit indirectly, i.e. adding theoretical views not thinking of *techno-regulation* nor *communication* but of what law means when constructed through technological means. It is the field of *intermediate institutional models* what might be worked out from the perspective of self-governance and socio-cognitive artificial systems.

9.3.4 PbD Strategies and CAPER Rules

Design also means *institutional design*. The notion of *institutional-Semantic Web Regulatory Model* (i-SWRM) leans on this assumption: a self-regulatory model embeds the dimension of PbD into a technological environment than can be represented as a social ecosystem (Casanovas 2015a, see below Sect. 9.4.2). In some domains —licensing, patents, or intellectual property rights— such an ecosystem entails an automated link between normative compliance and the effective legal action that implements some rights (Rodriguez-Doncel et al. 2016). OSINT platforms for massive surveillance require a different strategy. With some differences, the CAPER ecosystem is similar to the framework set by Koops et al. (2013) and specifically by Hoepman (2014).

Hoepman's PbD strategies model focuses on the inner structure of the modelling; signalling several points into a general framework for privacy (or data protection) closed managerial system. This is consistent with the idea of concentrating the effort on the interpretation of the law, leaving aside the specific problems, risk scenarios, and asymmetric multilayered governance of OSINT platforms, end-users and LEAs. Conversely, i-SWRMs and specifically the regulatory model designed to regulate the CAPER workflow system (CRM), are more focused on LEA's inner and outer relationships. The social ecosystem centred on the specific data that users are processing and "living by" can be outlined as a simple scheme described in advance (see Fig. 9.2).

To implement the rules contained in the CAPER final recommendations as guidelines to run and monitor the platform, we might place first the potential risks on its information flow (Fig. 9.3).[15] E.g. a rule like "The storage of the CAPER data should be implemented in a separate repository. No contact with ordinary criminal

---

[15] A Private Impact Assessment (PIA) was carried out all along the Project with LEA's analysts. Antoni Roig set the risks and rules in some non-public Deliverables (D7.2, D73. D7.5, and D7.6) and J. González-Conejero plotted them on the information workflow (Fig. 9.2). The CAPER Ethical Committee was composed by Ugo Pagallo, Giovanni Sartor, Danièle Bourcier, John Zeleznikow, and Josep Monserrat.

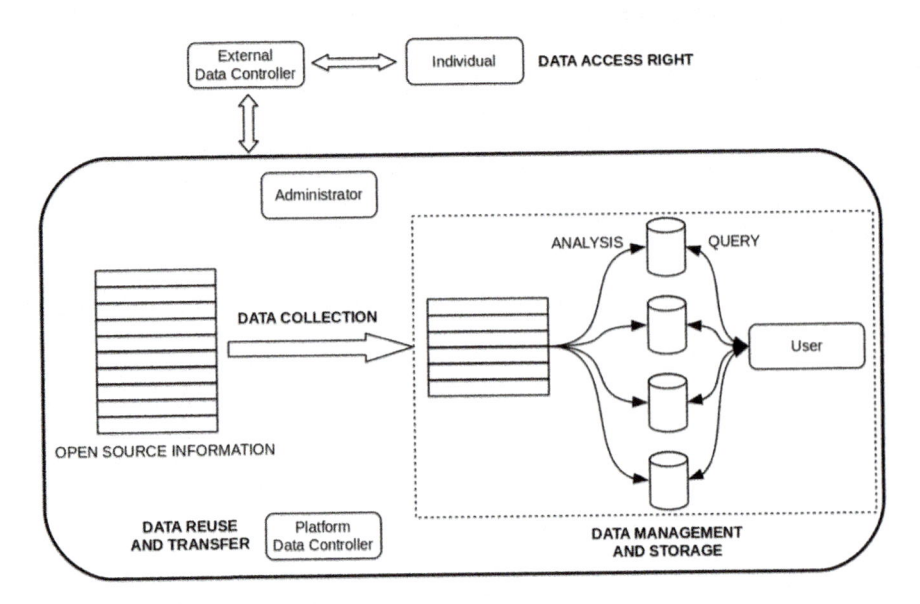

Fig. 9.2 CAPER regulatory scheme. Source: CAPER D7.8 (González-Conejero et al. 2014; Casanovas et al. 2014a, 2014b).

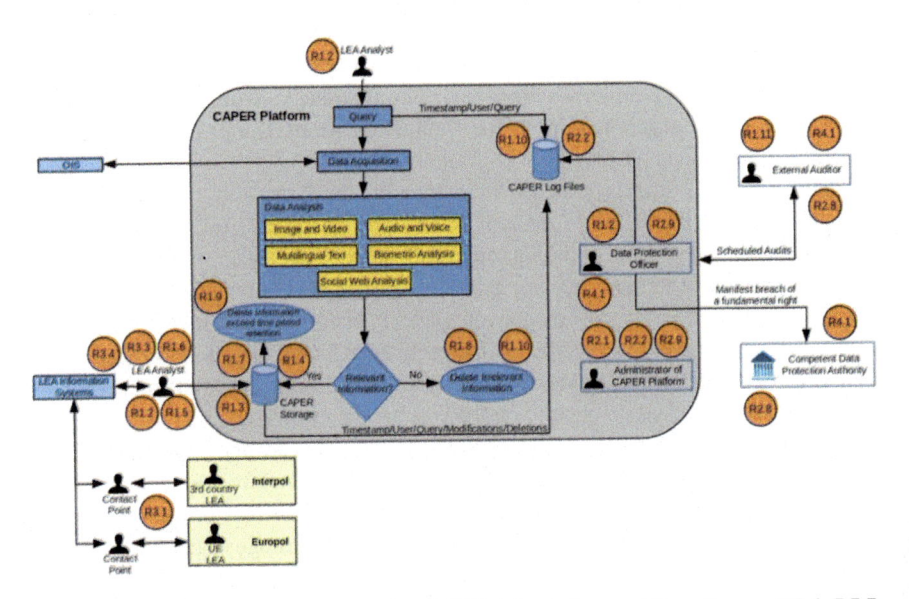

Fig. 9.3 Situated risks and rules onto the CAPER information workflow. Source: D7.6, D7.7, D7.8 CAPER (González-Conejero et al. 2014; Casanovas et al. 2014a, 2014b, http://www.fp7-caper.eu/results.html)

Table 9.1 Rules to regulate LEA's internal behaviour

I data collection and storage	R1.1 Each LEA should perform a specific Privacy Impact Assessment (PIA) according to the general framework offered by the CAPER Regulatory Model (CRM).
	R1.4 No automated classification of suspects, victims and witnesses can be inferred from CAPER results.
II Data management	R.2.3 Access to CAPER database should be granted for the purpose of prevention, detection or investigation of organized crime.
	R.2.4 Any other request of access for other purposes should be rejected.
	R.2.5 Non-authorised LEA and intelligence services or administrative bodies of authorized LEA should not have access to CAPER data.
	R.2.6 The use of system integrity tools should enable detection and reporting of changes applied on servers. In case of such an event the system should be able to notify specific users such as the creator of the query which results have been modified.
	R.2.7 Regular audits of the CAPER system should be performed by the external supervisor. The competent authority should be informed of the results, if necessary, according to national legislation, including the plans for enforcing recommendations.
III. Data reuse and transfer	R.3.2 No automated classification of suspects, victims and witnesses can be inferred from CAPER results.
IV Right of data access	R.4.2 The reasons to deny access should be clear and defined. Access can be denied when the access may jeopardise the fulfilment of the LEA tasks, or the rights and freedoms of third parties.
	R.4.2 The alleged reasons to deny access should be open to external supervision. The external supervisory authority should have free access to documents justifying the refusal. A short time-span of 3 months to give an answer to a previous request of access should be implemented.

Source: D7.8 Caper
Casanovas et al. (2014a, 2014b), http://www.fp7-caper.eu/results.html

data bases should be allowed" can be situated between the LEA analyst and repositories of raw data.

However, rules for the prevention of unauthorized access, data reuse and transfer, and for the protection of citizens' rights — e.g. the right to be notified when they notice that their personal data is being processed — cannot be plotted on the chart because they do not address solely the regulation of the information flow, but rights and obligations for citizens, third parties, controllers, and administrators. Regulations have a wider scope, covering functions, rules and players alike.

As stated, 9 out of the 23 designed rules cannot be situated because they don't apply to information processing, but to end-users' behaviour. R1.1, R1.4, R2.3, R2.4, R2.5, R2.6, R2.7, R3.2, R4.2, R4.3 are not plotted (Table 9.1). Under the new General Data Reform Package (GDRP) the scope of regulations is much wider. Thus, it is a hybrid model: some rules apply to the information processing flow, while others apply to the way how this information should be used, treated, protected and eventually deleted by analysts and internal controllers. This holds, e.g.

for a positive obligation such as "Each LEA should perform a specific Privacy Impact Assessment (PIA) according to the general framework offered by the CAPER Regulatory Model (CRM)", a general prohibition as "No automated classification of suspects, victims and witnesses can be inferred from CAPER results", and for specific obligations such as "The alleged reasons to deny access should be open to external supervision. The external supervisory authority should have free access to documents justifying the refusal. A short time-span of 3 months to give an answer to a previous request of access should be implemented".

9.4 The Regulation of OSINT Platforms

9.4.1 Rule and Meta-Rule of Law

What is the relationship between all these non-discrete categories? OSINT platforms raise a governance problem, because they should be as effective as possible to fulfil their goal to collect and store information, and at the same time they must comply with OSINT requirements, such as law compliance. Yet, organised crime has a transnational, extended dimension. There are about 3.600 organised crime organisations operating across Europe, with global connections. Therefore, what does "law compliance" mean?

Security, data, and privacy are the subject of quite different national regulations. Platforms and tools must respect statutes and regulations at the national level, attend the National and European Data Protection Agencies requirements, observe Human Rights case laws, and take into account LEA's professional best practices and culture. The notion of a *transnational rule of law* could be a good global strategy, since it grasps in a single concept all the legal sources while focusing on rights and regulations.

The rule of law implies that government officials and citizens are bound by and generally abide by the law, and legal pluralism refers to a context in which multiple legal forms coexist (Tamanaha 2011). Would it be possible to conceptualise and organise the relationship between rulers and ruled so that rulers themselves are subject to a rule of law in digital contexts, respecting pluralism?

In the absence of a transnational state, the notion of *Meta-level rule of law* has been coined at Elinor Ostrom's school to point to the tension between the "threat of chaos" and the "threat of tyranny" in the management of common perishable goods, such as water, wood or fisheries (Ostrom 2010; Aldrich 2010, Aligica and Boettke 2011). The "just right" solution lies in between. Rule of law scholars, such as Murkens (2007) and Palombella (2009, 2010) have recently pointed out the recoverable features of the classical model for transnational purposes as well (not bound by nation-state perspectives).

Michael Ritsch (2009) advances a "Virtual rule of law" for cyberspace. According to him, the "rule by law" [his spelling] must be: (1) non-arbitrary, (2) stable, (3) public, (4) non-discretionary, (5) comprehensible, (6) prospective, (7) attainable, (8) consistently enforced, (9) impartially applied, and (10) adjudicated in a factually neutral way. *"These indicia, however, do not include traditional elements of "liberal" rule of law, including democracy and personal rights"* (Risch 2009, 2).

Jonathan Zittrain's idea of "generative" Internet is relevant here. Zittrain's seeks to maintain the Internet use community-based, into the users' hands.[16] Following Zittrain (2008) and Lessig (2006), Mark Burdon (2010) provides examples of the importance of standards from a law and technology perspective. He distinguishes between first generation privacy laws, based on the application of information principles to interactions, and Web 2.0 second generation laws, which should take into account the collective and aggregated dimension of crowdsourcing and public intervention. E.g. Privacy invasive geo-mashups are unavoidable with the use of Googlemaps. The same situation is produced with the shared content needed in crisis mapping and the generalized use of local and personal images in disaster management (Poblet 2013). But potential solutions for the prevention and mitigation of privacy problems reside in the development of embedded technical and social standards, and not solely through "avenues of legal recourse founded on the concept of information privacy" (Burdon 2010, 50).

It is my contention that this problem should be tackled at different levels of abstraction: (i) the regulation-sourcing problem, in which the selection of legal sources is at play, should be treated both at the technical and regulatory dimension of the specific tool; (ii) the social regulatory ecosystem model set by Web Services and platforms (involving rights holders, managers and end-users); (iii) the conceptual meta-model drawn to design the regulatory system.

I will call *Meta-rule of law* the analytical management of rights and norms of the rule of law through computational models, information processing, and both virtual and physical organisations (Casanovas 2015a, 2015b). The CAPER Regulatory Model (CRM) is an example of this multi-layered and multi-dimensional approach.

9.4.2 CAPER Regulatory Model (CRM)

Figure 9.4 shows the emergence of institutional strengthening and trust from the two axes of the rule of law —binding power and social dialogue. The meta-model assumes that the degree of strength (*Macht, force*) exerted and the degree of non-cooperative behaviour is inversely proportional to the degree of dialogue and cooperation between rulers and ruled. In this way, the construction of a public space

---

[16] "The deciding factor in whether our current infrastructure can endure will be the sum of the perceptions and actions of its users. There are roles for traditional state sovereigns, pan-state organizations, and formal multistakeholder regimes to play. They can help reinforce the conditions necessary for generative blossoming, and they can also step in—with all the confusion and difficulty that notoriously attends regulation of a generative space—when mere generosity of spirit among people of goodwill cannot resolve conflict. But such generosity of spirit is a society's powerful first line of moderation." (Zittrain 2008, 246)

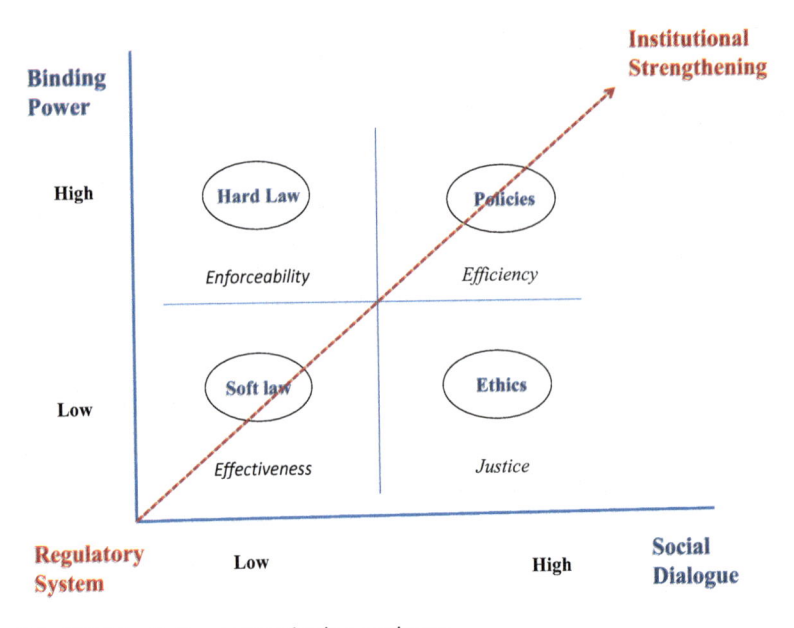

Fig. 9.4 CRM: Institutional strengthening *continuum*

depends on the combination of the regulatory components that can be ordered along the two axes to set up *intermediate institutions* such as the CAPER framework guidelines. This kind of institutions can be built to select legal sources and make them compatible with specific governance models.

Another way to describe them, drawing from classical legal theory, is to consider law application and implementation as the initial, framing point to instantiate the content of legal norms. Selection and interpretation of legal sources is a top-down semantic process (from some authorised bodies) as well as a pragmatic and dialectic one (stemming from the interaction of stakeholders, companies, citizens, consumers).

Figure 9.5 represents CRM dynamics. *Institutional strengthening* constitutes the third dimension emerging from the relationship between binding power and social dialogue, setting the regulatory system and fostering *trust*. Therefore, trust is a non-direct result coming out from a multilayered governance dynamics involving: (i) Courts (hard law), (ii) Agencies (policies), (iii) Experts (soft law), (iv) and Ethics (Committees balancing values, principles, and norms). It holds as a conceptual meta-model, framing the specific rules laid down for CAPER monitoring and managing. Enforceability, efficiency, effectiveness, and justice are first order properties, directly qualifying their space. *Validity* (of norms or rules) is a *second* order property, triggering *legality* — the final qualificatory stage of a behaviour, action or system. For a regulatory system be qualified as *legal* (or compliant with legality) it must be valid first, i.e. it must reach some threshold of compliance with first order properties. Both first and second order proprieties can be understood as graduated scales in a regulatory space. They are non-discrete categories (Casanovas 2013, Ciambra and Casanovas 2014).

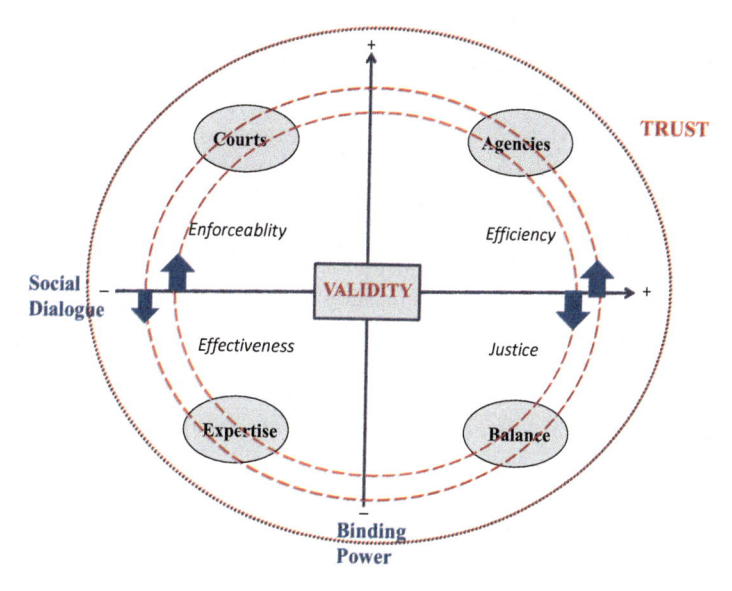

9 Cyber Warfare and Organised Crime. A Regulatory Model and Meta-Model… 155

Fig. 9.5 CRM: Validity and trust dynamics

9.4.3 Normative and Institutional Semantic Web Regulatory Models

Semantic Web Regulatory Models (SWRM) co-organise the conceptual architecture of (enforceable) hard law and policies, and (non-enforceable) soft law, and ethics (Casanovas 2015a). They operate at the *in-between* space (digital/real) described by Floridi. In this hybrid, augmented, semantically enriched reality, formal compliance with norms should not be limited to their content, but it should be expanded to the sustainable endurance and maintenance of their effects too. Thus, the validity of the intermediate system is shown through the emerging independent axis of *institutional strengthening*. This approach presents the additional advantage of being measurable.

I distinguish between *Institutional-SWRM* and *Normative-SWRM*, according to the focus and degree of automated execution. Again, this distinction is not absolute. Any regulatory model set for a specific ecosystem contains elements of both. *Normative-SWRMs* make use of RDF, RuleML, and computer versions of rights, duties and obligations. They may lean on the use and reuse of nested ontologies or ontology design patterns (ODP), and *Rights Expression Languages* (REL) representing legal licenses, intellectual property rights, or patents as data and metadata to be searched, tracked or managed. Rights Expression Languages (REL) are based on the instantiation of rules that can be semantically populated by means of extended vocabularies to express policies.[17] Therefore, *end-users and systems are linked*

---

[17] Cfr. https://www.w3.org/community/odrl/

through the same tool that is being used to manage and apply the modelled policy and legal knowledge.

Institutional-SWRM (i-SWRM) are focused on the relationship of end-users with their self-organised system. Inner coordination among electronic agents, outer interface with human (collective) agents, and their dynamic interaction within different types of scenarios and real settings are crucial. They can be applied to regulatory systems with multiple normative sources and human-machine interactions between organizations, companies and administrations. Thus, their conceptual scheme is linked with legal pluralism and with existing models of asymmetric multi-layered and networked governance. These are conceptual constructs compatible with Elinor Ostrom's social philosophy —polycentricism and social ecosystems— because their centre of gravity lies on their dynamic social bonds. *End-users and systems are connected through the social and legal bonds that externally link them through intermediate legal and governance institutions.*

In the CAPER example, dialogue with LEAs and security experts constitute a key point to understand where do the problems lie and why, and to let LEA's investigators participate into the regulatory process. At the same time, control is exerted because binding norms apply as well. E.g. the need for an internal and external DP controller (competent DP authorities) and the obligation to set a strict log file to keep records to sustain the accountability of the whole system.

So, the Caper Regulatory Model (CRM) lying behind the actions taken intends to bridge negotiations of social agents with the normative requirements and conditions of the rule of law (reinterpreted from this broader standpoint). This is why it can be implemented among LEA's organizations and embedded into the CAPER system to regulate the use of the platform. CRM is an example of i-SWRM (even if there are few nested automated rules into the system architecture).

9.5 Could CRM Be Applied to Cyberwarfare?

9.5.1 *Cyberwarfare*

The CRM is a regulatory programme that has been designed to furnish a solution to the OSINT regulatory puzzle. It remains still unclear whether it can be applied with success to related subjects. This chapter constitutes a first step in this direction. I have outlined (i) a specific model, (ii) a more general meta-model to be applied to the OSINT landscape, and (iii) some new concepts for institutional design around the notions of regulatory systems, i-SWRM and n-SWRM. Especially the notion of Meta-rule of law is deemed to be used in a wider political context.

Surveillance over the civil population, a rational response to cyber-attacks, and protections against terrorist threats constitute different objectives and entail different tasks that should be coordinated, albeit they can be analysed separately and regulated differently.

It could be taken into account that, a bit surprisingly, the fight against organised crime is a much more precise topic than cyberwarfare. Experts have already stressed that cyberwarfare constitutes a field in which conventional regulatory tools for warfare have to be carefully restructured to solve "the 3 R problems": rights, risks and responsibilities (Taddeo 2012). Broadly speaking, in addition to land, sea, air and space, information sets a 5th dimension of warfare. A transversal, not always violent domain, but with a great potentiality to cause harm to specific targets; be the target a state, a high-tech company, a conventional corporation, or a political community (Orend 2014). Floridi and Taddeo (2014) have turned this dimension into *Information Warfare*.

There is no general agreement yet about how this new space should be monitored and regulated. The *Tallinn Manual on International Law applied to Cyber Warfare*, issued by distinguished scholars in the field and edited by M. N. Schmitt (2013), does not address the subject directly. OSINT is never specifically mentioned. It would be possible to connect it to related notions such as "Active Cyber Defence" or "Supervisory Control and Data Acquisition", but regulatory models are still viewed as state devices, and customary public international law as an inter-state affair (Schmitt and Watts 2014).[18] This approach is not to be thrown. From a legal perspective, it constitutes a realistic standpoint to deal with. Many experts are still working under this umbrella, endorsing a legal understanding that defines *cyberattack* as "a trans-border cyber operation, whether offensive or defensive, that is reasonably expected to cause injury or death to persons, or damage or destruction to objects" (*Tallinn Manual*, Rule 30). Therefore, "cyberattacks are deemed to be simply another strategy or tactic of warfare, like armed drones and artillery barrages" (Solis 2014, 5).

Kilovaty (2014) distinguishes three conceptual perspectives: (i) Instrument-based approach, (ii) Target-based approach, (iii) Effects-based approach. He summarises the unsolved problems in the Tallinn definition: who would qualify as a combatant in the cyber context, what measures an attacked state can employ to repel a cyber-attack, and how the international law normally treats cyber espionage and the theft of intellectual property. As highlighted by Zeadally and Flowers (2014,15), the *Manual* definition of cyberwarfare excludes several relevant issues, e.g. cyber-operations designed to destabilize a nation-state's financial system since the attack does not directly result in death or physical destruction.

---

[18] *(…) as cyber activities become ever more central to the functioning of modern societies, the law is likely to adapt by affording them greater protection. It will impose obligations on states to act as responsible inhabitants of cyberspace, lower the point at which cyber operations violate the prohibition on the use of force, allow states to respond forcefully to some nondestructive cyber operations, and enhance the protection of cyber infrastructure, data, and activities during armed conflicts. These shifts will not be cost-free. They may, inter alia, prove expensive, affect privacy interests, extend to kinetic operations, and deprive battlefield commanders of options previously available to them. Ultimately, though, law reflects national interests. States will inescapably eventually find it in their interests to take such measures to protect their access to cyberspace and the goods it bestows.* (Schmitt 2014: 299)

On the contrary, the Geneva Centre for the Democratic Control of Armed Forces (DCAF) defines cyberwar as "warlike conduct conducted in virtual space". This is a more inclusive definition that allows the distinction between state-sponsored and non-state-sponsored cyberattacks. It also includes cyber-vandalism, cyber-crime, and cyber-espionage. This perspective links cyberwarfare to cyber-criminality and cyberspace, taking into account all players and intended and unintended effects on non-combatants, the industry and civil society.

CRM could be applied within this broader context as a tool to organise, monitor and control OSINT platforms through a multi-layered governance meta-model. A useful related approach, although more directly addressed to convert OSINT findings into legal evidence, can be found in Gottschalck (2009). Kornmaier and Jaouën (2014) also treat this subject and highlight the emerging role of the individual (2014, 142), trust and cooperative work. We do converge on this perspective:

> *Thus, establishing regulations in strategies and policies for the exchange of information and intelligence in [the above outlined] cyber context is the essential first step in the described process. It must be defined who shares what, with who, under what circumstances, how the information is handled, classified, processed and stored. These regulations are necessary, because on the one hand there exists no broadly accepted standard for sharing information or even intelligence across agencies or private companies. On the other hand – mentioned for completeness – trust is the key for increasing the sharing behaviour. Trust for the exchange occurs at the individual and organizational level. It is the degree of confidence to handle the information/ intelligence with the same sensitivity. Only then the exchange will take place. As well, cooperation between sovereign states is to be fostered for a better efficiency in cyber defence.* (Kornmaier and Jaouën 2014, 152).

9.5.2 *Expanding CRM to Cyberwarfare. Ten Preliminary Observations*

This is to be read as advance for future work. To broaden the scope of the kind of modelling proposed in this chapter I will depart from the following preliminary observations:

1. The advent of cyberspace has brought new features that make it a unique combat domain. Thus, it raises a governance problem that Schreier (2015, 91), among others, has stated clearly: *"in all states both the decision making apparatus for cyber-attack and the oversight mechanisms for it are inadequate today"*. He points out five distinguishing characteristics: (i) cyberspace has become a "global commons" existing everywhere and open to anyone; (ii) it provides an extended battle space with no real boundaries; (iii) ICT has demolished time and distance in a non-conventional convergence of technologies and infrastructures; (iv) cyberspace favours the attacker; (v) and, fifth, there is a permanent kaleidoscopic change of the components of cyberspace (Schreier 2015, 93).

2. Over 10 years ago, in the aftermath of September 11th in USA, Laura Donohue (2005–06, 1207–8) identified six possibilities in privacy protection to safeguard citizens from invasive ways of surveillance: (i) creating a property right in personal information, (ii) regulating the access, transfer and retention of data while providing remedies for violations, (iii) scaling back the existing powers, (iv) more narrowly defining "national security," (v) creating effective safeguards, (vi) and eliminating clauses that allow for such powers to be "temporary". I don't think things have changed that much. CRM attempts a seventh possibility: creating intermediate institutions to apply the safeguards of global ethics and the Rule of Law while easing security and intelligence services tasks.

3. Balancing cyber security risks against privacy concerns constitute a real challenge. Besides, as stated by Tene (2014, 392–93) *"this delicate balancing act must be performed against a backdrop of laws that are grounded in an obsolescent technological reality. Legal distinctions between communications content and metadata; interception and access to stored information; and foreign intelligence and domestic law enforcement — do not necessarily reflect the existing state of play of the Internet, where metadata may be more revealing than content, storage more harmful than interception, and foreign and domestic intelligence inseparable"* (Tene 2014: 392). Indeed, trying to focus on specific technological functionalities and affordances might facilitate the evaluating task. I do agree with Tene's conclusions: automated monitoring raises less privacy concerns than human observation. This, in turn, implies that the focal point for triggering legal protections should be the moment the system focuses on an individual suspect. This is consistent with Lin et al. suggestion (Lin et al. 2015, 57) to reframing the cybersecurity discussion closer to the individual-actor level.

4. There are striking differences among theoretical proposals to make such a balance. The notions of Data Protection by Design and by Default adopted by the European Union in the General Data Protection Regulation (GDPR) will be eventually enacted in 2016, and will come into force in 2 years. It constitutes an EU strong political bet that is at odds with the strictly liberal legal-market oriented perspective popular among American scholars, where it is assumed that public space is subordinated to private transactions.[19] A quick look to the last European Conferences on Data Protection leads to opposite results: PbD (or DPbD) is the only strategy widely accepted by all participants as a working alternative to generalized surveillance to protect consumers and citizens (Gutwirth et al. 2015).

5. The same difference can be found in legal articles about cloud services and cloud computing. The applicable body of law is separated in two tiers: *primary* privacy law, and *secondary* privacy law. The first one is created by the providers

---

[19] *"Privacy by design is a system designed not to work [...]. The market for consumer privacy has yet to be tested because "privacy by design" policies shift all of the transaction costs of privacy onto consumers. To discover what consumers make of privacy online, the transaction costs of privacy should be shifted from consumers to the owners of internet technology"* (Faioddt 2012, 104-5).

and users of services through privacy contracts, especially, privacy policies. The second one refers to statutes and policies.[20] From this perspective, public space is reduced, and clearly market-driven, *privatised*.

6. In everyday practice, fostering trust is as relevant as reshaping positive or customary international law to adapt norms to this changing digital environment. Therefore, the analytical proposal of a *Meta-rule of law* is entirely compatible with the idea of framing a Global Ethics to handle the fight against cybercrimes, cyberterrorism, and eventually cyberwarfare. This task is far from easy, because terrorist entities and organised crime frequently make use of websites that are beyond the reach of search engines in the so-called "deep" or "dark" Web. Non-indexed sites use a variety of methods to prevent detection from web crawlers —automated browsers that follow hyperlinks, indexing sites for later queries (Morishirna 2012).

7. Human-computer interface could be set as the landmark. Structuring data through second-generation Semantic Web tools starts and ends up into the pragmatic usage of structured information and knowledge in quite different scenarios and environments (Casanovas et al. 2016). Such knowledge can be *personalised*. Likewise, harm caused by worms and malware can reach individuals, groups, communities and political entities at different degrees and levels of granularity. But there is no way to separate casualties from targeted objectives. Civil society, as a whole, is affected. Therefore, any end-user, any member of cyberspace suffers a certain amount of harm. Adding regulations to protections, empowering and controlling LEAs and the military alike, adds normative complexity to the rule of law. Nothing prevents the application of global law to threats, conflicts and wars than can be private and public, virtual and physical at the same time, and most of all occur at the infra- and supra-state level. The Meta-rule of law seems adequate to handle the computer-language levels of contemporary environments, in real settings.

8. From a strictly military standpoint, John Arquilla (2013) recently summarised 20 years of *cyberwar*. Arquilla and Ronfeldt coined the term in a path breaking RAND paper that was published in 1993 in a scholarly journal. Comparing both papers the reader has the impression that Arquilla's original emphasis on the relevance of communications and *knowledge* has been framed later on into two broad and classical fields —war, and justice. *Just war* doctrines —*jus ad bellum, jus in bello*, (less) *jus post bellum*. In the middle, military doctrines of fast or attrition war, land vs. sea war, friction (Clausewitz) vs. geometrical (de Jomini) war (Arquilla and Nomura 2015). Thus, the power of nation-states still holds. Something is missing in between, because Arquilla thinks vertically (discourses on war / war of ideas) and horizontally (networked society, "swarm-

---

[20] "*The secondary privacy law, contained, for example, in statutes and regulations, is for the most part only applicable where no valid privacy contracts exist. This supremacy of privacy contracts over statutory and other secondary privacy law enables individualized privacy protection levels and commercial use of privacy rights according to the contracting parties' individual wish*" (Zimmeck 2012, 451).

ing") about *knowledge*,[21] without exploring its full potential to create self-regulatory institutional bodies, i.e. to reshape the legal landscape in which normative ethics operate. What it should be nuanced is not Arquilla's —and many others'— ideas of war and conflict, but his understanding of how law and regulations operate on and through the Web.

9. This is not saying that the ethical and legal problems pointed out by Arquilla —and many other ethicists— over the evolving cyberspace are not important.[22] They are. He realizes that the temporal sequence ante/and/post war does not represent what really happens in cyber-attacks, where this sequence does not hold. But ethics cannot be reduced to the principles endorsed by customary international law and the Geneva Convention. As I have shown, ethical principles can be embedded and nested on top of the CRM gradual scale to effectively regulate OSINT surveillance and to make the balance with citizens' rights. However, to embrace this point of view, a turning point is needed from a purely positive and normative understanding of law to an intermediate *computational* and *institutional* level. This is directly related to principles and values. E.g. *Fairness* is the tipping point in negotiation and ODR platforms (Casanovas and Zeleznikow 2014). *Accountability* and *transparency* seem to be key in OSINT cybercrime crawling (Casanovas et al. 2014a, 2014b). It is still unclear which ethical values count as the tipping point for justice in OSINT cyberwar crawling. The notion of Meta-rule of law could help thinking at different levels of abstraction (ground, models, meta-models) these ethical concerns, taking specific and situated human/computer interaction as a starting point (Casanovas 2013).

10. There is still room to improve cooperation between NATO and Europol, to put it gently (Rugger 2012). There are striking differences between USA and EU legal and policy strategies. The European Cybercrime Centre (EC3) commenced its activities in January 2013 at Europol.[23] Europol became an autonomous European agency in January 2010. Since then, cooperation among EU agencies is a fact (e.g. between Europol and Eurojust). But this does not hold for all state-members, especially in the counterterrorism area. Political and organizational barriers are still in place. The common USA-EU Safe Harbor Agreement has been replaced by the Privacy Shield policy on transatlantic

---

[21] "Swarming" means "simultaneous attack from many directions". See about the military development of cyberwarfare and the two competing paradigms of *"strategic information warfare as launching 'bolts from the blue' and cyberwar as doing better in battle strategic warfare"*, Arquilla (2011, 60).

[22] "(...) it seems that a kind of ethical 'bottom line' assessment might be discernible about cyberwar, in two parts. First, jus ad bellum *comes under great pressure in the key areas of right purpose, duly constituted authority, and last resort. However, the apparent benefits of waging preventive or pre-emptive war, concepts with a lineage dating from Thucydides and Francis Bacon,10 are largely illusory. Second, it seems that* jus in bello *considerations come off rather better in the areas of proportionality and non-combatant immunity although there is a bit of complexity in the parsing of notions of acceptable"* (Arquilla 2013, 85).

[23] https://www.europol.europa.eu/ec3

dataflows,[24] still under discussion.[25] This failure to cooperate presents several features: (i) disparities in the political, administrative and judicial frameworks of EU Member States obstruct effective information sharing and coordination; (ii) disparities between intelligence and police agencies arise because counter-terrorism issues are shared by organizations with mismatched interests; (iii) disparities in priorities and values; (iv) most important, disparities at the semantic interoperability level of the main legal and criminal concepts. This is an additional reason to think carefully and adopt the Meta-rule of law proposed in this chapter.

Acknowledgments This research has been funded by the F7 EU Project *Collaborative information, Acquisition, Processing, Exploitation and Reporting for the prevention of organised crime* (CAPER) —Grant Agreement 261712—; and by the National Project *Crowdsourcing*, DER2012-39492-C02-01 and the Australian project D2D CRC.

References

Albrecht, P. 2015. EU general data protection regulation: The outcome of the negotiations ("trilogues") and 10 key points. Lead European Parliament Committee: Committee on Civil Liberties, Justice and Home Affairs (LIBE). 17 December. http://www.janalbrecht.eu/fileadmin/material/Dokumente/20151217_Data_protection_10_key_points_EN.pdf Accessed 21 May 2016.

Aldrich, J.H. 2010. Elinor Ostrom and the "just right" solution. *Public Choice* 143: 269–273. doi:10.1007/s11127-010-9630-9.

Aligica, P.D., and P. Boettke. 2011. The two social philosophies of Ostroms' institutionalism. *The Policy Studies Journal* 39(1): 29–49. doi:10.1111/j.1541-0072.2010.0000395.x.

Aliprandi, C., J.A. Irujo, M. Cuadros, S. Maier, F. Melero, and M. Raffaelli. 2014. CAPER: Collaborative information, acquisition, processing, exploitation and reporting for the prevention of organised crime. *HCI* 26: 147–152.

Arquilla, J. 2011. From *blitzkrieg* to *bitskrieg*: The military encounter with computers. *Communications of the ACM* 54(10): 58–65. doi:10.1145/2001269.2001287.

Arquilla, J. 2013. Twenty years of cyberwar. *Journal of Military Ethics* 12(1): 80–87. doi:10.1080/15027570.2013.782632.

Arquilla, J., and D. Ronfeldt. 1993. Cyberwar is coming! *Comparative Strategy* 12(2): 141–165. Rand Corporation. http://www.rand.org/pubs/reprints/RP223.html. Accessed 21 May 2016.

Arquilla, J., and R. Nomura. 2015. Three wars of ideas about the idea of war. *Comparative Strategy* 34(2): 185–201.

Article 29 Working Party. 2014a. Joint statement of the European data protection authorities assembled in the Article 29 working party, November 25th (adopted on 26th). http://ec.europa.

---

[24] http://europa.eu/rapid/press-release_IP-16-216_en.htm

[25] Cfr. the Article 29 Data Protection Working Party *Opinion 01/2016 on the EU – U.S. Privacy Shield draft adequacy decision*, adopted on 13 April 2016: "*The check and controls of the adequacy requirements must be strictly performed, taking into account the fundamental rights to privacy and data protection and the number of individuals potentially affected by transfers. The Privacy Shield needs to be viewed in the current international context, such as the emergence of big data and the growing security needs*".

eu/justice/data-protection/article-29/documentation/opinion-recommendation/files/2014/wp227_en.pdf. Accessed 21 May 2016.
Article 29 Working Party. 2014b. Working document on surveillance of electronic communications for intelligence and national security purposes, adopted on December 4th, 2014. http://ec.europa.eu/justice/data-protection/article-29/documentation/opinion-recommendation/files/2014/wp228_en.pdf. Accessed 21 May 2016.
Article 29 Working Party. 2016. Article 29 Data protection working party, Opinion 01/2016 on the EU – U.S. Privacy Shield draft adequacy decision, adopted on 13 April 2016. http://ec.europa.eu/justice/data-protection/article-29/documentation/opinion-recommendation/files/2016/wp238_en.pdf Accessed 21 May 2016.
Backfried, G., C. Schmidt, M. Pfeiffer, G. Quirchmayr, M. Markus Glanzer, and K. Rainer. 2012. Open source intelligence in disaster management. 2012 European Intelligence and Security Informatics Conference, EISIC. *IEEE Computer Society* 254–258.
Best, C. 2008. Open source intelligence. In *Mining massive data sets for security: advances in data mining, search, social networks and text mining, and their applications to security*, ed. F. Fogelmann-Soulié et al. 19: 331–344. Amsterdam: IOS Press.
Best, R., and A. Cumming. 2008. Open Source Intelligence (OSINT): Issues for Congress, *CRS Report for Congress*, Order Code RL34270, Updated January 28 2008. https://www.fas.org/sgp/crs/intel/RL34270.pdf. Accessed 21 May 2016.
Bradbury, D. 2011. In plain view: Open source intelligence. *Computer Fraud & Security* 4: 5–9. Elsevier https://www.cse.msu.edu/~enbody/CFS_2011-04_Apr.pdf. Accessed 21 May 2016.
Burdon, M. 2010. Privacy Invasive Geo-mashups. Privacy 2.0 and the Limits of First Generation Privacy Law. *University of Illinois Journal of Law, Technology and Policy* 1: 1–50.
Burke, C. 2007. Freeing knowledge, telling secrets: Open source intelligence and development. *CEWCES Research Papers*. Paper 11. http://epublications.bond.edu.au/cewces_papers/11. Accessed 21 May 2016.
Cameron, K. 2005. The laws of identity …as of 5/11/2005. Microsoft Corporation, http://www.identityblog.com/stories/2005/05/13/TheLawsOfIdentity.pdf Accessed 21 May 2015.
Casanovas, P. 2013. Agreement and relational justice: A perspective from philosophy and sociology of law. In *Agreement Technologies*, ed. Sascha Ossowski, LGTS 8, Springer Verlag, 19–42, Dordrecht/Heidelberg: Springer. doi:10.1007/978-94-007-5583-3.
Casanovas, P. 2014. Open source intelligence, Open social intelligence, and privacy by design. European conference on social intelligence. *Proceedings of the European Conference on Social Intelligence (ECSI-2014)*, eds. Andreas Herzig and Emiliano Lorini, 174–185. Barcelona, Spain, November 35, 2014, CEUR http://ceur-ws.org/Vol-1283/ Accessed 21 May 2016.
Casanovas, P. 2015a. Semantic web regulatory models: Why ethics matter, special issue on information society and ethical inquiries. *Philosophy & Technology* 28(1): 33–55. doi:10.1007/s13347-014-0170-y.
Casanovas, P. 2015b. Conceptualisation of rights and meta-rule of law for the web of data. *Democracia Digital e Governo Eletrônico* 1(12): 18–41. http://buscalegis.ufsc.br/revistas/index.php/observatoriodoegov/article/view/34399. Accessed 21 May 2016. Reprinted in *Journal of Governance and Regulation* 4(4): 118–129.
Casanovas, P., Casellas, N., Tempich, C. et al. 2007. *Artificial Intelligence and Law* 15: 171. doi:10.1007/s10506-007-9036-2
Casanovas, P., and J. Zeleznikow. 2014. Online dispute resolution and models of relational law and justice: A table of ethical principles. In *AI approaches to the complexity of legal systems IV. social intelligence, models and applications for law and justice systems in the semantic web and legal reasoning*, ed. P. Casanovas et al., LNAI 8929, 55–69. Heidelberg/Berlin: Springer.
Casanovas, P., E. Teodoro, R. Varela, J. González-Conejero, and A. Roig, et al. 2014a. *D 7.8 EAG ethical report code. Final ethical audit on system development and deployment*, EU F7 CAPER, FP7-SECURITY-2010-1.2-1, 24/10/2014.
Casanovas, P., J. Arraiza, F. Melero, J. González-Conejero, G. Molcho, and M. Cuadros. 2014b. Fighting organized crime through open source intelligence: Regulatory strategies of the

CAPER project. In *Legal knowledge and information systems. JURIX 2014: The twenty-seventh annual conference*, Foundations on artificial intelligence, 271, ed. Rinke Hoekstra, 189–199, Amsterdam: IOS Press.

Casanovas, P., M. Palmirani, S. Peroni, T. van Engers, and F. Vitali. 2016. Special issue on the semantic web for the legal domain guest editors' editorial: The next step. *Semantic Web Journal* 7(3): 213–227. IOS Press. http://www.semantic-web-journal.net/system/files/swj1344.pdf. Accessed 21 May 2016.

Casellas, N., M. Blázquez, A. Kiryakov, P. Casanovas, M. Poblet, R. Benjamins. 2005. OPJK into PROTON: Legal domain ontology integration into an upper-level ontology. *On the Move to Meaningful Internet Systems 2005: OTM 2005 Workshops*, ed. R. Meersman et al., LNCS 3762, 846–855. Berlin/Heidelberg: Springer.

Castelfranchi, C. 2014. Minds as social institutions. *Phenomenology and Cognitive Science* 13(1): 121–143. doi:10.1007/s11097-013-9324-0.

Cavoukian, A. 2010. *Privacy by design. The 7 foundational principles. Implementation and mapping of fair information practices. Information and privacy commissioner*. Ontario, Canada. http://www.privacybydesign.ca/index.php/about-pbd/7-foundational-principles/. Accessed 21 May 2016.

Cavoukian, A. 2012. Privacy by design. *IEEE Technology and Society Magazine* 4: 18–19. doi:10.1109/MTS.2012.2225459. http://ieeexplore.ieee.org/stamp/stamp.jsp?tp=&arnumber=6387956. Accessed 21 May 2016.

Ciambra, A., and P. Casanovas. 2014. Drafting a composite indicator of validity for regulatory models and legal systems. In *AI approaches to the complexity of legal systems IV. Social intelligence, models and applications for law and justice systems in the semantic web and legal reasoning*, ed. P. Casanovas et al., 70–82. LNAI 8929, Heidelberg/Berlin: Springer.

Colesky, M., Hoepman, J. H., Hillen, C. A. 2016. Critical Analysis of Privacy Design Strategies. IEEE Symposium on Security and Privacy Workshops, 33–40.

de Hert, P. and Papakonstantinou, V., 2016. The new General Data Protection Regulation: Still a sound system for the protection of individuals?. *Computer Law & Security Review*, 32 (2): 179–194.

Donohue, L.K. 2005-2006. Anglo-American Privacy and Surveillance. *Journal of Criminal Law and Criminology* 93 (3): 1059–1208.

EGE. 2014a. Ethics of security and surveillance technologies, Opinion no. 28 of the European Group on Ethics in Science and new Technologies, Brussels, 20 May 2014, http://ec.europa.eu/bepa/european-group-ethics/docs/publications/ege_opinion_28_ethics_security_surveillance_technologies.pdf Accessed 21 May 2016.

EGE. 2014b. Press release on the EGE opinion 28, of 20 May 2014. http://ec.europa.eu/bepa/european-group-ethics/docs/publications/press_release_ege_opinion_28_.pdf. Accessed 21 May 2016.

EU Commission. 2014. Progress on EU data protection reform now irreversible following European Parliament vote European Commission – MEMO/14/186 12/03/2014, http://europa.eu/rapid/press-release_MEMO-14-186_en.htm. Accessed 21 May 2016.

Faioddt, J.T. 2012. Mixed reality: How the laws of virtual worlds govern everyday life. *Berckeley Technology Law Journal* 27 1/3: 55–116. doi:10.15779/Z38ST2W.

Fleisher, C. 2008. OSINT: Its implications for business/CompetitiveIntelligence analysis and analysts, OSINT: Its implications for business/competitive intelligence analysis and analysts. *Inteligencia y Seguridad* 4: 115–141. http://www.phibetaiota.net/wp-content/uploads/2013/02/2008-Fleisher-on-OSINT-English-and-Spanish.pdf. Accessed 21 May 2016.

Floridi, L., and M. Taddeo (eds.). 2014. *The ethics of information warfare*, LGTS 14. Heidelberg/Dordrecht: Springer.

Glassman, M., and M.J. Kang. 2012. Intelligence in the internet age: The emergence and evolution of Open Source Intelligence (OSINT). *Computers in Human Behavior* 28: 673–682. doi:10.1016/j.chb.2011.11.014.

González-Conejero, J., R. Varela-Figueroa, J. Muñoz-Gómez, and E. Teodoro. 2014. Organized crime structure modelling for European law enforcement agencies interoperability through ontologies. In *AI approaches to the complexity of legal systems. AICOL IV-V*, ed. P. Casanovas, U. Pagallo, M. Palmirani, and G. Sartor, 217–231. LNAI 8929. Heidelberg, Dordrecht: Springer. doi: 10.1007/978-3-662-45960-7_16.

Gottschalck, P. 2009. Information sources in police intelligence. *The Police Journal* 82: 149–170. doi:10.1350/pojo.2009.82.2.463.

Gutwirth, S., R. Leenes, P. de Hert (eds.). 2015. *Reforming European data protection law*, LGTS, Dordrecht/Heidelberg: Springer. doi:10.1007/978-94-017-9385-8.

Hoepman, J.H. 2014. Privacy design strategies (extended abstract). In ICT systems security and privacy protection. *29th IFIP TC 11 International Conference, SEC 2014, Marrakech, Morocco, June 2-4*, Proceedings. IFIP Advances in Information and Communication Technology, ed. N. Cuppens-Boulahia et al., 446–459. Heidelberg: Springer.

Jardines, E.A. 2015. Open source intelligence. In *The five disciplines of intelligence collection*, ed. Mark M. Lowenthal and Robert M. Clark, chapt.2, L.A., Washington: CQ Press.

Keane, J., and P. Bell. 2013. Confidence in the police: Balancing public image with community safety. A comparative review of the literature, *International Journal of Law, Crime and Justice* 41: 233–246. doi:10.1016/j.ijlcj.2013.06.003.

Kilovaty, I. 2014. Cyber warfare and the Jus Ad Bellum challenges: Evaluation in the light of the Tallinn manual on the international law applicable to cyber warfare. *National Security Law Brief* 5(1): 91–124. http://digitalcommons.wcl.american.edu/cgi/viewcontent.cgi?article=1066&context=nslb. Accessed 21 May 2016.

Kim, Y., M. Glassman, M. Bartholomew, and E.H. Hur. 2013. Creating an educational context for open source intelligence: The development of internet self-efficacy through a blogcentric course. *Computers & Education* 69: 332–342. doi:10.1016/j.compedu.2013.07.034.

Koops, B.-J., J.H. Hoepman, and R. Leenes. 2013. Open-source intelligence and privacy by design. *Computer Law & Security Review* 29: 676–688. doi:10.1016/j.clsr.2013.09.005.

Koops, B-J., and R. Leenes. 2014. Privacy regulation cannot be hardcoded. A critical comment on the 'privacy by design' provision in data-protection law. *International Review of Law, Computers & Technology* 28(2): 159–171. doi:10.1080/13600869.2013.801589.

Kornmaier, A., and F. Jaouën. 2014. Beyond technical data -a more comprehensive situational awareness fed by available intelligence information. *2014 6th International Conference on Cyber Conflict*, ed. P. Brangetto, M. Maybaum, and J. Stinissen, 139–156, NATO CCD COE Publications. https://ccdcoe.org/sites/default/files/multimedia/pdf/d0r0s2_kornmeier.pdf. Accessed 21 May 2016.

Kshetri, N. 2010. *The global cybercrime industry. Economic, institutional and strategic perspectives*. Heidelberg/Dordrecht: Springer.

Lessig, L. 2006. *Code* and *other laws of cyberspace* (2001), *Code 2.0* (2006). Crowdsourced version. http://codev2.cc/ Accessed 21 May 2016.

Lin, P., P. Allhoff, and K. Abney. 2015. Is warfare the right frame for the cyber debate? In *The ethics of information warfare*, ed. L. Floridi and R. Taddeo, 39–57. LGTS, Dordrecht/Heidelberg: Springer. doi:10.1007/978-3-319-04135-3.

McCary, M. 2013. Sun Tzu's battle for our footnotes: the emergent role of libraries in juridical warfare. *University of Miami National Security & Armed Conflict Law Review* 3(Fall): 46–103. http://www.m2lawpc.com/index_htm_files/McCary-Sun%20Tzu%20Footnotes-UM-NSAC%20L%20Rev-Vol-III-2013.pdf Accessed 21 May 2016.

Moshirnia, A.V. 2012. Valuing speech and open source intelligence in the face of judicial deference. *Harvard National Security Journal* 4(2012-3): 385–454. http://harvardnsj.org/wp-content/uploads/2013/05/Vo.4-Moshirnia-Final.pdf. Accessed 21 May 2016.

Murkens, J.E.M. 2007. The future of *Staatsrecht*: Dominance, demise or demystification? *The Modern Law Review* 70(5): 731–758. doi:10.1007/978-3-540-73810-7_2.

Noriega, P., and M. d'Inverno. 2014. Crowd-based socio-cognitive systems. In *Crowd intelligence: Foundations, methods and practices. European network for social intelligence*, ed. M. Poblet, P. Noriega, and E. Plaza, Barcelona, January 2014, http://ceur-ws.org/Vol-1148/CROWD2014 Accessed 21 May 2016.

Noriega, P., J. Padget, H. Verhagen, and M. d'Inverno. 2014. The challenge of artificial sociocognitive systems. In AMMAS 14' Proceedings. http://aamas2014.lip6.fr/proceedings/workshops/AAMAS2014-W22/p12.pdf. Accessed 21 May 2016.

O'Connor, T.R. 2006. Intelligence-led policing and transnational justice. *Journal of the Institute of Justice & International Studies* 6: 233–239.

Orend, B. 2014. Fog in the fifth dimension: The ethics of cyber-war. In *The ethics of information warfare*, ed. L. Floridi and R. Taddeo, 1–23. Dordrecht/Heidelberg: Springer. doi:10.1007/978-3-319-04135-3.

Ostrom, E. 2010. Institutional analysis and development. Micro *workshop in political theory and political analysis. Proceedings of the policy studies organization*, New series 9, 851–878 http://www.ipsonet.org/proceedings/category/volumes/2010/no-9/ Accessed 21 May 2016.

Palombella, G. 2009. The rule of law beyond the state: Failures, promises, and theory. *International Journal of Constitutional Law* 7(3): 442–467. doi:10.1093/icon/mop012.

Palombella, G. 2010. The rule of law as institutional ideal. *Comparative Sociology* 9: 4–39. doi:1 0.1163/156913210X12535202814315.

Phythian, M. 2009. Intelligence analysis today and tomorrow. *Security Challenges* 5(1): 69–85. doi:10.1080/13619462.2014.987530.

Poblet, M. (ed.). 2011. *Mobile technologies for conflict management. Online dispute resolution, governance, participation.* LGTS, Dordrecht/Heidelberg: Springer. doi:10.1007/978-94-007-1384-0.

Poblet, M. 2013. Visualizing the law: Crisis mapping as an open tool for legal practice. *Journal of Open Access to Law* 1. Ithaca, Cornell: https://ojs.law.cornell.edu/index.php/joal/article/viewFile/12/13 Accessed 21 May 2016.

Poblet, M., E. García-Cuesta, and P. Casanovas. 2014. Crowdsourcing tools for disaster management: A review of platforms and methods. In *AI approaches to the complexity of legal systems IV. Social intelligence, models and applications for law and justice systems in the semantic web and legal reasoning*, ed. P. Casanovas et al., 262–276. LNAI 8929, Dordrecht, Heidelberg: Springer. doi: 10.1007/978-3-662-45960-7_19.

Ratcliffe, J.H. 2003. Intelligence-Led policing. *Trends and issues in crime and criminal justice*, 248. Canberra: Australian Institute of Criminology.

Ratcliffe, J.H. 2008. *Intelligence-led policing*. Cullompton: Willan Publishing.

Risch, J.M. 2009. Virtual rule of law. *West Virginia Law Review* 112(1): 1–50.

Robson, T.A. 2009. A burning need to know: the use of open source intelligence in the Fire Service. Thesis. Monterrey: Naval School. http://citeseerx.ist.psu.edu/viewdoc/download?doi=10.1.1.3 87.6834&rep=rep1&type=pdf Accessed 21 May 2016.

Rodriguez-Doncel, V., C. Santos, P. Casanovas, and A. Gómez-Pérez. 2016. Legal aspects of linked data – The Europeanframework, Computer Law & Security Review: The International Journal of Technology Law and Practice (2016), doi: 10.1016/j.clsr.2016.07.005

Rugge, F. 2012. The case for NATO-EU cooperation in the protection of cyberspace. In *Cybersecurity Summit (WCS), 2012 Third Worldwide*, 1–10. IEEE.

Schmitt, M.N. (ed.). 2013. *Tallinn manual on international law applied to cyber warfare*. Cambridge: Cambridge University Press.

Schmitt, M.N. 2014. The law of cyber warfare: Quo Vadis. *Stanford Law and Policy Review* 25: 269–300.

Schmitt, M.N., and S. Watts. 2014. The decline of international humanitarian Law Opinio Juris and the law of cyber warfare. *Texas International Law Journal* 50: 189–231.

Schreier, F. 2015. On cyberwarfare. *DKAF Horizon* 2015, WP 7. http://docplayer.net/4159538-Dcaf-horizon-2015-working-paper-no-7-on-cyberwarfare-fred-schreier.html Acessed 21 May 2016.

Solis, G. 2014. Cyberwarfare. *Military law review*, 219(Spring): 1–52. http://www.loc.gov/rr/frd/Military_Law/Military_Law_Review/pdf-files/219-spring-2014.pdf Accessed 21 May 2016.

Steele, R.D. 2007. Open source intelligence. In *Handbook of intelligence studies*, ed. Loch Johnson, 129–147, New York: Routledge.
Taddeo, M. 2012. Information warfare: a philosophical perspective. *Philosophy & Technology* 25.1(2012): 105–120. doi: 10.1007/s13347-011-0040-9.
Tamanaha, B. 2011. The rule of law and legal pluralism in development. *Hague Journal on the Rule of Law* 3: 1–17. doi: http://dx.doi.org/10.1017/S1876404511100019.
Tene, O. 2014. A new Harm Matrix for cybersecurity surveillance. *Colorado Technology Law Journal* 12(2): 391–426.
Trottier, D. 2014. Police and user-led investigations on social media. *Journal of Law, Information and Science* 23: 75–96. AustLII: http://www.austlii.edu.au/au/journals/JlLawInfoSci/2014/4.html Accessed 21 May 2016.
Usmani, Z-ul-H. 2014. Predictive modeling to counter terrorist attacks. *Go-FigSolutions* An Interview with Max Ernst, Pranav Sharma, and Neil Singh, Providence, RI, 9 February.
Verfaille, K., and T.V.d. Beken. 2008. Proactive policing and the assessment of organised crime. *Policing. An International Journal of Police Strategy and Management* 31(4): 534–552. doi: 10.1108/13639510810910553.
Zeadally, S., and A. Flowers. 2014. Cyberwar: The what, when, why, and how [commentary]. *Technology and Society Magazine, IEEE* 33(3): 14–21. doi:10.1109/MTS.2014.2345196.
Zimmeck, S. 2012. The information privacy Law of Web applications and cloud computing. *Santa Clara Computer & High Technology Law Journal* 29: 451–487.
Zittrain, J.L. 2008. *The future of the internet – And how to stop It*. New Haven/London: Yale University Press & Penguin UK. Harvard University's DASH Repository. http://dash.harvard.edu/bitstream/handle/1/4455262/Zittrain_Future%20of%20the%20Internet.pdf?sequence=1. Accessed 21 May 2016.

Pompeu Casanovas is director of advanced research, professor of philosophy and sociology of law at the Autonomous University of Barcelona (Spain) and adjunct professor at Royal Melbourne Institute of Technology (RMIT, Australia). He is serving as head of the UAB Institute of Law and Technology as well. He is general coeditor of the Springer Law, Governance and Technology Series (Germany) and coeditor of the *Journal of Open Access to Law* at the University of Cornell (USA).

Chapter 10
A Model to Facilitate Discussions About Cyber Attacks

Jassim Happa and Graham Fairclough

Abstract The evolution of the Internet and digital systems is making it increasingly difficult to understand cyber attacks. Politicians, ethicists, lawyers, business owners and other stakeholders are all affected by them, yet many lack necessary technical background to make correct decisions in dealing with them. Conversely, cyber-security analysts have a better understanding about the technical aspects of cyber attacks, but many do not understand the repercussions of decisions made from their perspective alone. Both contextual (e.g. societal, political, legal, financial, reputational aspects etc.) as well as technical considerations must be taken into account in making decisions that relate to a cyber attack. A plethora of cyber-attack models exist today that aid (to some degree) understanding of attacks. Most of these however focus on delivering insight from a single perspective: technical detail or understanding of human-centric factors. These approaches do not outline how a discussion among expert-domain people of different backgrounds should be conducted to establish a basic situational awareness understanding, from which to make collective decisions. In this chapter, we present our efforts towards establishing such a model to enable a collective approach in discussing cyber attacks. In this paper, we propose a first version, but believe extensions should be made. We also acknowledge that testing and assessment in real environments is necessary.

Keywords Attack models • Mental models • Cyber attacks

J. Happa (✉)
Department of Computer Science, University of Oxford, Oxford, UK
e-mail: jassim.happa@cs.ox.ac.uk

G. Fairclough
Oxford Internet Institute, University of Oxford, Oxford, UK
e-mail: graham.fairclough@balliol.ox.ac.uk

© Springer International Publishing Switzerland 2017
M. Taddeo, L. Glorioso (eds.), *Ethics and Policies for Cyber Operations*,
Philosophical Studies Series 124, DOI 10.1007/978-3-319-45300-2_10

10.1 Introduction

The vocabulary of cyberspace is a much debated topic. The degree to which cyberspace, loosely defined as: the conceptual landscape where people and machines interact, actually exists and the extent to which it can be considered a new threat environment or simply a medium through which there is a means to an end are widely discussed. Several opinions exist, cyberspace can be described as: *"a global commons that has enhanced interaction, information exchange and productivity"* (INSA 2012); *"an operational domain framed by the use of electrons...to exploit information via interconnected systems and their associated technology"* (Nye 2011), or as Healey in his recent work on cyber conflict defines simply as *"interconnected information technology"* (Healey 2013).

Cyber attack is another conflicted term. Current definitions often fall within two principal categories: *process-driven* or *taxonomy/hierarchical* (Bishop 1995; Cohen 1997; Howard 1998; Lough 2001; Simmons et al. 2009; Hutchins et al. 2011, MITRE). Being able to precisely define a cyber attack is becoming increasingly difficult. The technical complexity of systems, the growing variety of exploitable attack vectors and perhaps most importantly the ubiquitous integration of Internet technology (into all aspects of our daily lives) compound this problem. This uncertainty is creating a disturbing trend: critical decisions concerning cyber attacks (whether it be incident response handling or new policy formulation (as a consequence of cyber attacks)) are dependent on a complete understanding of cyber attacks and how they relate to society or organisations, presently however, such decisions are dealt with by individuals who lack the necessary experience or who approach the issue from a narrow perspective. The failure to adopt a comprehensive approach to the problem is frequently the norm, leading to an incomplete understanding of the attack and a failure to provide an appropriate solution.

A mental model is a person's understanding of a real-world system (such as a cyber attack). Norman (1983) describes mental models as naturally evolving. They change over time as a person learns more about the system, and while they may not be complete or technically accurate, they must be functional. In other words, people must be able to act upon them.

While verbal communication limits our capability to share our own mental model, and erroneous information may be conveyed if attempted, mental models are useful in providing an up-to-date description of the behaviour of a system. **However, mental models are not perfect**, and **no two models will be perfectly aligned** because they are developed individually; they are often incomplete and become unstable over time as people forget details; they lack firm boundaries with similar devices and operations becoming confused and they are "unscientific" with people maintaining "superstitious" behave even when it is unnecessary.[1]

Even experts from the same domain disagree among each other in expert topics, and experts of differing domains are also likely to disagree even when their mental

---

[1] For further explanation of the issues surrounding mental models, see Norman (1983).

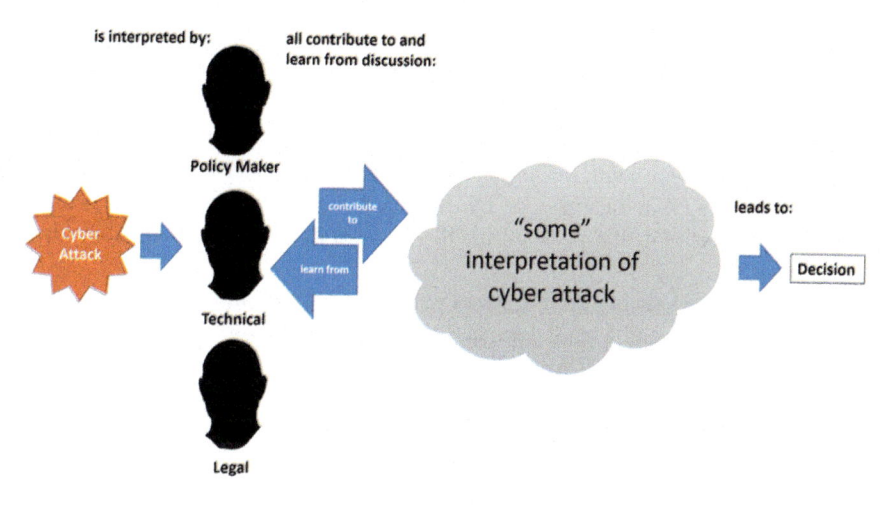

10 A Model to Facilitate Discussions About Cyber Attacks

Fig. 10.1 The cyber attack decision-making process

maps are closely aligned. These issues are of particular relevance when understanding the evolving cyber domain. Where people form mental models about *"what a cyber attack is"* and *"what a particular instance of a cyber attack has done"* regardless of any prior experience and background.

Our proposed approach, suggests that by aiming for **a common, shared reference point** will lead to a greater collective understanding of the same problem space, and a better ability to recognise their own gaps in knowledge. Allowing a "common" mental model of the cyber attack to be constructed. One that can be shared by all actors involved.

Figure 10.1 illustrates the response to a cyber attack decision-making process. Three different mental models of the attack, based upon the actor's role and/or previous experience are integrated to form the basis of "some" common interpretation of the event. This interpretation is then used to construct a decision on the response to the cyber attack. However, this interpretation is likely to lack the degree of detail and shared awareness that is required to produce the necessary level of understanding.

Only from this necessary level of understanding can the correct decision concerning any reaction to the cyber attack be made. We believe the **failure to move from "some" interpretation of the event to a single, shared level of understanding is a consequence of the lack of a common foundation or reference point from which each of the actors derives their individual mental models.** This point is shown in Fig. 10.1 where a policy maker, a technical analyst and a legal representative formulate a response – the decision. Based upon their own prior knowledge of cyber attacks and considered from their own perspectives loosely combined through "some" interpretation.

We argue that knowledge gaps occur because of the failure to move from "some" interpretation to the necessary level of understanding required to enable decision makers to take appropriate, compensatory actions i.e. seeking advice from those who do possess the requisite insight to fill in the respective gaps.

Our approach proposes a model that can be used to fill in these knowledge gaps. The outcome being a more appropriate decision. We envisage this transformation as being a weighting problem. For instance, if the decision maker has a non-technical background (e.g. law, business or ethics), more consideration (weight) will be required on the technical aspects during the decision making process to yield a balanced result.

We investigate existing definitions of cyber attacks with in order to understand their characteristics and what the impacts they have upon actors operating in cyberspace today. We use our investigation to propose our model to generate shared understanding of cyber attacks for decision makers. Our approach is a high-level (abstract), inclusive approach that enables decision makers from different backgrounds and roles to form greater understanding of cyber attacks. Our intent is to enable informed decisions on how to response to a cyber attack, which incorporate a number of perspectives to be made. We do not attempt to define concepts such as cyberspace, attacks or defences comprehensively believing such definitions are not necessary for our abstract approach and that adequate literature exists elsewhere.

The remainder of this chapter is as follows: Sect. 10.2 lists the related work in modelling cyber attacks. Section 10.3 outlines our model and how it can be used in relevant environments, while Sect. 10.4 presents a discussion about additional applications; how it might be extended for other cyber events. We also discuss means to test and assess the usefulness of our model and potential future work. Finally, in Sect. 10.5 we present our conclusion.

10.2 Related Work

The development of new technologies is frequently accompanied by the concurrent exploitation of identified and unidentified concerns within these systems. These weaknesses include design failures, network architecture or software challenges, those that occur during implementation and those that become apparent after adoption. An insider threat may use a tool or exploit social weakness that only becomes apparent after a period of time, while an *Advance Persistent Threat* (APT) may achieve success through agile exploitation of a chain of occurrences. Some of these occurrences exploit technical weaknesses while others exploit flaws in human behaviour. A large body of cyber security literature identifies a set of core components that constitute a cyber attack. We have framed these in a manner in which we assess most decision makers will be able to understand and make sense of; technology-centric models, social models and cyber situational models. We identify some of the more important models to illustrate existing threats and current academic focus.

10.2.1 Technology-Centric Models

Technology-centric models seek to define cyber attacks from the perspective of how an attack operates. These are typically described at a lower (detailed), technical level (e.g. how a piece of malware operates or how a vulnerability can be exploited), and frequently require a high degree of technical knowledge of the system and its constituent parts in order for sense to be made of a cyber attack. These models often remain difficult for decision makers lacking a technical background to comprehend the full implications of a cyber attack and through which to generate the necessary degree of understanding. Examples include:

- Bishop (Bishop 1995): Bishop presented a taxonomy that expresses attacks in the form of six axes: Nature of a flaw; Time of introduction; Exploitation gain; Effect domain; Minimum number necessary and the source of the identification of the vulnerability.
- Cohen (Cohen 1997): Cohen expressed network attacks based on a defined set of properties. These properties view attacks in terms of: Non-orthogonality; Correlation; Hardware Non-specificity; Description; Applicability and Incompleteness. This approach views cyber defence as representing a mirror model.
- Howard (Howard 1998): Howard's model is process-based, taking into account five stages of an attack: *Attackers, tools, access, results and objectives*. Events that occur in each of the process stages are used to derive understand of the nature of the attack.
- Lough (Lough 2001): Lough proposed VERDICT: the *Validation Exposure Randomness De-allocation Improper Conditions Taxonomy*. The model specifies four characteristics of network attacks: Improper validation (insufficient validation resulting in unauthorised access); improper exposure (a system or information is improperly exposed to attack); improper randomness: (insufficient randomness results in exposure to attack), improper de-allocation (information is not properly deleted after use and thus can be vulnerable to attack). VERDICT represents a deep technical approach to understanding cyber attacks.
- AVOIDIT (Simmons et al. 2009): The AVOIDIT methodology is a classification scheme outlining in a tree hierarchy what constitutes a network attack. Its focus is upon: Attack Vector; Operational Impact; Defence, Informational impact and Targets.
- Cyber Killchain (Hutchins et al. 2011): The Cyber Killchain is another process-based model for describing the stages of a cyber attack. These are Reconnaissance, Weaponization, Delivery, Exploit, Installation, Command & Control, and Act on Objectives.

10.2.2 Social and People-Centric Models

Social and people-centric models attempt to understand a cyber attack from a human behavioural perspective. Approaches focus on a wide spectrum of behaviour stretching from the impact of training, identification of non-trustworthy individuals who might represent a cyber-security risk to the discovery of how human behavioural failures can be exploited as part of the cyber-attack process (e.g. phishing). This approach is as technically demanding to the non-human behaviourist as the technical centric frameworks are to non-technicians. Examples include:

- Greitzer et al. (2009) describes an approach to predictive modelling for insider threat mitigation, and continued in 2013 (Greitzer et al. 2013) with a paper on methods and metrics for evaluating analytic insider threat tools.
- The Centre for the Protection of National Infrastructure (CPNI) presented guidance to help manage risk of employees' behaviour damaging businesses called *Holistic Management of Employee Risk* (HoMER).
- *Social engineering: The art of human hacking* (Hadnagy 2010) illustrates the methods by which individuals are "persuaded" to conduct actions that result in a negative impact upon them through subsequent loss in cyberspace.

10.2.3 Cyber Situational Awareness and Understanding Models

Cyber situational awareness and understanding models attempt to adopt a high-level, approach to considering cyber attack. Focus is on the environment in which the cyber attack occurs and the resultant impact upon different elements or layers within it. These models are differentiated by the nature of the elements or layers considered. The need for deep technical or behavioural knowledge is of less importance: the aim being to place cyber attacks into context as opposed to define a cause. Examples include:

- The UK Defence and Science Technology Laboratory (DSTL) (2012) showed a layered model for situational awareness in cyberspace. It consists of six layers of interaction: *Social; People; Persona; Information; Network; Real World*, and attacks can exist on any one or more of these layers.
- *NATO Cyber Security Framework Manual* (NATO 2012b). An interdisciplinary approach to its key activities, including: academic research related to the cyber domain from legal, policy, strategic, doctrinal and/or technical perspectives. The Framework supports the NATO Policy on Cyber Defence.
- *Mission Integration Framework* (Booz | Allen | Hamilton) provides a five pillar approach to considering the cyber security problem: Strategy and Policy; Management and Budgeting; Planning and Operations; People and Culture and Technology and Architecture.

- *The Corporate Insider Threat Detection* (CITD) conceptual model presented by Legg et al. (2013) describes a model for detecting insider threats by exploring hypotheses from measurements of the real world. The model suggests that any measurement is likely to fail on its own, but is likely to yield indicators of insider threats using machine learning and visual analytics.

10.3 Core Model to Discuss Cyber Attacks

To understand modern cyber attacks, it is no longer possible to only consider the technological level, but essential to consider all aspects of cyberspace and the threats generated within it. Our model is underpinned by three assumptions, the need to:

1. **Consider cyber attacks holistically** – attacks are no longer only technical, and other aspects need to be considered. Some of these other aspects will be measurable, but other aspects will be close to impossible to establish (such as attribution or motives).
2. **Involve a wide spectrum of stakeholders** when considering a cyber attack – attacks are best understood when stakeholders are able to communicate effectively among each other and make a decision collectively. While this may not always be the case, it is nevertheless important to involve all parties.
3. **Establish a shared understanding** – as attacks will consist of properties that relate to a variety of different fields (not just technical), it is important to establish a common perspective of what has happened. It is important to identify the knowledge gaps between each expert in order that they can be removed and thus enable better decision-making.

Our model is not intended to provide a solution to classifying cyber attacks. The model's purpose is to facilitate an informed discussion among actors of different backgrounds, including; ethicists, policy makers, lawyers, cyber security experts and boardroom executives (among others). By identifying aspects of an attack within a framework of relevant categories, each containing relevant sub-categories we believe a better comprehension of a cyber attack can be achieved regardless of prior knowledge and experience.

The model aims to enable stakeholders from a spectrum of backgrounds to communicate about cyber attacks in a manner that is enabled by a common position of understanding. From this common understanding, the most appropriate decisions can be made. Figure 10.1 showed this in the context of making such decisions within organisations. Our model has utility in regard to policy-making scenarios, incident response and education (more on this in Sect. 10.4). We believe a more complete understanding, and ability to make better decisions (as a result) is achieved through the use of abstraction and a common mental map of each category.

At the centre of this approach is our mental model process constructed around the construct of sensemaking. Sensemaking has been described as the process of *"how people make sense out of their experience in the world."* (Duffy 1995). It has

Fig. 10.2 Adopted sensemaking model from Xiao et al. (2006)

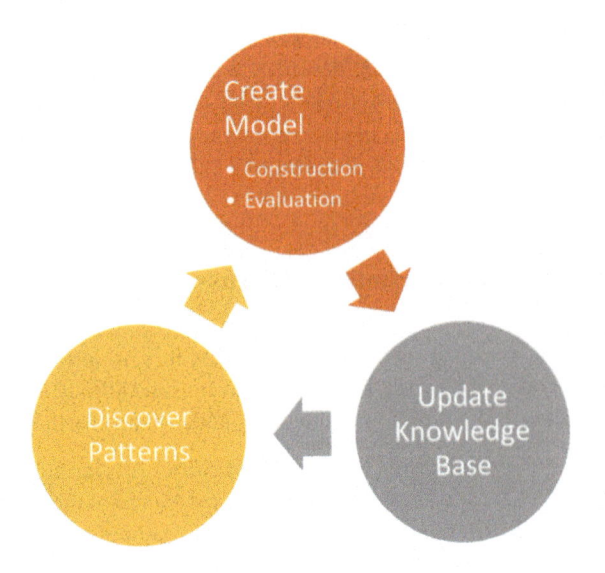

been used widely across a number of disciplines, including informational science (Dervin 1992) and organisational sensemaking (Wieck 1979) but has only been applied to cyber security in more recent years.

By developing an analytics model about network traffic Xiao et al. (2006) extrapolated meaning from activities occurring within the network. Alberts et al. (2002) also discussed the applications of sensemaking for cyber security in net-centric operations within cyberspace in understanding cyber information systems.

Figure 10.2 illustrates a high-level approach to sensemaking that can be used in making new discoveries and updating a mental model of an attack (adopted from Xiao et al. (2006)). The model is updated as more information about the attack becomes available. Xiao represents only one approach to consider sensemaking, and considerable numbers of work on sensemaking exist, other examples might be to use the sensemaking models proposed by Alberts et al. (2001), Pirolli and Card (2005) or Hutton (2008).

We propose an initial starting point of four core categories: Attributes of Attack; Values of Attack; Impact of Attack and Mitigation of Attack. These reflect the key pillars that underpin our sensemaking of a cyber attack. We do not see these categories as being exclusive, accepting that others might be added depending upon the context of any event and the purpose for which the model is being used. In this sense the model is both expandable and transformative driven by the nature of the cyber attack and the stakeholders who are responsible for subsequent decision-making.

We perceived our model as representing a holistic mental model that is intended to frame the construction of the necessary level of understanding from which appropriate decisions can be made. As new knowledge is obtained through an evaluation process it is incorporated into the mental model. In turn new patterns are discovered, which further aid in its updating. It is our intention to investigate this starting point further and refine our model in the future. Our approach is shown in Fig. 10.3.

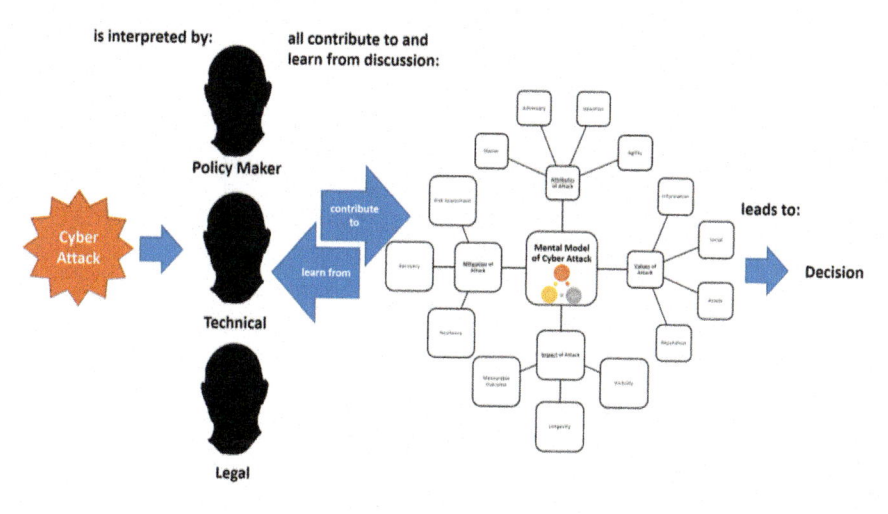

10 A Model to Facilitate Discussions About Cyber Attacks

Fig. 10.3 The cyber attack decision-making process using our model

Figure 10.4 shows an enlarged diagram of our model to understand a cyber attack. Below the four core categories or parent nodes previously identified sit sub-categories that represent activities or factors through which a cyber attack is defined. Xaio's approach can be seen to sit at the centre of this framework, reflecting that the application of sensemaking is a continuous process that builds upon constant evaluation and analysis of all aspects of the cyberspace environment. It is envisaged that the adoption of this approach will allow the identification of any interdependencies existing between the categories. Thus, leading to a more complete understanding of the threat and of those mitigation measures most likely to succeed.

Core categories and their sub-categories present in Fig. 10.4 are defined in the following section. Our concern is with the metadata that describes each of the respective core categories and their supporting sub-categories. Consideration of the measurement used to underpin each of our categories is the subject of separate work outside the scope of this chapter.

10.3.1 Attributes of Attack

Attributes of the attack describe the context in which the event occurred. Doing so allows the expert actor to place the attack into their own scheme of reference: into their own context. This is important. Context setting provides the platform from which the actor can consider the other three remaining core categories within the model from both their own perspective and in relation to those held by other stakeholders and subsequently make decisions. These values can be described through the application of existing models e.g. Howard or Cohen).

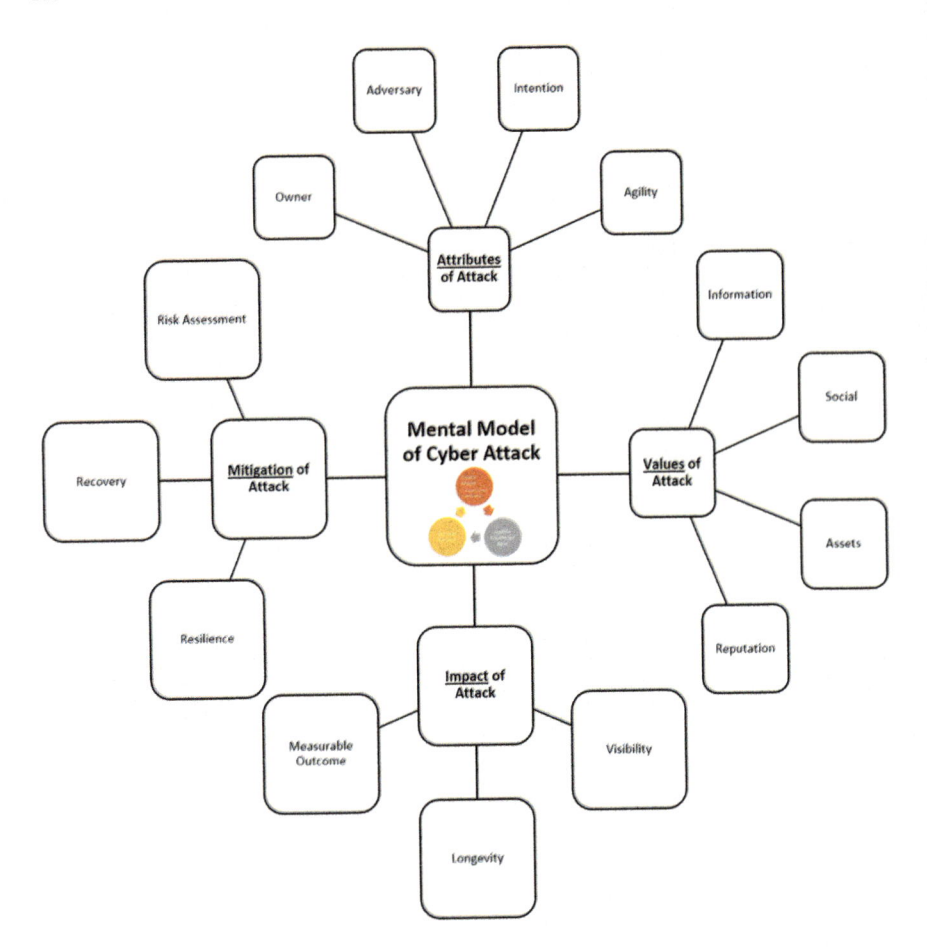

Fig. 10.4 Our cyber attack mental map model

Owner Relates to the owner of the information, asset or capability that is the subject of the attack.

Adversary Relates to the perpetrator of the attack. The Adversary is the entity who is behind the intent of the cyber attack. It should be noted that a separate actor may be responsible for the actual conduct of the attack.

Intention Intent describes the motive for attack. Motives and associated goals may not remain static, changing in response to external drivers or actions undertaken by the Owner of the asset under attack. Subsequently intent may be of short or long duration and may constitute an end in itself or form part of a complex objective. Examples include curiosity/challenge, political, financial, vengeance reasons (see "goals", from Howard's Taxonomy).

Agility We understand agility to be the capacity held by an acto to identify, select and exploit capabilities to achieve a desired intent. We consider all cyber attacks to consist of some form of agility and hence our model goes beyond the traditional, linear, kill-chain approach frequently referred to in the literature.

Duration Reflects a quantitative value that can be measured to capture the length of an attack defined by time. Measurement of duration can be used to define seriousness of the attack or level of intent of the Adversary.

Dimensions This attribute refers to the structure of the attack and is used to inform assessment of complexity and agility e.g. is the attack split into different stages/components. Sufficient knowledge of dimensions might be used as a form of attack profile or "fingerprint".

10.3.2 Values

Values describe the semantic meaning that is generated by the Owner following a cyber attack. These psychological perspectives will influence how the Owner interprets the attack and contextualises it in regard to their wider value system. Values will play an important role in constructing the Owner's "world view" of the event. They will subsequently inform decision making on the nature of any response.

Information Information is the real world meaning behind the data available. In practical terms we can compare this to the data stored on a database. Information can be the values put into systems (digital bits and bytes) or knowledge that is made available from this data, such IP (e.g. blueprints), or HR records at an organisation.

Social Reflects the people aspect of cyber attacks, i.e. whether a cyber attack can have an impact on the social dynamics of people in any way. Will it impact upon they way that they conduct their daily activities or with their relationships with other actors in either the physical world or the cyber domain.

Assets In this sub-category we cover the physical items in the organisation such as physical computers, *Supervisory Control And Data Acquisition* (SCADA) systems etc. that can be affected by cyber attacks.

Reputation The extent to which the outcome of a cyber attack can alter the way other actors perceive the Owner or the entity that they represent. Change is reflected in the impact that occurs in regard to the Owner's achieving their desired intent. While reputational damage is non-tangible it is most frequently reflected in a tangible impact such as loss of revenue.

10.3.3 Impact

Impact describes the consequence of a cyber attack in terms of the victim's ability to achieve their desired intent. Impact can be expressed in tangible or non-tangible terms e.g. online retail fraud or loss of custom through demise in reputation. Form and degree of impact will have considerable influence on the response that is made to a cyber attack. The response may differ in regard of time, allocation of effort and nature i.e. a technical response or a policy response.

Outcome Defines the nature of the impact of the attack. It identifies how and the extent to which the victim can no longer achieve their desired intent. We describe cyber attacks as events in cyberspace that can generate an outcome in the physical world or virtual domain. Outcome may be measurable in tangible or non-tangible terms. A prevented cyber attack is seen as having an unsuccessful outcome.

Depth Depth relates to the duration of an attack in terms of the period existing between its initial consequences being identified and the restoration of the normal, pre-event systems condition. Quantification of depth can assist in consideration of risk and the choice of appropriate mitigation.

Visibility Defines the extent to which the outcome of the cyber attack is observable, frequently reflected its impact on the victim's intent. This impact might not be visible to all actors who are impacted upon by the attack or visibility may only occur after the adversary's aim has been achieved e.g. State level espionage.

10.3.4 Mitigation

Mitigation represents those actions and capabilities that are available to the target in order to minimise the consequences of a cyber attack. These measures may be protective, existing prior to a hostile event occurring or be responsive, transpiring post event such as recovery software or operational redundancy. Mitigations may take the form of hard measures or soft measures or a combination of the two.

Recovery Recovery reflects the capacity of the Owner to regain pre-event capability post a cyber attack. Recovery can be measured in terms of time and/or use of other capability including manpower and technology. Recovery will be dependent on other attributes including Dimension and Depth.

Resilience Resilience reflects the Owner's ability to deal with the impact of a cyber attack as it is occurring without there being noticeable degradation of capability or asset. Resilience is influenced by Dimension, Depth and Recovery and the existence of prior measures of mitigation.

Risk Represents a formalised process through which the Owner considers the range of potential threats that may exist to their assets or capabilities.

Risk Assessment identifies those threats that the Owner can deal with and those which he cannot. This process provides the pivot/bridge between consideration of cyber attack and cyber defence.

10.4 Discussion

10.4.1 Applications of the Model

We have identified a number of other cyber-security challenges that could benefit from the application of our model. The first of these is in regard to cyber defence. Such usage envisages sense making of the same event but from an opposing perspective: a "mirroring" effect. This effect recognises that the attributes of cyber attack have equal relevance when considering their mitigation. In considering cyber defence, the Owner's intent is to fortify his cyber environment in order to protect assets and capabilities from a cyber attack. Achieving this intent requires the establishment of a robust cyber defensive posture founded upon a holistic understanding of the threat that takes consideration of not only the threat but in addition the wider context in which the entity to be defended exists. We see successful cyber defences as being as reliant on the existence of deterrent legal policy, as on strong firewalls and penetrating software based intruder detection systems.

Weick (1988) describes sensemaking in crisis conditions as being particularly difficult, and argues that commitment, capacity, and expectations affect sensemaking. We suspect that in instances where crucial decisions have to happen fast, our model in its form is too comprehensive to be appropriately used. Although, with suitable adaption it can still provide useful structure and guidance to decision makers.

The second area in which we see the model being able to contribute is that of cyber security policy generation. At the strategic level of activity, the establishment of shared situational awareness of the challenges present and the need to consider the perspective of all stakeholders is of equal or perhaps greater importance. The definition and implementation of meaningful cyber-security policy requires that understanding must exist across a wide spectrum, from the creation and enforcement of standards in respect of security measures to the establishment of appropriate authorities that allow Law and Order organisations to operate against potential adversaries proactively. In both of these cases technical input shares the same level of importance as that of a lawyer. We believe that our approach permits consideration of such wide ranging factors through the application of a simple conceptual model.

A third use of the model is in the field of cyber education. In this area, we consider that it has applicability in regard of those involved in the cyber security process, the expert-actors referred to previously and to the education of the users of cyber systems. In this context our model provides a means to identify, dependent upon the past experience and role of the actor those areas where benefit would be gained through the undertaking of education. As such our model provides an informed and targeted approach to education programme design.

Additionally, it provides a structure upon which to build an education programme. One that can be tailored dependent upon a particular subject of interest, with each of the categories being the basis for an individual module in the course timetable. A cyber attack based education programme based upon the graphical representation of our model shown at Fig. 10.2 would consist of at least 14 modules.

10.4.2 Future Work

We do not claim our cyber security mental model framework is complete. We believe that it represents only a first step and it will need continual refinement with experts from a wide variety of backgrounds who can contribute to its development. The first of these is in understanding more fully what are the important categories and sub-categories that must be included when considering a particular cyber security problem. Our illustrative model provides one set that we believe has importance in relation to cyber attack, while recognising that there may be others that also should be present. We would wish to examine in more depth the extent to which we have succeeded in identifying all of the relevant concepts necessary to underpin our model and the validity of the categories from which it is constructed. A second issue concerns the degree to which the categories and sub-categories need to be changed when the focus of the model alters. Is it valid to use the same sets in looking at cyber defence and cyber policy generation? If not, and this would seem the likely answer what new categories might be required and which ones could be removed. In regard to both of these areas we see the need for further analysis of the relevant literature and direct engagement with the cyber security expert community. The latter perhaps, in the long term, providing to be of the most use through the insight that communicating directly should bring.

In order to assess our model it will be necessary to put it in the hands of many different decision makers, obtain their feedback, based on their prior experiences and identify where this model might fit in, if at all. Validating our approach is not straightforward for real environments. For laboratory setting, one approach might be to introduce a group of participants to cyber attack scenarios before being introduced to the model, and make decisions, before introducing them to our model and present them with a different scenario in which the model should be used. Another approach to assessment would be to measure how well decisions maker are able to respond to cyber attacks, some with having been introduced to the model, and others not.

A third area of future work is identifying those cyber events that suit the application of the model. In the course of this chapter we have discussed a number of instances. Although of these we have only considered that of cyber attack in any detail. This strand of work will focus on identifying other cyber security events that would benefit from our work. Engagement with domain experts particularly as part of testing and assessment process referred to previously represent logical avenues of advance in regard to resolving this concern.

We judge however, that the more pressing area, which requires further work, concerns the identification of appropriate measures for our categories and how these measures can be obtained and validated. In some circumstances this challenge will be relatively easy to overcome while in others gaining a reliable metric may not even prove possible. Tangible values relating to loss of revenue or fall in customer numbers as a consequence of a cyber event against a business enterprise should be accessible as an element of an organisation's routine reporting process. Placing measurable values on the categories of Agility and Visibility are likely present challenges as their impacts may be non-tangible. Our approach in resolving these types of difficulties will be to consider how such non-tangible values are dealt with in other disciplines including risk assessment and medicine. In the latter case, inference is as a means to make an assessment. The applicability of this approach in respect of our model requires to be determined.

10.5 Conclusion

In this chapter we have proposed a conceptual model that we see as enabling a collective approach to understanding, discussion and subsequent decision-making about cyber attacks. The model is underpinned by three assumptions: the **need to consider cyber attacks holistically**, the requirement to **involve a wide spectrum of stakeholders** in discussing any cyber-attack problem, and finally, the critical need to **establish a shared understanding**.

The model is constructed from a set categories and sub-categories that reflect the diverse spectrum of factors that must be considered in dealing with a cyber attack. Our choice of assumptions and categories is based upon literature review and engagement with domain-experts. The chapter has also examined how the model might be applied more widely in the cyber security field. Consideration has also been given to the future work necessary in the development of our approach.

We do not claim that our approach to considering cyber security attacks and other cyber events is complete. We do however, believe that the approach offers a potential means to establish a common understanding and establishment of a starting point for collective discussion and decision making about cyber attacks.

References

Alberts, D.S., J.J. Garstka, R.E. Hayes, and D.A. Signori, 2001. Understanding information age warfare. Assistant secretary of defense (c3i/command control research program).

Bishop, M. 1995. A taxonomy of Unix system and network vulnerabilities. Technical Report CSE-95-10, Department of Computer Science, University of California at Davis.

Booz. Allen. Hamilton. 2012. Cybersecurity: Mission integration to protect your assets. https://www.boozallen.com/media/file/ Cybersecurity-Mission-integration-to-protect-your-assets-fs.pdf. Accessed 8 Sept 2014.

Cohen, F. 1997. Information system attacks: A preliminary classification scheme. *Computers & Security* 16(1): 29–46. Elsevier.

CPNI, Centre for Protection of National Infrastructure. http://www.cpni.gov.uk/advice/Personnel-security1/homer/ Accessed 7 Sept 2014.

Dervin, B. 1992. From the mind's eye of the user: The sense-making qualitative-quantitative methodology. In *Qualitative research in information management*, ed. J. Glazier and R.R. Powell, 61–84. Englewood: Libraries Unlimited.

DSTL Centre for Defence Enterprise (CDE) Cyber Situational Awareness Launch Presentation 2012. http://webarchive.nationalarchives.gov.uk/20140410091116/http://www.science.mod.uk/events/event_detail.aspx?eventID=184 Accessed 7 Sept 2014.

Duffy, M. 1995. Sensemaking in classroom conversations. In *Openness in research: The tension between self and other*, ed. I. Maso, 119–132. Assen: Van Gorcum.

Greitzer, F.L., and T.A. Ferryman, 2013. Methods and metrics for evaluating analytic insider threat tools. In *Proceedings of the 2013 IEEE Security and Privacy Workshops (SPW'13), California, USA*, 90–97. IEEE, May 2013.

Greitzer, F.L., P. Paulson, L. Kangas, T. Edgar, M.M. Zabriskie, L. Franklin, and D.A. Frincke, 2009. *Predictive modelling for insider threat mitigation*, Technical report PNNL-60737. Pacific Northwest National Laboratory.

Hadnagy, C. 2011. *Social engineering: The art of human hacking*. Indianapolis: Wiley.

Healey, J. 2013. A fierce domain: Conflict in cyber space. Cyber Conflict Studies Association.

Howard, J.D., and T.A. Longstaff. 1998. A common language for computer security incidents. Sandia National Laboratories.

Hutchins, E.M., M.J. Cloppert, and R.M. Amin. 2011. Intelligence-driven computer network defense informed by analysis of adversary campaigns and intrusion kill chains. Leading Issues in Information Warfare and Security Research.

Hutton, R., G. Klein, and S. Wiggins. 2008. Designing for sensemaking: A macrocognitive approach. In Sensemaking Workshop.

Intelligence National Security Alliance. 2012. Cyber Intelligence: setting the landscape for an emerging discipline. *Air and Space Power Journal*.

Legg, P., N. Moffat, J.R. Nurse, J. Happa, I. Agrafiotis, M. Goldsmith, and S. Creese. 2013. Towards a conceptual model and reasoning structure for insider threat detection. *Journal of Wireless Mobile Networks, Ubiquitous Computing, and Dependable Applications*. 4: 20–37.

Lough, D.L. 2001. *A taxonomy of computer attacks with applications to wireless networks*. Blacksburg: University Libraries, Virginia Polytechnic Institute and State University.

MITRE Corporation. 2012. Common Attack Pattern Enumeration and Classification (CAPEC). https://capec.mitre.org Accessed 7 Sept 2014.

NATO. 2012a. NATO policy on cyber defence. NATO Cooperative Cyber Defence Centre of Excellence. https://web.archive.org/web/20120310083820/http://www.nato.int/nato_static/assets/pdf/pdf_2011_09/20111004_110914-policy-cyberdefence.pdf Accessed 7 Sept 2014.

NATO. 2012b. NATO cyber security framework manual. NATO Cooperative Cyber Defence Centre of Excellence. https://www.ccdcoe.org/publications/books/NationalCyberSecurityFrameworkManual.pdf Accessed 7 Sept 2014.

Norman, D.A. 1983. Some observations on mental models. In *Mental models*, ed. D. Gentner and A.L. Stevens, 7–14. Hillsdale: Lawrence Erlbaum Associates Inc.

Nye, J. 2011. *The future of power*. New York: Public Affairs.

Pirolli, P., and S. Card. 2005. The sensemaking process and leverage points for analyst technology as identified through cognitive task analysis. In *Proceedings of International conference on intelligence analysis*.

Simmons, C., C. Ellis, S. Shiva, D. Dasgupta, and Q. Wu. 2009. *AVOIDIT: A cyber attack taxonomy*, Technical report: CS-09-003. University of Memphis (August 2009).

Weick, K. 1979. *The social psychology of organizing*. New York: McGraw-Hill.

Weick, K.E. 1988. Enacted sensemaking in crisis situations. Journal of Management Studies 25(4): 305–317. Wiley.

Xiao, L., J. Gerth, and P. Hanrahan, 2006. Enhancing visual analysis of network traffic using a knowledge representation. In *Proceedings of the IEEE Symposium on Visual Analytics Science and Technology*. IEEE.

Jassim Happa is a postdoctoral researcher in cybersecurity at the University of Oxford. His research interests include computer graphics, cybersecurity, rendering, visualization, human visual perception and human-computer interaction. He obtained his BSc (Hons) in computing science at the University of East Anglia in 2006, after which he worked as an intrusion detection system analyst. In October 2007 he began his PhD in engineering at the University of Warwick where he developed a number of novel computer graphics techniques to appropriately document and reconstruct real-world heritage sites based on evidence available today. He defended his PhD in January 2012 and has since December 2011 worked at Oxford. In more recent years, he has spent his research efforts on visual analytics and cybersecurity research challenges. He is responsible for creating, implementing and assessing novel visualization approaches, but also investigates the topic of threat modelling.

Graham Fairclough is a DPhil candidate in cybersecurity at the University of Oxford. Prior to commencing his studies, he was career soldier in the British Army, reaching the rank of colonel. Throughout his time serving, he was employed on intelligence and security duties, including operational tours in Northern Ireland, Belize, the Balkans, Iraq and Cyprus. Senior appointments included 3 years in the United Kingdom's Permanent Joint Headquarters (PJHQ), responsible for the delivery of intelligence architecture and capability in Iraq and Afghanistan in the period 2007–2010, and as the chief of staff to the United Kingdom's Defence Intelligence between 2010 and 2013. He has served with the Government Communications Headquarters (GCHQ) and has worked closely in a number of roles with other elements of the United Kingdom's Intelligence Community. He possesses an MSc in knowledge management systems and an MA in defence studies. His current research interests include the establishment of an appropriate level of understanding on how cyber power can be used to deliver military effect in support of national objectives, the impact of the emerging operational cyber domain on the future character of war, how cybersecurity incidents are understood by decision makers and the potential empowerment of non-state actors through the low barriers of entry that exist for cyberspace.

Chapter 11
Strategies of Cyber Crisis Management: Lessons from the Approaches of Estonia and the United Kingdom

Jamie Collier

Abstract This chapter compares the cyber crisis management strategies of Estonia and the United Kingdom—two leading nations in the field of cyber security. The two countries' strategies differ significantly. The most important variables influencing these differences are history, size (both demographic and material resources), political philosophy, digital dependence, and the nature of the threats and adversaries each country faces in the cyber domain. Given the importance of these factors in determining Estonia's and the United Kingdom's cyber crisis strategies, it is difficult to draw from these two cases any generalisable recommendations that apply to other states; rather, the main significance of this study is another: it draws attention to the important role that political, historical, and cultural variables play in the definition of a nation's cyber crisis strategy—and, consequently, the need to fit specific policy approaches within the bounds set by these factors. The chapter seeks to demonstrate that while cyber attacks may be highly technical in nature, organisational responses to them have crucial political and social determinants that may supersede the significance of technical factors.

Keywords Cyber Security Strategy • Critical Infrastrucutre • UK • Estonia

11.1 Introduction

Cyberspace has played a decisive role in improving and expanding the operations of critical infrastructures in modern society. This is particularly apparent in economically developed and technologically advanced nations, where information technologies are constantly integrated into a growing range of core services across a number

---

This publication is funded by the European Social Fund and the Estonian Government

J. Collier (✉)
Department of Politics and International Relations and Centre for Doctoral Training in Cyber Security, University of Oxford, Oxford, UK
e-mail: Jamie.collier@cybersecurity.ox.ac.uk

© Springer International Publishing Switzerland 2017
M. Taddeo, L. Glorioso (eds.), *Ethics and Policies for Cyber Operations*, Philosophical Studies Series 124, DOI 10.1007/978-3-319-45300-2_11

of public and private industries. For example, "smart" power grids utilising cloud computing can deliver significant cost savings and maximise energy efficiency (Clastres 2011), while electronic health records can optimise hospital services and give impetus to innovative medical research (Kotz et al. 2009).

Yet alongside these benefits are considerable risks. The increasing reliance of modern society on complex digital systems—particularly in the operations of critical national infrastructures—increases the risks of disruptive or destructive attack in two significant ways. First, digital systems contain a theoretically limitless number of software (and other) vulnerabilities; hence, the number of possible attack sequences available to an aggressor is also limitless. Second, the wide diffusion of cyberspace empowers new threat actors. Compared to conventional security and defence domains, the cyber domain has low barriers to entry; not just states but also non-traditional and militarily weak actors can use cyberspace to cause significant harm (Nye 2011 Ch. 5). Cyber security, in short, has become a crucial component of national and economic security strategy; it is a central and necessary feature of a modern nation's crisis management strategy.

This chapter focuses on cyber crisis management, defined as the cyber security aspects of a crisis situation, which in turn is defined as a situation where a state's citizens are subject to significant danger. Specifically, the chapter examines three domains of crisis management: the organisation of the governmental institutions responsible for crisis situations; the role of non-governmental stakeholders; and international cooperative efforts, with a special emphasis on the countries' use of traditional international security mechanisms (such as NATO) to manage crises. The study seeks to identify similarities and differences in the cyber crisis management strategies of two nations—Estonia and the United Kingdom—and to explore the main explanatory variables that shape and constrain each. This comparative study aims to contribute to cyber security scholarship and practice by identifying broad lessons and insights that might be applicable in other national contexts—even if specific national factors are not replicable. The study aims to create new knowledge by demonstrating the analytical usefulness of a comparative approach to the formulation of cyber security policy, while uncovering the broad political, cultural, and historical factors that influence the national strategies of Estonia and the United Kingdom. Why Estonia and the United Kingdom? These two states were chosen for comparison because they are often regarded as international leaders in cyber security (for example, both are members of the so-called Digital 5—or D5—network of leading digital governments) (Williams-Grut 2014). It is important to note that there are significant similarities and differences between the two states: both are developed European economies as well as European Union and NATO member states and both feature a high degree of dependence on digital systems; yet at the same time they are notably different in size, threat landscape, and other factors.

The chapter argues that Estonia and the United Kingdom differ, at times significantly, in their strategic responses to cyber crisis situations because of a combination of political, historical, and cultural variables. These variables include history, size (referring to both demographic and material resources), political philosophy, digital dependence, and the threats and adversaries that each nation faces in the cyber domain. Both case studies demonstrate the importance of non-technical variables in the development of national cyber strategy. Indeed, if Estonia and the United Kingdom differ in their cyber crisis strategies, then other more disparate nations are likely to differ even more significantly—thus highlighting again the importance of a customised and context-specific approach to the study of national strategies. Thus although states caught in a cyber crisis situation are likely to face many similar policy and technical challenges, responses to these challenges at the *strategic* level may vary enormously. Therein lies an important insight for scholarship and practice: because non-technical variables can significantly affect a state's cyber crisis management strategy, these variables require serious attention in assessing the adequateness of national responses. Of course, some features of national strategy—such as the need to define clear and coherent defensive priorities—can be applied broadly. But given the importance of non-technical and country-specific factors in shaping a cyber crisis management strategies, such generalisable lessons are likely to be limited. A strategy or policy that is successful in Estonia might not work effectively in the United Kingdom and vice-versa.

In sum, this chapter will seek to demonstrate that given the importance of political, historical, and cultural variables, there can be no "one-size-fits-all" approach to cyber crisis management. Rather, the best crisis response strategy is that which most effectively integrates the strengths and limitations of each nation's respective political and cultural circumstances.

The remainder of the chapter is organised into five sections. Section 11.2 describes the scope of analysis and research methods. Section 11.3 examines the cyber crisis management strategy of Estonia. Section 11.4 reviews the strategy of the United Kingdom. Section 11.5 identifies the main explanatory variables that shape and constrain the two nations' strategies. Section 11.6 discusses the broader policy implications of the study.

11.2 Scope and Methods

A few words on the chapter's scope and key terms are in order. In considering responses to cyber threats, the chapter focuses on indirect cyber attacks—specifically, attacks on critical national infrastructure. Many security analysts have claimed that cyber attacks are only indirect in nature; that is, they do not produce direct effects on a target (Rid 2013). Others, however, have argued that direct forms of cyber attack are possible (for example, digital manipulation of electronic pacemakers) and should not be dismissed. In addition, as new types of devices increasingly become connected online, the possibility of direct damage by cyber attacks will only increase. But it is true that the vast majority of cyber attacks are indirect; they damage digital systems that run vital services or critical functions that reside outside cyberspace, with the intent of indirectly harming, or at least disrupting, these services. By focusing on indirect attacks, the chapter does not mean to downplay the danger of direct attacks. But because academic and policy understandings of such direct attacks is still rudimentary, policy reaction to them has been limited; consequently, it is difficult to meaningfully examine the national strategies dealing with direct attacks—but for this same reason, the development of such strategies should be a key government priority. Third, the chapter focuses on strategic—not technical—responses to a crisis. Although cyber attacks are highly technical in nature, the response required in a crisis often has clear political and other non-technical components, for example the question of how government departments work with civilian stakeholders and foreign allies. Non-technical factors may, in fact, be the more important in determining the nature, scope, and success of a crisis response.

Methodologically, the chapter draws on a diverse range of sources that includes the secondary academic literature and primary data such as publicly available governmental and business reports. Fieldwork was conducted in both Estonia and the United Kingdom and involved a number of semi-structured field interviews with leading practitioners in the field of cyber security and defence, including officials at the NATO Cooperative Cyber Defence Centre of Excellence (CCDCOE). The analysis also draws from a number of conversations with individuals working within critical national infrastructure as well as academics and researchers specialising in relevant fields. The chapter encountered a few notable methodological obstacles. For one thing, the chapter is unable to identify interview participants owing to the sensitivity of the topic and in order to elicit open and frank responses from them. For another, Estonia's and the United Kingdom's cyber crisis strategies are largely hypothetical. That is, the paucity of cyber crises available for empirical investigation renders—inevitably—any analysis of this subject speculative to a degree. It cannot be guaranteed that during an actual crisis situation a state will execute its intended crisis strategy; indeed, as discussed below, crisis situations have previously led to extemporaneous responses. Moreover, important procedural details of Estonian and UK strategies remain unclear or unknown because neither country has outlined its cyber crisis management strategy in one single document or because the data remains at least partly classified. This may be because governments are naturally reluctant to disclose the full details of their crisis management, since disclosure may reveal weaknesses that an opportunistic assailant could exploit. Nevertheless, the qualitative data and sources of this study have made possible a rich empirical analysis on which future studies may build.

11.3 Estonia

Estonia's digital infrastructure has grown significantly since the nation regained independence in 1991. It has progressed from a country where only half the population had telephone lines to one that is a recognised global leader in information technology and e-government. In 1992, Estonia decided to orient its developmental strategy to the development of superior digital systems (Laar 2002). Starting with a blank slate, it was able to leapfrog previous types of infrastructure—for example, Estonia was able to develop a sophisticated digital land registry system soon after it's re-independence. The state also launched an innovative digital ID-card system. (The Economist 2013). Technology has therefore played a large role in Estonia's attempt to overcome the legacy of Soviet occupation. With the limited resources and budget constraints faced by a small state, technological solutions and innovative approaches in its public sector services have been vital to the country's development. This trend has continued through 2015, with Estonia becoming one of the most connected nations in the world: Estonians pay taxes and vote online; their health records are stored digitally; and concerned parents can access their children's exam results, attendance, and homework assignments via the Internet (Mansel 2013). Notably, and uniquely, Estonia created the "X-road," a decentralised data exchange environment that connects the government's various e-services databases (Kotka et al. 2015).

Such high levels of digital dependence make Estonia potentially vulnerable to cyber threats. This vulnerability was demonstrated in 2007. Following the removal of a Soviet war memorial from Tallinn's city centre, a number of Estonian e-services and websites were overloaded by a distributed denial of service (DDOS) attack, believed to be of Russian origin (Landler and Markoff 2007). The attack degraded the availability and functioning of services crucial to Estonia's information society, including government websites and online banking systems (Schmidt 2013). The attack, however, inflicted no long-term property damage, loss of life, or substantial financial loss (Cardash et al. 2013). Nevertheless, the attack illustrated the potential for cyber attacks to threaten Estonia's national and economic security. Since this incident, there has been notable change in Estonia's strategic posture: cyber security is now regarded as a higher political priority. A number of initiatives have since been launched, including the publication of an official Estonian cyber security strategy, the creation of a formalised cyber reserve force, and the establishment of the NATO Cooperative Cyber Defence Centre of Excellence in Tallinn (Jackson 2013). What follows is a review of Estonia's current cyber crisis management strategy.

11.3.1 Domestic Institutional Organisation

In Estonia, there is a clear centralised leadership structure for response to cyber crisis situations. The responsibility for overall cyber security strategy and policy coordination lies with the Ministry of Economic Affairs and Communication; this responsibility was relinquished by the Ministry of Defence in 2011 (Osula 2015b). Within the Ministry of Economic Affairs and Communication, the Estonian Information Systems Authority (*Riigi Infosüsteemi Amet*—RIA) serves as the central cyber security competence and coordination centre, supervising the application of cyber security measures for systems that provide vital services to the Estonian state. The Ministry of Interior also acts as a coordinating body during crisis situations.

The 2009 Emergency Act requires providers of vital services to immediately notify the relevant national government department about attacks, and they must also provide information to other bodies upon request. Governmental powers can be further escalated through the 1996 State of Emergency Act if a threat emerges that threatens the constitutional order of Estonia (Osula 2015b). This Act significantly centralises the crisis management response by concentrating rights, duties, and liabilities in the Prime Minister, who becomes the chief authority during a state of emergency. In serious crisis situations, the Prime Minister is able to restrict certain rights and freedoms in the interest of national security and public order. Therefore, while certain cyber crisis management responsibilities lie with individual government departments, if a situation becomes serious and significantly threatens Estonia, there is a very clear process for the escalation of powers to central bodies such as the Estonian Information Systems Authority and the Prime Minister.

The Estonian Computer Emergency Response Team (CERT-EE) provides a support function in cyber crisis situations. It is responsible for handling security incidents within Estonian networks, providing warning of potential security incidents, and analysing the spread of malware and incidents that have taken place in Estonian computer networks (Osula 2015b). With the high level of e-government services, it is clear that CERT-EE has an important role in safeguarding Estonian infrastructure. Moreover, the government utilises its embassies abroad to maintain secure cloud systems—"data embassies"—that can sustain e-services in the event of a cyber attack on Estonia's domestic infrastructure. Estonia's "digital continuity" project aims to ensure that in the event of any cyber crisis situation, crucial e-government services will continue to function (Kotka et al. 2015). Thus the idea is that if Estonian-based systems are under attack or services become compromised in the event of a national emergency, data embassies abroad can assist in the recovery process.

11.3.2 Stakeholder Mobilisation

The implications of cyber crisis situations reach beyond governmental institutions. For any response to be successful, governments must interact with other stakeholders. The Estonian government has adopted a comprehensive and inclusive approach with non-governmental actors playing important roles in cyber crisis situations. In Estonia there has been a particular emphasis on mobilising civil society. The current Estonian Deputy Director for National Security and Defence Coordination, Kristjan Prikk, has claimed that while other states have sought to overcome cyber threats by allocating vast budgets to the problem, Estonia has instead strived to create a nation of citizens alert to online threats (Jackson 2013). In Estonia, there is an emphasis on educating citizens about risks and promoting an understanding of cyber security (Grobler et al. 2013). According to Prikk, Estonia has adopted "not just a whole of government approach, but a whole of nation approach" (Jackson 2013).

This "whole of nation" approach was clearly evident during the 2007 DDOS attack, during which a number of stakeholders collaborated in order to defend against and mitigate the effects of the suspected Russian action. Three days after the initial attacks occurred, a number of Estonian experts came together, including Internet service providers, banks, police, governmental envoys, mobile telecommunication firms, and envoys from the government's Security and Information Boards (Schmidt 2013). This informal alliance was able to organise assistance in a number of ways, for example, by supplying relevant appliances and hardware, filtering malevolent traffic, and providing information on the threat's scope, nature, and technical detail (Schmidt 2013). By creating opportunities for collaboration, this informal alliance enabled different actors to work closely together in order to help overcome the unfolding crisis. Principles of loose governance and trust-based information-sharing facilitated this ad hoc collaborative framework.

Yet despite the effectiveness of the Estonian technical community's response, the attack also marked the end of that community's autonomy from state interference and regulation (Schmidt 2013). In the aftermath of the attack, the Estonian government woke up to the danger of cyber crisis situations and increasingly attempted to formalise the "whole of nation" strategy by creating institutions and regulations to oversee and guide what was previously an informal and largely self-driven approach. This formalising process accelerated after 2007, as demonstrated by one of the most significant legacies of the attacks—the establishment of the Defence League Cyber Unit (popularly referred to as the Cyber Defence League, or CDL) within the Estonian Defence League, the reserve military force (Cardash et al. 2013). By creating the CDL the Estonian government has been able to harness the desire of civil society to contribute to defensive efforts in the midst of a crisis situation (Jackson 2013). The CDL consists of volunteers who possess some form of expertise in cyber security and, indeed, many of its volunteers were part of the informal 2007 alliance. In crisis situations, members of the CDL can be assigned to CERT-EE to help protect critical national infrastructure. The CDL has also acted pre-emptively in other instances: in 2011, the CDL was on standby during the country's parliamentary elections: with many Estonians voting online, the voting system was an obvious target for cyber disruption. In 2012, the CDL organised a cyber crisis simulation exercise for Cabinet Ministers, fostering interest and preparedness in cyber crisis situations among the highest levels of government (Cardash et al. 2013).

Formation of the CDL confers a number of advantages to Estonia's crisis management capacity. First, it allows the government to effectively borrow civilian cyber security expertise as and when it is required. CDL members can work in the private sector yet still contribute meaningfully to national crisis situations. The 2007 informal alliance relied on a preexisting network of civil actors with high levels of trust between certain individuals; however, there were inevitable gaps in this network because it lacked a parent organisational structure. By contrast, the CDL does not rely on individual and personal relationships; it has a structure that renders it more coherent and sustainable in the long term (Pernik and Tuohy 2013). Second, the CDL gives the government access to a cost-effective, highly skilled, and specialised reserve force. CDL members are volunteers who only get paid when they are formally called up. This is a much more efficient allocation of Estonian resources than the employment of cyber security professionals on a permanent basis. Third, as part of the military reserve, CDL members can perform a variety of functions including planned missions, training, exercises, planning, etc. This means that the CDL can play a meaningful role in any type of cyber crisis response. Unlike traditional Estonian reserve units, which protect military assets, the CDL protects non-military assets, including civilian critical national infrastructure. The CDL is therefore viewed favourably among the general public and the cyber security community in Estonia, generating a sense of goodwill in the popular perception that can motivate citizens to join its ranks (Cardash et al. 2013). In addition to the CDL, the Estonian government has developed a close relationship with private sector firms that own and manage critical national infrastructure. Given the high degree of economic liberalism in the state, maintaining such relationships remains highly important. These close industry links often emerge naturally owing to Estonia's small population and geographic size. And the links are in some respects formalised. Providers of vital provide valuable input to government departments (Pernik and Tuohy 2013). For example, the Committee on the Protection of Critical Infrastructure was set up in 2011 and also includes private sector IT managers and risk management specialists (Pernik and Tuohy 2013).

Establishment of the CDL was the most significant step the Estonian government has taken to formalise interaction with multiple stakeholders in dealing with crisis situations. It provides an established institution and forum for such interaction to take place. In addition, the presence of critical national infrastructure firms in government initiatives such as the Committee on the Protection of Critical Infrastructure shows that the government regards the private sector as an important voice. The initiatives discussed above, however, rely on volunteers or firms agreeing to voluntary guidelines. In addition to these largely voluntary initiatives, the government also enforces strict regulation of owners of critical national infrastructure subsystems. This has enabled the Estonian government to enforce governmental guidelines in cyber crisis situations (Osula 2015b).

11.3.3 International Engagement

Turning to Estonia's interaction with the international community, it is clear that there is a strong emphasis on international engagement in the country's cyber crisis management strategy. In addition to its inclusive approach to domestic stakeholders, Estonia is also active in international fora. It acts as a strong advocate for greater international cooperation within the cyber security domain—and it has even encouraged citizens in other nations to become "e-residents" in Estonia, allowing them to establish Estonian digital identities as part of its vision of a "country without borders" (Kotka 2014). While much of Estonia's international engagement relates to broader initiatives of this sort and long-term cyber security capacity-building, there is also a strong specific emphasis on cyber crisis management. This can be observed in the country's of various international security mechanisms, including bilateral, regional and multilateral institutions.

On a bilateral level, Estonia enjoys a close relationship with the United States. A letter signed on December 3, 2013 by U.S. Secretary of State John Kerry and Estonian Minister of Foreign Affairs Urmas Paet pledged to create stronger ties between the two states (Paet and Kerry 2013). This included not only an effort to enhance ties between each state's central cyber security coordination bodies but also expressed an aspiration to build relationships between specific government departments (such as between the U.S. Department of Energy and the Estonian Ministry of Interior). There is also a clear agreement to cooperate at a cyber crisis management level: closer links were pledged between U.S. institutions, such as the National Cyber Security and Communications Center and the U.S. Computer Emergency Readiness Team (US-CERT), and Estonian institutions, such as the Estonian Information Systems Authority and CERT-EE. Both states also agreed to exchange information on instances of best practice in the protection of critical national infrastructure. This cooperation has extended to a working level encompassing the CDL, which has partnered with the 175th Network Warfare Squadron of the Maryland Air National Guard (Cardash et al. 2013). Estonia also places emphasis on regional (and bilateral) cooperation with other Nordic-Baltic countries. This cooperation includes a variety of functions such as sharing technical information, offers of emergency assistance, pooling of resources, and specific inter-agency cooperation (Areng 2014). The closeness of such regional cooperation is not surprising: because the Nordic-Baltic states are small nations, they are—as a practical matter—often compelled to work closely together on security affairs.

On the multilateral stage, Estonia is a highly active and leading proponent of intergovernmental cooperation in cyber security through fora such as North Atlantic Treaty Organisation (NATO), the European Union (EU), the United Nations, the Council of Europe, the Organisation for Security and Co-operation in Europe, the International Telecommunication Union, and other bodies (Osula 2015b). In particular, Estonia has been a strong proponent of NATO's role in cyber security. The country has hosted cyber-related exercises such as the NATO Cyber Coalition and has also offered the Estonian Defence Force's cyber range for NATO alliance training (Pernik and Tuohy 2013). Perhaps most notably, Estonia also hosts the NATO Cooperative Cyber Defence Centre of Excellence in Tallinn, whose simulation exercises give the centre some crisis management relevance despite its primary mission being research and analysis.

11.4 United Kingdom

Cyber security is an acute security concern for the United Kingdom. In 2014, 90% of all UK households had a broadband connection, ranking fourth highest in the EU (Osula 2015a). Many services vital to the UK economy are reliant on digital technologies—law, financial services, aerospace design, etc. In these sectors, a reliable ability to store information and communicate digitally is vital. As the government continues to implement a "Digital by Default" strategy by which government services are increasingly run online, the United Kingdom's digital dependence will only grow larger, meaning that cyber crisis management will become both more complex but also more important.

The first UK Cyber Security Strategy was published in 2009 and then updated in 2010. This led to the inauguration of cyber crisis management bodies such as the Office of Cyber Security, located in the Cabinet Office, and the Cyber Security Operations Centre, housed by the UK Government Communication Headquarters (GCHQ) (UK Government 2011). Yet it was not until the 2011 Cyber Security Strategy that the issue was supported with significant resources over a longer period with spending plans outlined until 2015. The Conservative-Liberal Democrat coalition signalled their commitment to the issue by allocating £650 million in 2011 with a further £210 million investment in 2013, while at the same time, cutting government spending in other areas significantly (Maude 2013). Although the strategy includes a number of broader objectives (such as education and training and long-term capacity-building), cyber crisis management is a main priority. Indeed, one of the four core objectives of the 2011 strategy is for the United Kingdom to be more cyber resilient. The protection of critical national infrastructure is at the heart of the document. In addition, the 2010 National Security Strategy ranked cyber attacks as a Tier One threat—above threats posed by nuclear attacks and organised crime.

11.4.1 Domestic Institutional Organisation

In many respects, the government's internal response to cyber crisis management echoes its stance on traditional crisis management situations. The UK Cabinet Office Briefing Room (COBRA), chaired by the Prime Minister, acts as the central crisis management response system for crisis situations, including those that are cyber-related. Similar to the United Kingdom's conventional response to crisis situations, however, the power and responsibility for a cyber crisis situation rests with the relevant government departments. Although the 2009 National Cyber Security Strategy created national entities to deal with cyber security, this trend has been reversed. The 2011 Cyber Security Strategy led to caution in escalating responsibility to centralised bodies (UK Government 2009; Osula 2015a). Instead, such responsibilities "are provisional and can be adjusted if experience suggests that a different mix of inputs will produce better results" (Osula 2015a). This has resulted in a highly decentralised approach where a number of different government departments function in a multi-layered and dynamic system of coordination.

The emphasis in the United Kingdom's response is therefore on decentralisation. This means that much of the responsibility in cyber crisis situations lies with the specific government departments affected. Departments are allocated responsibility for sectors relevant to them. For example, the Department of Health would lead a response against any attacks on hospitals, while the Department of Transport would lead against attacks affecting rail or aviation networks. This decentralised strategy is intended to facilitate a formalised, yet flexible approach: the government can bring in different departments as and when the situation requires. As responsibility for each sector have been allocated to a government department, however, the decision-making process on which departments should handle a particular cyber crisis situation is intended to remain predictable and clear.

Within this decentralised approach, central government bodies have only limited authority; they act merely in a coordinating capacity seeking to deliver a coherent government response. The Cabinet Office provides the lead for this coordination role. Historically, responsibility for cyber security was vested in the Home Office, but in 2009 this role moved to the Cabinet Office in order to enable a response that focused on cyber security in a broader sense, that is, with a focus in areas traditionally outside the purview of the Home Office such as education and training (Osula 2015a). Within the Cabinet Office specifically, the Office of Cyber Security was formed in 2009, later becoming the Office of Cyber Security and Information Assurance in 2011. And although CERT-UK has technical staff and expertise it can offer to government departments, use of this capability is reserved for severe emergencies on the theory that departments will normally be able to address problems themselves. In this regard, while CERT-UK does provide a technical perspective in cyber crisis situations, it only does so at a high level and the technical assistance it offers is deliberately limited. The main focus of the Centre for Protection of National Infrastructure (CPNI) is to prevent cyber crisis situations occurring by recommending security controls and offering advice on how to avoid crisis situations occurring. While necessary, the preventative emphasis of the CPNI means that the organisation has a limited role when a crisis situation is occurring (Harrop and Matteson 2014). Within this role, however, the CPNI has achieved a strong network of security-cleared contacts that can be utilised in the event of a crisis situation.

As regards the defence sector, the United Kingdom's cyber security strategy clearly emphasises a demilitarised approach. A large degree of the security responsibility is handed down to civilian departments. The military's primary responsibility is to protect its *own* networks; its remit does not extend to other government departments. Neither the 2011 National Strategy for Defence nor the Ministry of Defence (MoD) Defence Plan 2010–2014 mentions cyber attacks as a threat. In addition, the MoD has no jurisdiction to develop policy to protect critical national infrastructure and maintains only a limited advisory role to the CPNI and intelligence agencies such as GCHQ (Osula 2015a).

The United Kingdom's decentralised strategy has the advantage of facilitating a flexible response to a crisis situation, but it comes with a number of shortfalls. From an institutional perspective, cyber crisis management has become highly crowded; inevitably, there is overlap and inefficiency between different departments. For example, both Her Majesty's Revenue and Customs (HMRC) and the Department for Work and Pensions run their own security operation centres, despite such a facility also being provided by CERT-UK. This results in unnecessary redundancy and an increasingly costly institutional response. There also concerns that with responsibilities divided among central government departments including the Home Office, the Foreign and Commonwealth Office, and the Cabinet Office, that the institutional response is overly complex (Osula 2015a). In addition, such a devolved approach can foster dysfunctional relationships between departments. Government departments are not necessarily incentivised to cooperate with one another. Each department sees itself as "sovereign" and therefore prioritises its own survival and growth on a stand-alone basis. Different departments are potentially faced with conflicting objectives. It is reported that cultural gaps exist between organisations that seek to prevent cyber-attacks (e.g., law enforcement) and those that exist to respond to cyber attacks and minimise the damage (e.g., CERT-UK) (Caldwell 2014). There is a real danger that the lack of cohesion creates a climate of mistrust, whose consequences for policy effectiveness during a crisis are potentially severe.

11.4.2 Stakeholder Mobilisation

From a cyber crisis management perspective, the UK government's interaction with other stakeholders is focused predominantly on privately owned critical national infrastructure. Liberal economic policies and a belief in market mechanisms have resulted in a large proportion of the United Kingdom's critical national infrastructure remaining in the private sector. Indeed, with 80% of the United Kingdom's critical national infrastructure owned and managed by the private sector, it dominates the critical national infrastructure landscape. These principles of economic liberalism and faith in market mechanisms are reflected in a cyber crisis management capacity. With the exception of some caveats (discussed below), the UK government treats the owners of private critical national infrastructure as responsible for managing cyber attacks against their own computer systems. There is a degree of trust upon which firms recognise that it is in their own interest to implement effective cyber security and cyber crisis management procedures. The government has created its own certificate in Cyber Essentials and Cyber Essentials Plus that firms can advertise if they adhere to a set of government cyber security standards. Although now mandatory for government contractors, the principle idea behind the certification scheme was to help private sector firms turn cyber security into a competitive advantage by exhibiting their competence in this area.

Government organisations play only a supporting role for an already sophisticated private sector, recognising that the private sector possesses many of the technical skills required and the government can make a contribution at a higher level in a coordination capacity. The government aims to use CERT-UK as a body to bring together private and public sector expertise in order to create a coordinated and cohesive response to cyber crisis situations. In this regard, CERT-UK takes a high-level, strategic view of cyber crisis situations, with less emphasis on providing a sophisticated technical capability. For example, CERT-UK established the Cyber Security Information Sharing Partnership (CISP) in order to create a portal where both governmental organisations and private sector firms can share threat intelligence. Indeed, in crisis situations, CERT-UK are able to direct owners of critical national infrastructure to trusted security vendors rather than provide technical assistance itself.

From a cyber crisis management perspective, the UK government takes a special interest in privately owned critical national infrastructure assets. Both CESG (the cyber security branch of GCHQ) and the Centre for the Protection of National Infrastructure (CPNI) perform risk assessments and audit security arrangements for critical national infrastructure, producing risk and vulnerability reports. The CPNI has also made 20 specific control recommendations for owners of critical national infrastructure to implement (Harrop and Matteson 2014). Privately owned critical national infrastructure is, however, predominantly managed independently. There is a clear absence of any central point of control or of laws that require operators of critical national infrastructure to conform to minimum security standard (Ashford 2011). The UK government acknowledges that regulation would be a blunt instrument in the field of cyber security. The government is aware that critical national infrastructure exists in a variety of sectors that should be managed and regulated differently. This is because of the specific challenges faced in certain industries, given the highly specialised nature of the technology used. For example, within the energy sector, infrastructure such as the power grid needs to be compatible with both antiquated physical structures such as power stations, installed over 50 years ago before cyber security was a priority, and highly modern systems and technologies. This is highly complex from a regulatory perspective: the rate of technological change is likely to outpace regulatory timelines. In addition, the skills required to regulate in this area are highly specific, requiring a combination of legal skills and a deep technical understanding of complex, unique, and highly sophisticated systems.

The government's hands-off role and emphasis on self-regulation is not merely a matter of its own policy choices. There is a desire within government to implement stricter regulation and laws over particularly vital services such as the power grid and water supply. There is especially a concern within government about potential cyber crisis situations that would negatively affect the public to a significantly greater extent than an individual firm. For example, within the National Grid, a large-scale power outage would have a much greater effect on the nation (with loss of life and civil unrest possible) compared to the impact on National Grid itself (a small number of staff might be affected). This creates potential for misaligned priorities between the owners of privately operated critical national infrastructure and the public as a whole.

Despite the government recognising the need for further private sector regulation, the process remains difficult to implement. The private sector is protected from government interference through strong property rights that significantly constrain the government's role, making the enforcement of mandatory guidelines in a cyber crisis situation far from straightforward. Furthermore, difficulty in interpreting preexisting regulation has provided a challenge for government organisations. Traditional crisis management regulation, such as that which was focused on business continuity, was written before cyber security was considered an important issue. Yet there is nothing in such regulation that precludes its application in the domain of cyber security. Government departments are therefore currently in a process of understanding how preexisting regulation might apply in a cyber security context. For example, the financial services sector is an example where progress has been made. There is a strong link between industry and the Bank of England, which has established strong regulatory power, having already overseen simulated cyber attack exercises with banks.

Outside of the conventional defence structures, the United Kingdom operates a cyber reserve force in the form of the Joint Cyber Unit (JCU). Although the JCU comprises civilian reservists (Morbin 2013), civil society as a whole is reluctant to participate in securing the United Kingdom. JCU reservists receive payment for their involvement; there would be significantly less interest if reservists were asked to volunteer for free services. In addition, the remit of the JCU is narrow, covering only military networks. In short, civilian and volunteer groups do not play a significant role in securing the digital systems of UK critical national infrastructure.

11.4.3 International Engagement

The United Kingdom's international engagement in cyber crisis situations largely reflects preexisting security and intelligence partnerships. In this regard, the United Kingdom's first point of contact would be trusted allies that would be consulted on almost any security issue. The United Kingdom predominantly utilises partnerships with historical allies such as the United States. Close multilateral alliances are also used. For example, the "Five Eyes" intelligence alliance between Australia, Canada, New Zealand, the United Kingdom and the United States is seen as one of the main multilateral channels to engage with trusted allies on the issue of cyber security. The United Kingdom is most willing to cooperate with foreign powers within a preexisting alliance structure that represents a track record of trusted exchange and shared language (Areng 2013).

Apart from cooperating with its closest trusted allies, in the event of a cyber crisis, two other avenues might be followed by the United Kingdom on a more ad hoc basis. First, the United Kingdom would seek to engage with the state where the source of the attack is located. While this does not assume origin or intent of attack, the United Kingdom would typically request assistance from this state. However, the United Kingdom's response will largely depend on the domestic character of the state in question and its ability and willingness to cooperate. The Foreign and Commonwealth Office is likely to provide assistance in opening up a dialogue between the United Kingdom and other such states. Second, the United Kingdom would also seek to establish links and cooperation with other victims of the attack. If another country is simultaneously being attacked, the potential for collaboration will increase; if the attack has already taken place, previous victims of the perpetrator may also be consulted.

Although the United Kingdom is actively engaged in international cooperative efforts on both bilateral and multilateral levels, within both cyber crime and cyber security capacity-building initiatives, the country appears less willing to cooperate with international partners on the cyber crisis management plane. This is partly because crisis partnerships are typically less publically visible; however, the United Kingdom appears comparatively reluctant to engage with others on this front. For example, the United Kingdom is considerably less vocal in supporting NATO's transition into the cyber domain, questioning the applicability of collective defence to the management of cyber conflict. Compared to Estonia, the United Kingdom is reluctant to extend Article Five of the NATO Washington Treaty (the concept of self-defence and the notion that a cyber attack on one NATO state would be interpreted as an attack on all NATO members) to the cyber domain. In addition, CERT-UK did not participate in Locked Shields, a NATO CCDCOE cyber simulation exercise in 2015 as well as previous years. Instead, the UK has focused the majority of if its international efforts in cyber security to non-crisis related issues such as building cyber security capacity and developing cyber norms.

Overall, the United Kingdom's approach to cyber crisis management is not highly reliant on international cooperative mechanisms. Although London does regard international engagement as important, the domestic initiatives represent a significant proportion of the United Kingdom's crisis control capability. Outside case-by-case cooperation during specific incidents, trusted allies and preexisting intelligence-sharing mechanisms such as the Five Eyes are the main locus of cooperative efforts, underlining the importance that the United Kingdom places on trust and historical ties in the management of its international security relationships.

11.5 Determinants of National Cyber Crisis Management Strategies

Having examined the Estonian and the UK cyber crisis management strategies individually in the preceding sections, this section compares them and reviews their similarities and differences. In particular, the analysis identifies explanatory variables that may account for major differences in the two national models. The section also draws lessons and insights for policy and strategy.

11.5.1 Comparison Between Estonia and the United Kingdom

Estonia and the United Kingdom have several similarities in their organisation of government institutions in relation to responding to cyber crisis situations. For both countries, there is a clear focus on civilian-led organisations that manage and coordinate cyber crisis situations, with the responsibility held by the Estonian Ministry of Economic Affairs and Communication and the UK Cabinet Office respectively. Both states have sought to demilitarise the issue, raising questions more broadly on the role national militaries can and should play in cyber security and cyber crisis situations. Moreover, both states have, to an extent, delegated cyber crisis situations to individual government departments; thus, specific departments typically find themselves responsible when cyber attacks are directed at, or affect, sectors relevant to them. But while both states have adopted similar approaches in this regard, the strategies, outcomes, and implications are quite different.

In Estonia, there remain very clear procedures to escalate responsibility to coordination bodies such as the Ministry of the Interior, the Estonian Information Systems Authority, and even the Prime Minster's Office. By contrast, given the United Kingdom's much larger governmental structure, the decentralisation process is more complex, with a multi-layered approach that makes it more difficult to escalate responsibility to higher coordination bodies. This also reflects constraints on the UK Prime Minister's power, given the existing limits on any rapid assumption of centralised control in crisis situations. For these reasons, central bodies such as the Office for Cyber Security and Information Assurance, CERT-UK, and CPNI largely have coordination roles with less decision-making authority or hands-on technical duties. Although perhaps inevitable given the larger size of the UK government, this reality makes for a less efficient allocation of resources with instances of repeated spending and the duplication of similar products and services.

Both states have also recognised that the government alone cannot respond effectively to cyber crisis situations. Their approaches in interaction with non-governmental stakeholders have been quite different, however. Estonia has taken a very wide approach to mobilising multiple stakeholders in a "whole of nation" approach that is consistent with the security strategy historically adopted by low-population states. This strategy is often referred to as a Total Defence Model, where it is recognised that given a small population, a full-time military force is alone insufficient to provide national security (Laaneots 1999; Jundzus 1996; Davis 2008). In a cyber security context within Estonia, the participation of civil society is reflected through the workings of the Cyber Defence League. Although it is part of a military organisation, the Cyber Defence League is clearly civilian in nature: it comprises volunteers from civil society who contribute in their spare time and are available in the event of a crisis situation. Furthermore, the remit of the CDL extends far beyond military infrastructure to the protection of national critical infrastructure in non-military departments. Estonia has backed up voluntary institutions with relatively strict regulation applicable to privately owned critical national infrastructure; this creates a sense that the Estonian government is the clear source of control and leadership in crisis situations.

By contrast, within the United Kingdom there is a clear emphasis on a private sector–led approach, given that businesses own and manage the majority of critical infrastructure. The UK government's hands-off role has developed within the backdrop of a state that embraces economic liberalism and market mechanisms. Although largely deliberate, the lack of government interference can also be explained due to difficulties in interpreting existing regulation. In addition, unlike Estonia, the United Kingdom has been unable to mobilise other stakeholders such as civil society more broadly, operating a significantly more limited cyber reserve force. Given the United Kingdom's comparatively larger population—and by extension larger military presence—British citizens do not feel as compelled to participate in the provision of national security when compared to Estonia as there is a recognition that national security is sufficiently provided by full-time military personnel.

In terms of international engagement, Estonia is a much more active participant at the bilateral, regional, and multilateral levels. Although well organised and advanced at the policy level, Estonia simply does not have the same level of financial resources and government personnel as other states, meaning international engagement is crucial. That is not to say that the United Kingdom regards international engagement as unimportant. It should also be noted that the UK has made significant contributions to international cooperation in broader aspects of cyber security, including capacity building and the formulation of cyber norms. In a cyber crisis management context, engagement with trusted UK allies and as ad hoc cooperation with other potential victims, is taken seriously and seen as a useful way to avoid or mitigate a crisis situation. The United Kingdom, however, is not as reliant on international mechanisms, given the level of internal financial resources and government personnel that exist within the United Kingdom. This difference is clearly reflected in each state's stance towards NATO: Estonia has been an active proponent for extending NATO's remit to include cyber security and to apply Article Five to cyber attacks; the United Kingdom appears sceptical of NATO involvement in the cyber domain, arguing that its remit should extend to only the protection of NATO's own digital infrastructure and networks.

In sum, when comparing Estonia and the United Kingdom, it is difficult—perhaps impossible—to determine which state has the superior cyber crisis management strategy, because both states have developed tailored policies specific to their own characteristics and needs. The focus of analysis now turns to an examination of the factors that explain the divergence in the two countries' strategies.

11.5.2 *Explanatory Variables*

This section identifies the most significant variables that influence the two state's cyber crisis management strategies and explains how the different approaches emerged, emphasising the important role played by the nations' size, history, threat landscape, political philosophy, and digital dependence.

11.5.2.1 Size

Size—both in relation to population size and level of government resources—has a significant effect on cyber crisis management strategy. Estonia demonstrates the advantages that a small-population state enjoys: it has an inherently small government that allows flexibility in adapting to new challenges. It is no coincidence that a number of historically low-tech states such as Sweden, Denmark, Finland, Austria, and Ireland, have gained leading positions in new industries like nanotechnology, biotechnology, telecommunications, and cyber security (Areng 2014). With shorter communication links in a society that fosters trust and flexibility, it becomes easier to organise a cohesive cross-governmental response to cyber crisis situations (Areng 2014). In addition, the Estonian government has been able to apply its Total Defence model to cyber security, initially with an informal alliance that has now progressed into a more formalised framework within the Cyber Defence League. On cyber power terms, then, Estonia's size has a curious reversible property. As noted earlier, a process of power diffusion is taking place in the cyber domain: states are struggling to adapt to cyber security challenges in part because low barriers to entry and the affordable cost of information technologies have empowered non-traditional actors such as private sector firms and civilians (Nye 2011 Ch.5). Yet, by virtue of being small, Estonia has expanded its Total Defence model into the cyber security context; therefore, the country is better equipped to mobilise non-governmental actors in a manner larger states have struggled to replicate.

Estonia's strategy is consistent with the historical behaviour of "small powers" in the international system. According to Rothstein, "a small power is a state which recognises that it cannot obtain security primarily by the use of its own capabilities, and that it must rely fundamentally on the aid of other states, institutions, processes, or developments to do so" (Areng 2014). Although the Estonian cyber crisis management strategy is run efficiently, Estonia does not have the financial resources or government personnel of larger states and therefore looks to international partners to complement its smaller, albeit impressive, capability (Burton 2013). Small states such as Estonia understand that maintaining their "soft power" and good standing with other states is vital in order to garner international support in crisis situations (Areng 2014). It is therefore important for Estonia to meet its international legal and political commitments; indeed, Estonia is the only country in Europe to meet the rules of every international organisation it belongs to—including the Eurozone's target on debt, inflation and government deficit, and NATO's minimum defence spending threshold (2 % of GDP) (Areng 2014).

By contrast, the United Kingdom's larger population has resulted in a larger government administration with a multi-layered approach. Although necessary given the population of the United Kingdom, it is arguably too complex, fostering institutional rivalry between departments and potentially resulting in inefficient allocation and duplication of resources. The United Kingdom is in many respects only a medium-sized country and it is likely that the institutional makeup would be even more complex (and potentially inefficient) in a country such as the United States that has an even larger population and federal government structure. Although its strong focus on the private sector appears deliberate (Bada et al. 2014), the United Kingdom would be unable to mobilise civil society during a crisis in the way that Estonia can. While there are many reasons for this, it is most significant that civilians are unlikely to feel the same need for society-led security initiatives in a state with greater resources. In addition, as a larger state with greater resources, the United Kingdom is by definition less reliant on international cooperation.

11.5.2.2 History

Historical ties and institutional dependencies are highly influential in determining the allies and international partners each state prefers to work with. Many geopolitical alliances, such as NATO and regional Nordic-Baltic alliances, are still being employed despite the fact that cyber security is essentially borderless and cyber allies can theoretically exist anywhere in the world. Yet historical ties, institutional dependencies, and trust have been prioritised over cooperation with more disparate and non-traditional allies. Estonia has a strong relationship with other Nordic and Baltic states and has long been a proponent of NATO, given the geopolitical situation. These links demonstrate how important trust is in building a network of state partnerships. Likewise, the US has deemed cooperation with Estonia a priority given their long-held presence and interest in Eastern Europe after the Cold War. While cyber security may be very different from conventional security challenges, Estonia has nonetheless relied upon historical allies and relationships.

The Estonian case also demonstrates the importance of historical factors in other ways. As a newly independent nation, within Estonia there is a sense of unity and the need for society to act as one in order for the nation to survive and grow. Moreover, many of Estonia's current cyber security policies were implemented after the 2007 attack—before 2007, Estonia did not even have a formal cyber security strategy (albeit like many states at the time including the UK). Thus, although long-rooted historical factors are important determinants of domestic and international partnerships, an actual recent cyber attack is significant because it concentrates resources and political capital to the issue in a way that would otherwise appear unjustified. The latter situation prevails in the United Kingdom: the British public are relatively unconcerned with cyber security in comparison to other security issues; the topic, for instance, barely figured in the 2015 general election campaign.

11.5.2.3 Threat Landscape

Both Estonia and the United Kingdom have different areas of concern regarding their most immediate threats in the cyber domain as reflected in their cyber crisis management strategies. Estonia, a newly (re)independent state, faces a clear threat from its neighbour Russia that is significantly more powerful in a military context (Park 1995). This highlights the relevance of geopolitical factors (Aalto 2000). Although cyber attacks are in effect borderless, when a state such as Estonia is concerned about a physical attack from a local neighbour, this concern inevitably spills over into the cyber domain given that aggressor states increasingly regard cyber attacks as a valid use of force—either in isolation or in tandem with a conventional attack.

Given the importance to the domestic economy of sectors such as financial services and aerospace design, it is important for the United Kingdom to have access to reliable online services and the ability to store intellectual property securely.

Therefore, with cyber crime reportedly costing the UK economy £27 billion a year (Raywood 2011), it may be that cyber crime and fraud are deemed as greater priorities in the United Kingdom than in Estonia. Although cyber crisis situations are undoubtedly still regarded as a significant threat to the United Kingdom, the importance of other cyber security issues may explain why a smaller proportion of resources are dedicated to cyber crisis management in the United Kingdom. The focus on a greater number of cyber security issues also reflects the availability of relatively more resources in the United Kingdom.

11.5.2.4 Political Philosophy

Political philosophies, specifically the views held on the role of the state, are also significant determining variables. Within the United Kingdom, there is a strong sense that the private sector is the most efficient provider for a number of critical services. Therefore, even seemingly unrelated policies such as the reduced or minimal state agenda that goes back to the period of Margaret Thatcher have implications for cyber security in 2015. There is also a belief from a crisis management perspective that private sector firms are commercially incentivised to implement their own security. Although there are concerns about the potential of market failure, trust in the private sector remains a key feature of the UK strategy. Similar issues have played out in Estonia where market mechanisms have likewise been embraced by policymakers.

The United Kingdom's decentralised strategy within government, where centralised cyber security bodies operate a limited hands-off coordination function with weak authority, also largely reflects the structure of the country's political system, where much of the power for managing departments lies with the relevant cabinet minister as opposed to the Prime Minster. By contrast, in Estonia there is more emphasis on the need for a wide group of stakeholders to contribute to security. This factor relates to variables discussed above, such as Estonia's size and history. Estonia also enforces compulsory military service for all male citizens, meaning that many citizens have prior experience in contributing to state security. In addition, compared to other states, the use of technology is more deeply embedded in the day-to-day life of its citizens. Because of the preeminent role that technology has played in Estonia's transition from Soviet occupation, technology and digital systems are deeply entrenched in notions of security and, indeed, in the very concept of statehood (as exemplified by new initiatives such as "e-residency") to a far greater extent than in the United Kingdom.

11.5.2.5 Digital Dependence

Digital dependence clearly shapes the relationship between technology and security in the two countries. As Estonia has invested in technology in order to overcome many of the transitional challenges it has faced in recent history, the country has become highly reliant on many of these digital systems. This level of dependence clearly has implications from a cyber crisis management perspective. Generally, as states increase their reliance on cyber systems, the consequences of attack against those systems are more severe, meriting a greater proportion of resources being dedicated to cyber security.

The United Kingdom also has a high level of reliance on technology. As critical infrastructure becomes increasingly connected to digital systems, there are very clear consequences in the event of an attack. At the moment, however, the level of digital dependence is lower that in Estonia, with fewer systems digitalised; therefore, the issue of cyber crisis management is further down the political agenda. In addition, the United Kingdom's main economic sectors make other cyber security issues, such as cyber crime and financial fraud, bigger priorities for the UK cyber crisis management system.

***

The explanatory variables discussed above are by no means exhaustive: it may be that other factors are more significant where other states are concerned. This chapter is unable to suggest which variables are most significant; there is, however, potential for further study where more variables are controlled. For example, by looking at all the Baltic and Nordic states (which are more similar in a number of the explanatory variables discussed), the importance of specific independent variables may be clearer. Strategic thinking and policies on cyber crisis management and cyber security more broadly are still rudimentary; it may therefore take considerable more time before states have fully developed their strategies and can be compared in a more meaningful manner by future scholars and analysts.

11.5.3 Lessons for Strategy and Policy

Having examined the strategies of Estonia and the United Kingdom, this chapter now considers policy lessons and conclusions—both for Estonia and the United Kingdom as well as for other states more broadly.

Estonia deserves credit for its response to the cyber security challenge. Given the lack of government resources of a smaller state, it is vital for Estonia to maintain its efforts to embed cyber security into its national culture and ethos and garner the expertise of civil society in the provision of state-level cyber security. Russia's invasion of Eastern Ukraine demonstrates that the Kremlin remains a viable threat to Estonian interests and provides ample incentives for Estonians to remain active participants in their state's security. Estonia has also sought to make up for its lack of internal resources by engaging with international partners. As previously discussed, there are undoubted benefits to this cooperation. There is, however, a danger that Estonia will grow overly reliant on international assistance. As Russian invasions into Georgia in 2008 and Ukraine in 2013 have shown, Eastern European states hoping for international protection against Russian threats are likely to end up disappointed. Although Estonia is a NATO member, it should never take collective defence for granted—especially in the cyber domain, where precedents for the application of this doctrine (and other inherited security mechanisms) are still being established (Kello 2013).

The United Kingdom, meanwhile, should develop and improve its organisation of cyber crisis management *within* government. Given its large size, it is appropriate to decentralise cyber security responsibility to individual departments. A centralised strategy such as Estonia's would likely not work in the United Kingdom; it could merely fracture government institutional relationships further. Nonetheless, the disadvantages of a decentralised strategy require remedy: the government should consider policies that incentivise collaboration between departments and provide clearer guidelines over institutional remits and it should develop further oversight and coordination efforts in order to reduce unnecessary overlap and repeated spending by different departments on the same services, thereby making the implementation of cyber crisis management policy more cost-effective.

Although UK policymakers may be tempted to replicate the successes of the Estonian cyber reserve model (Blair 2015), such an approach would not necessarily succeed in the United Kingdom. Crucially, with the absence of a clear and more powerful threat (as Estonia experiences with Russia) and as a larger state, the United Kingdom will struggle to motivate civil society to participate in security as willingly as their Estonian counterparts do. The UK government will have to continue to provide strong incentives for participation, including paying members for the training and exercises they participate in. The United Kingdom can take inspiration from the broad remit of the Estonian Cyber Defence League. The responsibility of the UK Joint Cyber Unit is restricted to strictly military assets but should expand to protect other government assets in the future. Although private-sector firms may be reluctant to grant a reserve force access to their systems, there are a number of government functions and systems that would benefit from the skills and expertise of a reserve force. On the international front, the United Kingdom is making encouraging steps forward in its cooperative efforts. These steps should continue and are likely to do so naturally as institutions such as CERT-UK continue to develop their resources and organisational cultures. Although international cooperation alone is by no means a definitive solution to cyber security problems, it can offer a useful supplement to other measures that involve, for example, information sharing across diverse actors and sectors in multiple national jurisdictions.

The lessons drawn from this comparative study may apply, in a limited fashion, more broadly in other national contexts. The most important lesson to take away from the Estonia-United Kingdom comparison concerns the importance of a *relative* approach to the formulation of cyber crisis management policy—that is, one that builds on nation-state characteristics. States whose policy directions in the cyber domain represent an abandonment of national identity will face difficulties in implementing policy, whether because it will encounter problems of domestic legitimacy or problems of institutional disorientation. Policymakers should not regard this reality as a restriction on their ability to write policy; they should instead realise the opportunity that exists to exploit the natural advantages that each nation possesses. For example, a small state is naturally poised to exploit fast communication links and low levels of bureaucracy; a state with a sophisticated private sector or academic institutions is well placed to utilise knowledge and expertise outside of government; a state with strong international partnerships should ensure cooperation is extended into the cyber domain; and so forth. There is in cyber security problems a source of optimism for policymakers: while the related policy challenges are largely novel, preexisting policies and partnerships are often more germane than they may first appear. For example, both the UK and Estonia have successfully extended historical international security partnerships into the cyber domain and aspects of traditional security strategy are reflected in the cyber security strategy. Nonetheless, policymakers should look toward other states for inspiration: a limited but still important number of policies and strategic decisions may be transferrable from one state to another. Furthermore, learning from states with more mature cyber security strategies and policies can be a valuable tool for decision-makers operating in less developed policy contexts—there are possible advantages to being late in the game. But in the end it is imperative that specific political, historical, and cultural factors are considered to ensure the implementation of policy remains viable.

11.6 Conclusion

States are faced with a number of options in framing strategic responses to cyber crisis situations. As the above discussion has shown, there is no "one-size-fits-all" policy: no single strategy has proved to be inherently more effective than its alternatives. A good strategy, rather, is that which is tailored to the specific circumstances of the state in question—both to optimise the best possible response from the state as a whole and to create a strategy that accurately reflects the challenges and threats that the state actually faces. Inevitably, tensions exist among the possible strategic choices that states can make. For example, within the institutional organisation of the government, it is important for the whole of government to assume involvement and responsibility in order to deliver a broad and comprehensive defence. It is also important, however, to have a cohesive strategy as well as a clear control and leadership structure. Thus there is a tension between a bottom-up approach that builds inclusion and a top-down focus on concentration of policy direction and control. Furthermore, a number of dilemmas surface when states have to address the extent to which they should rely on a wide range of stakeholders. Crucially, Estonia has shown how integrating voluntary domestic networks and international partners into the crisis control strategy can bolster the strength of the response; however, voluntary agreements are not binding, making it difficult to predict and guarantee the actual availability of assistance during crisis situations (Ottis 2012). Conversely, government regulation can secure a specific level of support by compelling a given stakeholder response, but may be off-putting to the affected parties if they perceive it a blunt instrument.

With these strategic dilemmas in mind, no one state has an approach that represents the superior way to manage cyber crisis situations. Neither Estonia nor the United Kingdom has a naturally superior strategy. In fact, in many respects, the strategy of each state has distinct advantages and disadvantages, linked to each country's particular characteristics. For example, as a small state, Estonia benefits from a clear and efficient governmental process for managing crisis situations; at the same time, because of its fewer resources, Estonia is more reliant on international assistance. By contrast, as a larger state, the United Kingdom is more self-reliant—but this comes at the price of a more complex and disparate governmental approach to crisis management.

Increased reliance on cyberspace has led to a hugely challenging security environment. Cyber attacks have the potential to destabilise critical national infrastructure; a viable response mechanisms to the possibility of such attacks is a crucial part of a nation's crisis management strategy. Although states are faced with novel technical challenges, in many respects cyber security reflects traditional security challenges. For example, geopolitical issues remain relevant in determining the nature of cyber threats that a state faces—as illustrated by Estonia's concerns over future Russian attacks. International cooperation in cyber crisis situations shows that historical ties and institutional dependencies are influential in alliance formation. This relationship, however, may be inappropriate in situations where historical allies have not adapted to cyber security challenges in the same way or at the same rate as advanced cyber nations such as Estonia. Historical ties and institutional dependencies are important in 2015, primarily because of the rudimentary state of cyber security strategy; in consequence, this chapter predicts that future alliances will increasingly form on the basis of ideological similarities as opposed to historical factors or geographical proximity (though ideological similarities will often arise where states are in proximity to each other or share historical links). Given the largely borderless nature of the cyber domain, there is also a real prospect that states will form non-traditional and geographically disparate alliances.

The role of the state in cyber security also differs from other security domains. Given the diffuse nature of the issues, it may be extremely difficult for the military to adapt to and confront cyber threats. By contrast, the private sector is increasingly relevant because of the lack of relevant skills and resources possessed by governments and because in some states it and manages the majority of critical digital systems. In countries such as Estonia and the United Kingdom, where there is a very large private-sector involvement in the provision of cyber security, the government faces significant constraints in its ability to manage crisis situations; it is therefore unclear what role the government will be able to play in the long term. Less government control in the cyber domain does not necessarily equate to a weaker cyber crisis management responder for the country concerned, however. Indeed, the ability to mobilise wider stakeholders is an important aspect of the strategies of both Estonia and the United Kingdom. In both countries, non-state actors such as certain private firms and civilian groups are accorded a privileged role in the conduct of state security, highlighting the increasingly blurry distinction between state and non-state agency in the cyber domain.

This chapter has identified a number of political, historical, and cultural factors that significantly shape and constraint states' cyber crisis management strategies. While many states share similar technical challenges, their responses to cyber crises may be quite different at the strategic level, given the importance of non-technical factors. For this reason, significant strategic differences exist between Estonia and the United Kingdom—two technologically advanced states. Therefore, there are likely to be even greater differences between developed and developing states. As a result, there are significant limitations to the generic advice and recommendations that can appropriately be applied to all states based on these two case studies. Instead, the analysis of cyber crisis response strategies requires an emphasis on nation- and context-specific factors. While particular features of Estonia's or the United Kingdom's strategies merit general emulation, there is no single ideal approach that is universally applicable.

Bibliography

Aalto, Pami. 2000. Beyond restoration. *Cooperation and Conflict* 35(1): 1–24.

Areng, Liina. 2013. International cyber crisis management and conflict resolution mechanisms. In *Peacetime regime for state activities in cyberspace*, ed. K. Ziolkowski, 565–588. Tallinn: NATO Cooperative Cyber Defence Centre of Excellence.

Areng, Liina. 2014. Lilliputian states in digital affairs and cyber security. *The Tallinn Papers*: 1–15.

Ashford, Warwick. 2011. Is UK critical national infrastructure properly protected? *Computer Weekly*. March 3.

Bada, Maria, Sadie Creese, Michael Goldsmith, Chris Mitchell, and Elizabeth Phillips. 2014. Computer Security Incident Response Teams (CSIRTs) An overview. *Global Cyber Security Capacity Centre*: 1–23.

Blair, David. 2015. Estonia recruits volunteer army of "Cyber Warriors." *The Telegraph*. April 26.

Burton, Joe. 2013. Small states and cyber security: The case of New Zealand. *Political Science* 65: 216–238.

Caldwell, Tracey. 2014. Call the digital fire brigade. *Network Security* March 2014: 5–8.

Cardash, Sharon L, Frank J Cilluffo, and Rain Ottis. 2013. Estonia's cyber defence league: A model for the United States? *Studies in Conflict & Terrorism* 36: 777–787.

Clastres, Cedric. 2011. Smart grids: Another step towards competition, energy security and climate change objectives. *Energy Policy* 39(2): 5399–5540.

Cyber Emergency Response Team Launched by UK. *BBC News*. 31 March. http://www.bbc.co.uk/news/technology-26818747. Accessed 12 July 2015.

Davis, Milton Paul. 2008. An historical and political overview of the reserve and guard forces in the Nordic countries. *Baltic Security and Defence Review* 10: 171–201.

Grobler, Marthie, Joey Jansen van Vuuren, and Jannie Zaaiman. 2013. Changing the face of cyber warfare with international cyber defense collaboration. In *Case studies in information warfare and security*, 38–54. Reading: ACPI.

Harrop, Wayne, and Ashley Matteson. 2014. Cyber resilience: A review of critical national infrastructure and cyber security protection measures applied in the UK and USA. *Journal of Business Continuity Emergency Planning* 7: 149–162.

How did Estonia become a leader in technology? 2013. How did Estonia become a leader in technology? *The Economist*, July 30.

Jackson, Camile Marie. 2013. Estonian cyber policy after the 2007 attacks: Drivers of change and factors for success. *New Voices in Public Policy* 7: 1–15.

Jundzis, Talavs. 1996. Defence models and strategies of Baltic states. *The International Spectator: Italian Journal of International Affairs* 31(1): 25–37.

Kello, Lucas. 2013. The meaning of the cyber revolution: Perils to theory and statecraft. *International Security* 38(2): 7–40.

Kotka, Taavi. 2014. 10 million e-Estonians"by 2025! https://taavikotka.wordpress.com/. Accessed 4 Dec 2015.

Kotka, Taavi and Innar Liiv. 2015. Concept of Estonian Government cloud and data embassies. In *Electronic government and the information systems perspective: Proceedings of the international conference EGOVIS 2015*, ed. Andrea Kő and Enrico Francesconi, Valencia: Springer International Publishing, 1–3 September.

Kotz, David, Sasikanth Avancha, and Amit Baxi. 2009. A privacy framework for mobile health and home-care systems. In *Proceedings of the first ACM workshop on security and privacy in medical and home-care systems, SPIMACS*, 1–12.

Laaneots, Ants. 1999. The Estonian defence forces – 2000. *Baltic Defence Review* 10: 1–7.

Laar, Mart. 2002. *Estonia: Little country that could*. London: Centre for Research into Post-Communist Economies.

Landler, Mark, and John Markoff. 2007. Digital fears emerge after data Siege in Estonia. *The New York Times*. May 29.

Mansel, Tim. 2013. How Estonia became E-stonia. *BBC News*. May 16. http://www.bbc.co.uk/news/business-22317297. Accessed 5 July 2015.

Maude, Francis. 2013. UK cyber security strategy: Statement on progress 2 years on. *Presented as a written statement to Parliament*. December 12.

Morbin, Tony. 2013. Cyber reserves call on private sector. *SC Magazine*. October 4.

Nye Jr., Joseph S. 2011. *The future of power*. New York: Public Affairs.

Osula, Anna-Maria. 2015a. *National cyber security organisation: United Kingdom*. Tallinn: NATO Cooperative Cyber Defence Centre of Excellence.

Osula, Anna-Maria. 2015b. *National cyber security organisation: Estonia*. Tallinn: NATO Cooperative Cyber Defence Centre of Excellence.

Ottis, Rain. 2012. *Lessons identified in the development of volunteer cyber defence units in Estonia and Latvia*. Tallinn: NATO Cooperative Cyber Defence Centre of Excellence.

Paet, Urmas, and John F Kerry. 2013. *US-Estonian Cyber Parternship Statement*. 3 December.

Park, Andrus. 1995. Russia and Estonian security dilemmas. *Europe-Asia Studies* 47: 27–45.

Pernik, Piret, and Emmet Tuohy. 2013. *Cyber space in Estonia: Greater security, greater challenges*. Tallinn: International Centre for Defence Studies.

Raywood, Dan. 2011. Cost of cyber crime in UK estimated £27 billion. *SC Magazine*. February 17.

Rid, Thomas. 2013. *Cyber War will Not take place*. London: Hurst.

Schmidt, Andreas. 2013. The Estonian cyberattacks. In *A fierce domain: Conflict in cyberspace, 1986 to 2012*, ed. Jason Healey. Arlington: Cyber Conflict Studies Association.

UK Government. 2009. *Cyber security strategy of the United Kingdom*. London: Cabinet Office.

UK Government. 2011. *Cyber security in the UK*. 2011. London: Parliamentary Office of Science and Technology.

Williams-Grut, Oscar. 2014. London launch for "D5" alliance of digital nations. *The Independent*, December 8.

Jamie Collier is a DPhil candidate in cybersecurity at the University of Oxford Centre for Doctoral Training. Having completed his undergraduate studies in international relations at the University of Nottingham, he is predominantly interested in the political aspects of cybersecurity. He has recently completed an internship at the NATO Cooperative Cyber Defence Centre of Excellence in Tallinn and has prior experience working with PwC.

Chapter 12
Lessons from Stuxnet and the Realm of Cyber and Nuclear Security: Implications for Ethics in Cyber Warfare

Caroline Baylon

Abstract This commentary examines Stuxnet and the realm of cyber and nuclear security more broadly, looking at what the sector can teach us about ethics in cyber warfare. It considers five key ethical questions: First, it evaluates whether a cyber weapon can be ethical if it is used for a purpose that the majority of the international community would consider "good". Second, it looks at whether it is ethical for companies (or individuals) to sell zero-day or other vulnerabilities and exploits which can be used for cyber warfare. Third, it assesses whether it is ethical to freely share (or sell) tools designed for "good" purposes that also make it easier to engage in cyber attacks on critical infrastructure and other targets. Fourth, it deliberates whether profit-focused entities, like nuclear facilities or equipment vendors, can be trusted to invest sufficiently in cyber defence. Fifth, it discusses whether governments are at times too focused on short-term benefits—including trade deals and gaining electorate support—without contemplating the long-term consequences for cyber security.

Keywords Stuxnet • Nuclear • Cyber Weapon • Zero-day vulnerabilities • Ethics

At Chatham House, I ran an 18-month project on cyber and nuclear security, which examined the challenges that information and communication technologies (ICTs) pose for the nuclear sector. Of course, concerns surrounding cyber warfare in the nuclear industry—including Stuxnet and its implications—were an important component of the project. The nuclear industry is a particularly useful case study for cyber warfare since the Stuxnet worm that infected Iranian nuclear facilities in 2010 is considered by many to have been the world's first true cyber weapon. That is,

---

C. Baylon (✉)
Research Associate in Science, Technology and Cyber Security,
The Royal Institute of International Affairs (Chatham House), London, UK
e-mail: caroline@csdr.org

it was the first publicly known malware designed to cause significant physical damage to such a major infrastructure target. Many of the project findings thus have applications for the realm of ethics in cyber warfare, both in the nuclear industry and beyond.

12.1 Background: About the Project on Cyber and Nuclear Security

Funded by a grant from the MacArthur Foundation, the Chatham House project on cyber and nuclear security's objective was to assess the cyber security threats for nuclear facilities and the wider nuclear industry as well as to identify potential policy measures to enhance cyber security in the sector. The project focused on the nuclear sector because—although nuclear plants may be less likely targets as they are more difficult to attack than other types of critical infrastructure and for a country to do so would open a veritable Pandora's box—the potential consequences of such an attack could be far more catastrophic. While it would require a highly sophisticated state actor, the deployment of a cyber weapon against a nuclear facility could conceivably be used to trigger a release of radioactive material, causing thousands of deaths and environmental degradation both in the targeted country and in neighbouring countries. Even a cyber weapon only impacting the functioning of a specific facility could trigger widespread panic among the population at large.

As part of the project, I conducted interviews with a mix of cyber security and nuclear specialists—some thirty people in all. They included nuclear plant managers; nuclear engineers; directors of cyber security companies; IT practitioners; government officials, including from an energy ministry; and academics. The interviewees were from a range of countries—notably the United States, United Kingdom, Canada, France, Japan, Ukraine, and Russia, as well as from major international organizations, including the IAEA (International Atomic Energy Agency) and ENISA (European Union Agency for Network and Information Security). I also organised three roundtable meetings at Chatham House bringing together cyber security and nuclear specialists from industry, government, and academia to discuss the challenges that ICTs pose for the nuclear industry. The academic literature on cyber security challenges for the nuclear industry is limited, so I also drew on a wide range of sources from news articles to industry reports. The project culminated with a final report that I authored on cyber threats to the nuclear sector.[1]

---

[1] https://www.chathamhouse.org/publication/cyber-security-civil-nuclear-facilities-understanding-risks

12.2 Ethics of the Use of Offensive Cyber Weapons in the Nuclear Sector

12.2.1 Can a Cyber Weapon Be Ethical If It is Used for a Purpose That the Majority of the International Community Would Consider "Good"?

12.2.1.1 Stuxnet

Stuxnet was allegedly developed by the US and Israel and specifically targeted Iranian nuclear facilities with the aim of sabotaging Iran's ability to enrich uranium. It worked by causing the centrifuges used to enrich uranium to spin too fast and break apart, while sending false information to the control room to make it appear that the centrifuges were functioning normally. The worm is thought to have damaged some 1000 centrifuges at the country's Natanz facility.

Perhaps the most interesting aspect of the world's first cyber weapon is that—in contrast to prevailing fears that countries might employ cyber weapons to attack other countries' critical infrastructure in ways that could cause civilian deaths, like by causing trains to collide—Stuxnet was designed to achieve an aim that the majority of the international community would consider "good": *preventing* Iran from acquiring nuclear weapons. Thus, some might argue that Stuxnet is an ethical cyber weapon since it can be seen as improving global security. In forestalling a scenario in which a nuclear Iran could unleash nuclear weapons, Stuxnet might have saved a large number of lives. Stuxnet can also be seen as a safer way of impeding Iran's nuclear programme since airstrikes on a nuclear facility can result in civilian casualties and are also dangerous due to the risk of radioactive fallout: Stuxnet was thus "aimed at slowing down the Iranian nuclear effort without having to resort to risky airstrikes."[2]

12.2.1.2 Ethical Issues

An Attack on a Sovereign State Although there are compelling arguments for Stuxnet to be considered "good", it still cannot be viewed as ethical for several reasons. First, Stuxnet constituted an attack on a sovereign state. Iran has a right to develop a peaceful civilian nuclear energy programme and, while the US, Israel, and a number of other members of the international community suspect Iran of seeking to develop nuclear weapons, they have no actual proof that Iran is trying to do so. This means that Iran has a legitimate grievance against the US and Israel and would be fully entitled to retaliate by cyber or other means.

---

[2] http://www.theregister.co.uk/2013/07/08/snowden_us_israel_stuxnet/

Escalation The worm also had a number of unanticipated consequences that have been highly detrimental. Stuxnet's designers had thought that its existence would never be known; when it was discovered, Stuxnet touched off a cyber arms race—increasing the risk of cyber warfare. In the case of Iran, the country accelerated its own cyber weapons programme following the worm's discovery. It is somewhat ironic that while the United States and Israel intended to curb Iran's nuclear capabilities, the result led to an increase in the country's offensive cyber capabilities instead and, in the long run, probably did not have a significant impact on restricting Iran's ability to develop nuclear weapons. At best, experts' estimate that Stuxnet may have postponed Iran's ability to develop nuclear weapons by 2 years, which is not a sizeable achievement.

Stuxnet also pushed a large number of additional countries to develop their own offensive cyber capabilities. The worm's unprecedented sophistication drew their attention to the level of advancement of states like the US and Israel in the cyber domain, and they have rushed to develop their own cyber programs accordingly. As a participant at one of the Chatham House roundtable meetings commented:

> Stuxnet and others have created an arms race where every state in the world wants to attack in cyberspace and will be busy having a team of people writing and compiling capability.

Collateral Damage Despite having been 'a marksman's job' specifically targeted at Iranian nuclear facilities, Stuxnet infected those beyond Iran. According to Eugene Kaspersky, CEO of Kaspersky Lab, a Russian nuclear plant was also affected by the worm.[3] The nuclear sector is even more reluctant than other industries to disclose cyber attacks when they experience them, so it is possible—and even likely—that additional nuclear facilities in other countries were unintentionally infected as well, posing a security risk to these facilities and causing them financial loss. Thus, even highly targeted attempts to inflict damage on a country like Iran have the potential to cause harm to other countries as well. In a worst-case scenario, this could draw additional countries into a conflict, further contributing to escalation. Kaspersky commented that,

> Unfortunately, it's very possible that other nations which are not in a conflict will be victims of cyber attacks on critical infrastructure. It's cyber space. [There are] no borders, [and many facilities share the] same systems.[4]

Repurposing by Cybercriminals The worm is also alleged to have bolstered cybercriminals' capabilities. As soon as Stuxnet was discovered, security researchers began analysing and publicly sharing information about how its code was designed. Cybercriminals were able to learn from the way Stuxnet functioned, incorporating some of its sophisticated features into malicious software used to steal credit card information or engage in other forms of theft. For example, malware produced by

---

[3] http://www.independent.co.uk/life-style/gadgets-and-tech/news/russian-nuclear-power-plant-infected-by-stuxnet-malware-says-cybersecurity-expert-8935529.html

[4] http://www.theregister.co.uk/2013/11/11/kaspersky_nuclear_plant_infected_stuxnet/

cybercriminals increasingly uses stolen certificates to evade antivirus software—an innovative technique implemented first by Stuxnet. The worm has enabled cybercriminals to copy some of the techniques that the world's most advanced cyber powers have spent millions to develop for warfare in order to commit online theft:

> the techniques used in sophisticated, state-backed malware are trickling down to less-skilled programmers who target regular web users and their online accounts or credit card details.[5]

Inspiring Terrorists Another concern about the use of a high-profile cyber weapon like Stuxnet is that terrorists, seeing their potential for destruction and fear-mongering, may be inspired to engage in similar practices. Speaking at the IAEA's first ever International Conference on Computer Security in a Nuclear World in June 2015, Mikko Hyponnen of F-Secure commented that terrorist groups like ISIS (the Islamic State in Iraq and Syria) are increasingly developing cyber capabilities and may be interested in targeting nuclear facilities. In October 2015, a US Department for Homeland Security official revealed that ISIS has been attempting to hack the US power grid.[6] They currently do not have the advanced capabilities that would be needed to cause significant damage—but they are steadily improving and have strong intent.[7] It is also conceivable that—if they sought to do so—terrorist groups like ISIS could purchase these capabilities from private hacking-for-hire companies in the future.

This raises an important question surrounding the ethics of hacking-for-hire companies: would some of them be willing to work for terrorists? It is certainly a possibility. One roundtable participant commented that:

> Some private organisations in parts of the world that are effectively private hacking-for-hire companies have capabilities of a similar standard to a state; they sometimes work for states and sometimes for other individuals.

The participant added that:

> I am not necessarily suggesting that there may be an imminent change where they might hack for terrorists. However, if people are hacking with that capability for money, is it not inconceivable that a well-funded terrorist organisation might find their way in the future to the capabilities that might give them that access and vicariously mount an attack like that?

Many hacking-for-hire companies work for a wide range of governments, including repressive and brutal governments, so they are not overly troubled with ethical issues. (The ethics of shadowy companies selling vulnerabilities and exploits is further explored in a later section.)

---

[5] http://www.technologyreview.com/news/429173/stuxnet-tricks-copied-by-computer-criminals/
[6] http://money.cnn.com/2015/10/15/technology/isis-energy-grid/
[7] http://www.express.co.uk/news/uk/625470/Britain-power-grid-risk-ISIS-terror-attack-kill-thousands

12.2.1.3 Ethical Issues Beyond Stuxnet

Planting Cyber Weapons During Peacetime A major ethical concern involving cyber weapons more generally is that the nature of these weapons push countries to plant dormant malware in other countries' critical infrastructure—including in nuclear facilities—during peacetime so that they can be activated in case of a conflict. As Richard Clarke explains in his book on cyber warfare, cyber weapons take advantage of vulnerabilities in software—including in the various SCADA (Supervisory Control And Data Acquisition) systems that run the majority of critical infrastructure—that have not yet been discovered and patched. The shelf life of vulnerabilities thus puts pressure on countries to exploit them at present; if they wait until an outbreak of hostilities, the vulnerability might have been closed by then. This has led to a situation in which countries are actively infiltrating other countries' systems, planting logic bombs that execute malicious functions when certain conditions are met or trapdoors that grant access a system later. China, Russia, and other countries are known to have been planting malicious software in the electric grid of the United States since at least 2009[8]—and the US has done the same in other countries' infrastructure, justifying this under the doctrine of 'preparing the battlefield'. This certainly includes nuclear facilities; as one roundtable participant remarked,

> to suggest this is not happening in the nuclear sector as well is naïve.

Yet, these actions are intrusive in peacetime and constitute a violation of other countries' sovereignty. They also contribute to a related ethical concern: the danger that a country might launch a cyber weapon by mistake—for instance, by accidentally triggering a logic bomb it has planted in another country's systems. The situation also contributes to the threat of escalation. It is considered so hazardous that in their paper on cyber risks to nuclear assets the EastWest Institute called for countries to agree that the nuclear sector be off limits for cyber warfare.[9] The August 2015 UN (United Nations) GGE (Governmental Group of Experts) report proposing norms in cyberspace goes further and recommends that countries agree not to cause harm to critical infrastructure of any type.[10]

---

[8] http://www.wsj.com/articles/SB123914805204099085
[9] https://www.eastwest.ngo/sites/default/files/A%20Measure%20of%20Restraint%20in%20Cyberspace.pdf
[10] http://www.un.org/ga/search/view_doc.asp?symbol=A/70/174

12.3 Ethics Surrounding Companies That Sell or Share Information or Tools That Could Be Used to Harm Critical Infrastructure

12.3.1 Is It Ethical for Companies to Sell Zero-Day or Other Vulnerabilities and Exploits That Can Be Used for Cyber Warfare?

12.3.1.1 Companies Selling Zero-Day Vulnerabilities

There are a growing number of companies selling vulnerabilities and exploits (computer code written to take advantage of a vulnerability) to governments and other paying customers that can be used to carry out cyber attacks on critical infrastructure and other forms of cyber warfare. The most prized vulnerabilities and exploits are known as "zero-days" because only the person or company who found them knows of their existence (i.e. they have been known about for zero days); thus, there are no patches against them. One of these companies is Gleg, which is based in Russia and sells SCADA zero-days, including for nuclear power plants. ReVuln, located in Malta, specializes in zero-day exploits for industrial control systems, such as those running power generators. Exodus Intelligence, in the United States, is also a seller of zero-days, including for SCADA systems. Although such companies are not illegal—they are unregulated and thus operate in a grey area—they do raise profound ethical questions.

12.3.1.2 Ethical Issues

A Thin Pretence of "Good"? In order to maintain a positive public image, some companies selling vulnerabilities and exploits to private buyers assert that their products are for "good" uses. Gleg, for instance, packages its vulnerabilities and exploits into Canvas, a penetration testing tool. Thus, the company claims that it helps organizations find vulnerabilities in their systems so that they can fix them: according to the CEO of Gleg as reported in Forbes:

> The goal...is simply to 'illustrate' vulnerabilities and their risk.[11]

However, such arguments of being used for "good" in order to make the Internet safer are pretence. Although these companies' client lists are generally secret, leaked information as to who their clients include indicates that their products appear to be primarily intended for governments to make use of in cyber warfare. ReVuln customers, for instance, are rumoured to include both the NSA

---

[11] http://www.forbes.com/sites/thomasbrewster/2015/10/21/scada-zero-day-exploit-sales/#6d149a77d96c

(National Security Agency) and the Iranian Revolutionary Guards, who are some of the leading cyber warfare actors (and, who are actively engaged in thwarting each other in cyberspace).

Not Disclosing Vulnerabilities Above all, the business models of companies selling vulnerabilities and exploits to private buyers—especially governments—are unethical at their core because they rely on not disclosing vulnerabilities to vendors. Traditionally, companies and individuals would disclose any vulnerabilities they found to vendors so that they could develop patches and thus increase the security of their products. But companies that sell vulnerabilities and exploits to governments and other private buyers have an incentive to ensure that the vulnerabilities they uncover remain unpatched so that any exploits that make use of them remain effective—including for use in cyber warfare.

A related issue is that the emergence of such companies has also caused many individuals who previously disclosed the vulnerabilities that they discovered to vendors or the public to instead sell them to these companies for profit. While some vendors are willing to pay bounties to individuals who find vulnerabilities, a number of industry insiders have argued that such companies (and their customers) have driven up prices to beyond vendors' willingness or ability to pay, further diminishing security.

Of course, there are a range of business practices among the companies engaged in selling vulnerabilities and exploits to private buyers, some being more ethical than others. For instance, Exodus Intelligence first provides information on the zero-days that it finds to its customers—and then it informs the vendors. Naturally, the vendors receive the information after Exodus Intelligence's government customers have likely already used it to gain access to critical infrastructure targets that use the vendor's products and installed backdoor access—but it does nevertheless help improve the security of the vendor's remaining products.

ReVuln, by contrast, does not inform vendors of any vulnerabilities that it has found. One of the co-founders commented in an interview with TechTarget that:

> We have our own team focused on researching vulnerabilities, and we don't report our findings to vendors in order to respect both the investments of our customers, and the other companies that follow a business model similar to ours.[12]

In a Reuters interview with the ReVuln founders, they displayed a significant lack of ethical concern:

> Asked if they would be troubled if some of their programs were used in attacks that caused death or destruction, they said: 'We don't sell weapons, we sell information. This question would be worth asking to vendors leaving security holes in their products.'[13]

Such a flippant response indicates considerable disregard for security.

---

[12] http://searchsecurity.techtarget.com/feature/Private-market-growing-for-zero-day-exploits-and-vulnerabilities

[13] http://www.reuters.com/article/us-usa-cyberweapons-specialreport-idUSBRE9490EL20130510

12.3.2 Is It Ethical to Freely Share Tools Designed for "Good" Purposes That Also Make It Easier to Engage in Cyber Attacks on Critical Infrastructure and Other Targets?

12.3.2.1 Open Source Tools

Automated Exploit Tools A related issue is whether or not freely sharing certain tools is ethical if they also make it easier for hackers to attack critical infrastructure and other targets. Several of the cyber security specialists interviewed for the project raised concerns about the growing popularity of automated exploit tools like the Metasploit Framework. The tool appears to have been designed for an uncontestably "good" purpose: to automate the process of penetration testing. The Metasploit Framework enables users to execute any exploit (the delivery vehicle for a payload) combined with any payload in order to test for vulnerabilities in computer systems. This allows companies to find vulnerabilities in their systems so that they can fix them and strengthen security. The tool is available for free, although there are also higher-functionality commercial versions of the product.

Yet the Metasploit Framework can also be misused. By replacing the payload with a malicious one, hackers can use these same exploits to attack a system. The Metasploit Framework can thus also make it easier for less skilled hackers to engage in cyber attacks.[14]

Search Engines for Internet-Connected Industrial Control Systems Similarly, free to use specialized search engines that allow users to search for critical infrastructure that is connected to the Internet, such as Shodan and ERIPP (Every Routable IP Project), are also concerning. Such search engines also appear to have been designed for a "good" purpose—or at the very least for a purpose that was not malicious. For instance, Shodan was originally created as a market research tool in order to enable companies to better understand how their customers are using their products, whether their customers are patching them, etc.[15] Currently, many cyber security researchers use the tool to find a company's devices that are Internet-connected and then inform the company so that it can take those devices offline. Law enforcement also makes frequent use of the tool. The basic version of the search engine is available for free, and there is also a paid version that offers more features.

However, the drawback is that hackers can use these specialized search engines too.[16] One interviewee, a cyber security director at a large security company, explained that,

---

[14] http://www.wired.com/2012/01/scada-exploits/

[15] http://thirdcertainty.com/qa/shodan-search-engine-exposes-built-vulnerabilties/

[16] http://www.csoonline.com/article/2867407/network-security/shodan-exposes-iot-vulnerabilities.html

we did research in which we used Shodan and found all of the nuclear plants in France that are connected to the Internet. If a user knows what s/he is looking for, s/he could easily find this information.

For example, with Shodan's geolocation capability, it is possible to view SCADA systems' positions on a map. Since the location of nuclear plants in France is public knowledge, a hacker can—by making some educated guesses—determine which of these SCADA systems are at nuclear facilities. Once a hacker has identified a facility's Internet-connected systems, they may be able to take advantage of default settings to gain access. The default passwords for standard equipment made by major manufacturers is widely shared on hacker websites, and many nuclear facilities do not change the default passwords on equipment. Shodan can thus provide a route of entry for hackers.

12.3.2.2 Mitigating Factors

Making Vulnerabilities Public By contrast with companies that sell zero-day and exploits to private buyers and do not disclose the vulnerabilities they find, these open source tools—automated exploit tools and specialized search engines—are less of an ethical challenge since they raise awareness of vulnerabilities and contribute to their public disclosure. From a moral standpoint, companies or individuals which are aware of vulnerabilities should disclose them, so that they can be fixed and the internet made more secure. In making the existence of particular vulnerabilities known and publicized, open source tools put pressure on both vendors to develop patches and on companies to install those patches.

Restrictions on Use To further improve their ethical position, these open source tools can ensure that technical restrictions are in place to limit their ability to be misused. For instance, Shodan has limited its search results to 10 at a time for those using the search engine without an account. Those that wish to see more results must create an account, enabling them to view up to 50 results at a time. They must also provide information as to what they are looking for as well as make a payment, revealing their identity in the process.[17] Additional measures of this nature, both by Shodan and by other companies, can help further mitigate the ethical challenges.

---

[17] http://money.cnn.com/2013/04/08/technology/security/shodan/index.html

12.4 Ethics Surrounding Companies and Governments That Give Precedence to Financial Considerations Over Cyber Defence Needs

12.4.1 Can Profit-Focused Entities Like Nuclear Facilities or Equipment Vendors Be Trusted to Invest Sufficiently in Cyber Defence?

12.4.1.1 Nuclear Facilities

An important ethical question is whether business desires to maximize profits are compatible with ensuring adequate cyber defence. A consultant to the IAEA that I interviewed raised the issue of whether it is possible to trust a profit-motivated environment to handle a difficult security problem like cyber security, pointing out that nuclear facilities are fundamentally businesses that are driven by the need for return on investment. The project found a number of instances in which business interests are trumping cyber security interests at nuclear facilities.

12.4.1.2 Ethical Issues

Connecting to the Public Internet In particular, nuclear facilities have made a number of business decisions involving equipment that lowers costs but are highly detrimental to cyber security. For instance, nuclear facilities are increasingly connecting to the public Internet in order to enhance efficiency—yet doing so also provides a route that hackers can use to access the facilities. In a number of cases, nuclear facilities have even implemented virtual private networks (VPNs)—which allow vendors and other contractors to provide remote monitoring and support to a facility through an encrypted Internet channel—as this offers considerable cost savings. However, VPNs also provide an entry point for hackers.

Use of Commercial Off-the-Shelf Systems Nuclear facilities are also increasingly turning to commercial off-the-shelf systems due to the cost savings that they provide; however, these systems are easier to hack.[18] Historically, systems were to a large extent customized for a particular facility, and hacking them required specialized knowledge of how they were built. Commercial off-the-shelf products, by contrast, provide a high payoff for hackers since a large number of industries across the globe make use of a relatively small number of products. Many hackers have therefore already found vulnerabilities in these systems and written exploits that they can use, making commercial off-the-shelf products more susceptible to cyber attacks.

---

[18] http://edocs.nps.edu/npspubs/institutional/newsletters/strategic%20insight/2011/SI-v10-I1_Kesler.pdf

Use of Foreign Components A related concern is that most commercial products make use of foreign components, yet these run the risk of having been compromised by those countries' intelligence agencies; they may have intentionally introduced vulnerabilities in those components in order to be able to exploit them for espionage or cyber warfare (the ethics surrounding the behaviour of intelligence agencies is discussed in further detail in a later section). The IAEA consultant interviewed explained that:

> the digitalization of existing systems and the design of the new systems has this fundamental dilemma about either building everything bespoke from the bottom up out of components that one can trust or using the components that are commercially available at a sensible price.

While ensuring that all of a product's components are sourced nationally would be best from a cyber security perspective, the globalized nature of the supply chain means that this is no longer economically feasible. Most countries do not have domestic manufacturers that can make all of a product's components, and developing their own components from scratch would be very costly. This provides yet another example of how business considerations are further eroding cyber defence.

Insufficient Spending on Cyber Security Beyond making decisions involving business equipment that have negative consequences for cyber security, nuclear facilities are not investing sufficiently in cyber security overall. One cyber security practitioner interviewed commented that:

> the owner-operators of nuclear facilities are not terribly interested in spending money on cyber security, such as upgrading or replacing existing process control systems. Cyber security is not something that gets you more production, more efficiency.

Another interviewee, a shift manager at a US nuclear facility, shared his view that

> nuclear facilities are always trying to cut to the bare minimum, simply because it costs so much money to operate a plant. They will always deny that and say they are working smarter, but that is not true.

Nuclear facilities often try to spend the minimum possible for cyber defense.

12.4.1.3 Vendors and Ethical Issues

Equipment vendors, too, tend to focus on maximizing profits at the expense of cyber security. They often need to rush to get products to market quickly in order to gain commercial advantage. The result is that in some cases they roll out products before they are ready and that therefore contain a number of vulnerabilities.

Furthermore, even when vendors are made aware of a vulnerability in their products, many do not necessarily develop patches for them—or if they do, they do not do so rapidly. One cyber security practitioner that I interviewed commented that,

> We find that if we tell the vendor about a vulnerability, they do not necessarily take action.

From the perspective of many vendors, developing patches costs them time and money without bringing them any direct financial gain. This, too, impedes cyber defence.

12.4.2 Is There a Tendency for Governments To Prioritise Short-Term Benefits—Including Trade Deals and Gaining Electorate Support—Without Contemplating the Long-Term Consequences for Cyber Security?

12.4.2.1 Governments

The tendency of nuclear facilities and vendors to give precedence to financial considerations over measures that would increase cyber security indicate that there may be a need for government to intervene, both in terms of raising these businesses' awareness of cyber security concerns and if necessary by legislating on certain issues. However, at times governments give rise to ethical questions, too, surrounding whether their priorities—which include garnering electorate support—are always compatible with ensuring cyber defence. The project found a number of instances in which governments prioritised short-term benefits that suited their own interests over ones that would strengthen their countries' cyber defence posture in the long-term.

For instance, some governments take a permissive stance when it comes to allowing politically sensitive countries to invest in critical infrastructure—including in nuclear facilities. Lured by the influx of capital that foreign investment brings—and the corresponding boost in public opinion that comes from securing trade deals—a number have pursued economic agreements without concern as to how it may impact the country's ability to defend against cyber attacks. As one roundtable participant explained,

> future power plants might be owned by the Chinese, Japanese, or even by the Russians.

Discussions with Russia The United Kingdom, for instance, has engaged in protracted discussions with Russia regarding the country's interest in building a nuclear power plant in the UK. In September 2013, the UK Department of Energy and Climate Change and Rosatom, the Russian state nuclear company, signed a Memorandum of Understanding that agreed on a programme of co-operation "designed to be the most effective means of enabling Rosatom to prepare for entry into the United Kingdom civil nuclear market."[19] As part of the plan, Rosatom was given access to the UK government's watchdogs, the Office for Nuclear Regulation

---

[19] http://www.theguardian.com/environment/2014/mar/11/russian-nuclear-firm-build-power-station-uk

and the Environment Agency, so that it could understand British regulatory and licensing requirements, and the Department for Business, Innovation and Skills agreed to "have detailed discussions with Rosatom to facilitate commercial links with the United Kingdom's industry." These discussions were ongoing at least as late as April 2014—at the height of tensions between Europe and Russia over Crimea. Although the plans are on hold for the moment due to sanctions against the country, The Guardian reports that Rosatom "hop[es to be able] to revive the plan". It is therefore possible that the project will resurface in the future.[20]

A Potential Chinese Deal Similarly, the UK government has actively sought Chinese investment in several new nuclear power plants to be built in the country. This is part of a project led by French company Electricite de France (EDF) which involves plans to build a new nuclear power plant at Hinkley Point. EDF was not willing to move ahead on the project until the Chinese agreed to partner in order to help shoulder the high costs of the project, and it was UK Chancellor of the Exchequer George Osborne who secured the Chinese investment agreement during his September 2015 visit to the country.[21] EDF therefore signed the Strategic Investment Agreement with the China General Nuclear Power Corporation (CGN), a Chinese state-owned consortium, in October 2015—to coincide with Chinese President Xi Jinping's visit to the UK—for the construction and operation of Hinkley Point; the plans call for EDF to have a 66.5% stake in the £18 billion project, with CGN holding the remaining 33.5%. At the time, President Xi issued a joint statement together with UK Prime Minister David Cameron to promote the project.[22]

The UK has further plans to collaborate with China as well: EDF and CGN have also agreed on Heads of Terms to jointly develop two more new nuclear power plants that will be built subsequently at the Sizewell and Bradwell sites. The plans for the Bradwell plant provide for an even larger role for China: The proposal features a Chinese reactor design and CGN is planning to take a majority stake—66.5%—in the project, leaving EDF with a minor role.[23]

12.4.2.2 Ethical Issues

Access to Confidential Information These projects have numerous security concerns from a cyber warfare standpoint. For one, they would give China an advantage in cyber warfare since it will have access to confidential information about how UK

---

[20] http://www.theguardian.com/uk-news/2016/may/13/russias-state-owned-nuclear-group-keen-to-break-into-uk-market

[21] http://www.theguardian.com/politics/2015/sep/25/george-osborne-presses-on-with-hinkley-power-station-despite-criticism

[22] http://www.theguardian.com/world/2015/oct/21/xi-jinping-poised-to-sign-nuclear-deal-as-uk-seeks-to-clinch-30bn-of-contracts

[23] http://www.bbc.com/news/business-34587650

plants are constructed and thus where the vulnerabilities are. A Ukrainian cyber security expert that I interviewed explained that Ukrainians are concerned about the safety of their nuclear facilities because they were built during Soviet times. As a result, Russia has a great deal of information about how those facilities are built, making it easier for them to design cyber weapons to target them.

Active Cyber Warfare Participants Moreover, the UK's choice of partners—China and Russia—are perhaps the countries that are the most actively engaged in carrying out covert cyber warfare activities against the UK and other countries. As discussed earlier, China and Russia have been planting dormant malware in the power grids of other states. China is also the leading country in the use of cyber espionage for the stealing of trade secrets. For the UK to partner or to consider partnering with these countries in such a sensitive area is therefore highly detrimental when viewed in a cyber warfare context.

Of course, advocates of the deal argue that if a country like China has a large enough investment in a country such as the UK—and especially in its critical infrastructure—its desire to protect its assets may make it less likely to engage in cyber attacks against the UK. However, this reasoning overlooks the position of relative weakness that such investment puts the UK in relative to its relationship with China in the event of a conflict—the UK would be highly vulnerable. This may also affect its negotiating ability with China.

It is important to note that different countries have taken a wide range of approaches to the tradeoff between profit and cyber security. In the United Kingdom and United States, for example, business efficiency appears to have taken priority over cyber security considerations in the above instances. By contrast, Russia appears to be more cautious about ensuring that it does not let business considerations override its need to protect its critical infrastructure from cyber attack. For instance, Russia has a clear national directive prohibiting nuclear power plants from connecting to the Internet. It also urges the use of national components to the maximum extent possible. Some countries therefore display more ethical considerations than others when it comes to these issues.

12.4.2.3 Cyber Commands/Intelligence Agencies and Ethical Issues

Another ethical concern surrounds whether the activities of countries' cyber commands and intelligence agencies are necessarily compatible with cyber defence. In fact, at times the opposite appears to be the case. Since patching a vulnerability closes it not just for the systems in one country but for those in an adversary's country as well, it is difficult to simultaneously engage in both offense and defence in cyberspace: cyber commands often have to decide whether to keep a vulnerability open for everyone so that they can exploit it for cyber warfare or else to close it for everyone and make the Internet safer.

Meanwhile, intelligence agencies have a clear vested interest in keeping vulnerabilities open so that they can make use of them for espionage. The Snowden leaks have revealed that the US Cyber Command and the US National Security Agency (NSA) typically chose to engage in offense rather than in defence, and that in some instances they have intentionally built vulnerabilities into commercial off-the-shelf products for use in cyber warfare and surveillance. Yet this means that they are contributing to making systems—including critical infrastructure—more vulnerable to cyber attack.

12.5 Conclusions

The nuclear industry provides a useful case study for the rise of ICTs and their implications for cyber warfare. This commentary explores five ethical questions in particular:

- Can a cyber weapon be ethical if it used for a purpose that the majority of the international community would consider "good"?
- Is it ethical for companies to sell zero-day or other vulnerabilities and exploits that can be used for cyber warfare?
- Is it ethical to freely share tools designed for "good" purposes that also make it easier to engage in cyber attacks on critical infrastructure and other targets?
- Can profit-focused entities like nuclear facilities or equipment vendors be trusted to invest sufficiently in cyber defence?
- Is there a tendency for governments to prioritise short-term benefits—including trade deals and gaining electorate support—without contemplating the long-term consequences for cyber security?

The paper finds that even a cyber weapon like Stuxnet, which was ostensibly employed for a "good" purpose, presents major ethical concerns since it represented an attack on a sovereign state and led to an escalation of cyber capabilities among countries. It also resulted in collateral damage, with at least one nuclear facility in a country that was not targeted being harmed as well, and had unintended consequences as cybercriminals repurposed ideas in the code, enabling them to more effectively steal funds. Moreover, cyberweapons also carry the danger of inspiring terrorist groups. Beyond Stuxnet, ethical concerns surround the planting of logic bombs during peacetime and the possibility of accidentally launching a cyber weapon.

It also finds that, despite their protestations to the contrary, companies selling zero-day vulnerabilities and exploits are highly unethical because their business models rely on not disclosing vulnerabilities, which harms cyber defence. Open source tools like the Metasploit Framework and Shodan can be exploited by cybercriminals, but they pose less of an ethical challenge since they raise awareness of vulnerabilities and contribute to public disclosure. Further technical restrictions on these open source tools can also limit their ability for misuse and improve their ethical posture.

Finally, the paper indicates that business interests are trumping cyber security interests at nuclear facilities in a number of instances. For example, nuclear facilities have made certain business decisions involving equipment that lower costs but are detrimental to cyber security, like connecting to the Internet and making increased use of commercial off-the-shelf products. Vendors also tend to put commercial interests first, and the rush to get products to market quickly means that they often contain vulnerabilities. Governments are not immune either and at times prioritise their own interests over policies that would strengthen cyber defence. For instance, the UK government has been open to Russian and Chinese foreign investment in nuclear power plants despite the cyber security concerns. Cyber commands and intelligence agencies are also not compatible with cyber defence since these organisations have a vested interest in keeping vulnerabilities open in order to use them for cyber warfare and espionage. The complexity of these issues indicates that there is a key role for ethics studies to play in dealing with the cyber warfare challenge.

Caroline Baylon at the time her chapter was written, was the lead researcher on cybersecurity at Chatham House (the Royal Institute of International Affairs) in London, United Kingdom. While there, Caroline's work focused on critical infrastructure protection, notably on cybersecurity challenges for nuclear facilities and on cybersecurity threats to satellites. She was also the editor in chief of the *Journal of Cyber Policy*, a peer-reviewed academic journal published by Routledge, Taylor & Francis. Prior to joining Chatham House, Caroline was the vice president of the Center for Strategic Decision Research in Paris, France, where she still serves as senior advisor. Caroline holds a Master of Science in social science of the Internet from Balliol College, University of Oxford, and a Bachelor of Arts in economics from Stanford University. She is currently carrying out two research projects funded by the UK government: one on prospects for curbing the proliferation of cyberweapons and another on the countries' use of proxy groups in conducting cyberattacks.

Chapter 13
NATO CCD COE Workshop on 'Ethics and Policies for Cyber Warfare' – A Report

Corinne N.J. Cath, Ludovica Glorioso, and Mariarosaria Taddeo

Abstract The NATO Cooperative Cyber Defence Centre of Excellence (NATO CCD COE) workshop on 'Ethics and Policies for Cyber Warfare' took place on 11–12 November 2014 at Magdalen College, Oxford. It brought together 10 distinguished experts from the United Kingdom, the United States, Germany, Spain, Italy, the Netherlands, Norway, and Estonia, gathering representatives from academia and from international organisations such as the European Union, the United Nations Institute of Disarmament Research, the Cyber Security Centre in Oxford, and Oxford University. The workshop was chaired by Dr Mariarosaria Taddeo and Captain Ludovica Glorioso and was the second of its kind organised by the NATO CCD COE.

Keywords Cyber warfare • Cybersecurity • Ethics • Politics law • Ethical guidelines • NATO

The first workshop on ethics of cyber conflict was held in Italy at the Centre for High Defence Studies (CASD), in November 2013. The proceedings of the presentations are available at NATO CCD COE's website (https://ccdcoe.org/multimedia/workshop-ethics-cyber-conflict-proceedings) and reflect ideas related to the ethical aspects of a number of issues such as the 'Just War Theory' in cyber conflict, cyber warfare, cyber espionage and the status of cyber combatants, and the ethical basis of law.

Consistent with the approach of the previous event, to have an arena of discussion on the legal, ethical, and technical issues of cyber warfare, the workshop in Oxford was organised to allow speakers from three interacting fields – politics, law and cyber security – to develop their views about the existing regulatory gaps in cyber warfare and its ethical underpinnings. Each of the speakers dealt with the

---

C.N.J. Cath (✉)
Oxford Internet Institute/Alan Turing Institute, University of Oxford, Oxford, UK
e-mail: corinnecath@gmail.com

L. Glorioso
Security Force Assistance Centre in Cesano, Rome, Italy

M. Taddeo
Oxford Internet Institute, University of Oxford, Oxford, UK

© Springer International Publishing Switzerland 2017
M. Taddeo, L. Glorioso (eds.), *Ethics and Policies for Cyber Operations*,
Philosophical Studies Series 124, DOI 10.1007/978-3-319-45300-2_13

issues surrounding the definition of cyber warfare and the practical consequences of their definition for filling the regulatory gaps. This report will present the views as given by those experts and identify the recommendations made.

The first part, Politics, began with a discussion of the extent to which current international structures are able to develop cyber security norms, and continued focusing on the possible ways in which these issues can be addressed by international bodies, considering their limited mandates for addressing the international security aspects of cyber interaction. In general, the work of United Nations Group of Governmental Experts (UNGGE) and the conclusions of the NATO CCD COE in the form of the Tallinn Manual on the International Law Applicable to Cyber Warfare were acknowledged and regarded as a confirmation of the intent of nations to remain bound to the existing legal, political and ethical frameworks for warfare. The discussion moved to assessing in what way exactly the international legal norms of warfare can be applied to cyber warfare. This illustrated the lack of consensus amongst the participants, some holding that the current international legal framework is able to regulate cyberspace efficiently as we develop better fitting systems, while others argued that cyber warfare is a radical and new phenomenon which brings a need for new frameworks. Two specific political obstacles to filling the current regulatory gap were highlighted: (i) the fact that cyber does not fit our post-Cold War structured world of nation-states and international bodies, and (ii) the hyper-connectivity associated with cyber, which prompts the need for policy-makers to focus more on the interconnectedness of the different issues at stake instead of their compartmentalised uniqueness, as well as to adapt to the fact that policies constantly lag behind technological developments.

The discussion focused on efforts to find a set of common norms or a shared approach towards cyber warfare in the international community. The experts debated the difficulties of developing a set of shared ethical principles when dealing with diverse political actors and entities. They also discussed the problems arising from the lack of a commonly accepted definition of cyber warfare and the limited ability of the international community to contain any potential forms of cyber warfare.

The central conclusion focused on the need to develop a consensus on a unified set of norms from which new principles can be derived, and into which old ones can be integrated, so as to manage cyber warfare. In particular, the need was emphasised for nation-states and international bodies to engage with a wide range of actors seeking consensus on what are acceptable cyber warfare practices. The need to develop an adequate definition of 'cyber weapon' was also identified, and it was argued that this definition could help advance international understanding of what constitutes cyber warfare and to what rules it should adhere.

The second part of the workshop focused on the law and examined the applicability of current legal mechanisms of warfare to cyberspace, looking specifically at the issues of deterrence, proportionality, perfidy, and casus belli. Improved transparency and attribution mechanisms were also identified as important aspects in developing morally sound and effective norms for cyber warfare. The speakers agreed on the need to work with existing legal and regulatory structures until a better fit, or an entirely new set of rules, could be developed (if at all). They had different perspectives, some speakers arguing that cyberspace is pre-political, and therefore pre-ethical, and thus prior to the application of public policy discourse to it, cyber space needs to be constituted as a political space. Deterrence was offered as a possibility for guiding this process.

The analysis of perfidy and cyber attacks as a casus belli focused on the implications of the non-physicality of cyber warfare in applying concepts like perfidy and others of Just War Theory to the case of cyber warfare. The speakers focused on the distinctive nature of cyber warfare, distinguishing it from regular warfare and concluding that we should re-evaluate how we understand cyber warfare strategies and assess their consequences.

The third section focused on the role and contribution of experts from academia in filling the regulatory gaps in cyber warfare. This question was examined using a multi- disciplinary approach, bringing together the various fields of expertise of the speakers. As with the previous discussions, much of the focus lay with the lack of a coherent definition of cyberspace. The speakers agreed on the need for such a definition to advance the discussion, but differed in their opinions on how to formulate it. The delegates looked at different mechanisms for developing ethical norms, universal principles that could be applied to cyber warfare and gave their views on the usefulness of Just War Theory, information warfare and a meta-level rule of law to develop ethical norms to regulate cyberspace. One possible solution proposed was to extend the scope of the moral scenario to include informational objects, as they have an increasingly important role in the functioning of our societies. The experts stressed the need for an inclusive approach, encompassing the different stakeholders affected by cyber warfare and accounting for the networked nature of the cybersphere, to fill the current ethical regulatory gap surrounding this phenomenon.

13.1 Political Section – Summary

In this section the discussion focused on the difficulties of defining cyber warfare within the international community and the subsequent issues surrounding regulation and norm development for cyber warfare.

When discussing the regulatory gap concerning cyber warfare and the ethical problems underpinning it, it was argued that many of the issues spring from the divergent views on cyber warfare held by different international actors. This lack of consensus on an international level, caused by diverging needs, interests and cultures, complicates our ability to define cyber warfare and to justify the ethical norms that should guide it.

At the same time, the nature of cyberspace itself complicates any attempts to find a common definition. Cyberspace is a fast-changing environment that brings together a multitude of actors beyond those traditionally involved in setting the regulatory agenda for warfare. The technological development of cyberspace is also affected by the constant strain of physical geo-political events, which also influence the power dynamics in cyberspace. It is becoming increasingly difficult to define cyber warfare, and the ethical principles that should guide it, through consensus between sovereign entities that act in different international and multi-national bodies.

Although the debate over the definition of cyber warfare was not settled at this conference, and the experts' opinions on the nature of cyber warfare differed, there was general consensus that cyber is sufficiently different to warrant a definition that goes beyond putting the prefix 'cyber' before the word 'warfare'. One of the definitions that pointed out the unique nature of this type of warfare focused on its ability to render transparent what was once opaque. Another definition given during the conference focused on the difference between cyberspace as a location compared to physical space, emphasising that cyberspace is man-made, privately owned, and mainly used by civilians. These definitions led to an interesting debate on the nature of the cyberspace.

It was also suggested that we move beyond the term 'cyber warfare' as it is invokes an image of trenches and bunkers, which makes cyber seem like an isolated phenomenon. This was explained as being both untrue and unhelpful, because it misses the crucial point that one of the defining elements of cyberspace is its interconnectivity. According to one expert, this meant that we are all now just as strong as our weakest link. This scenario poses the need for an approach to the regulation of international relations different from the one developed during the post-Cold War period. Another possible solution presented by the experts focused on the need for the international community to interact with a wider range of stakeholders, including those that would normally not have been part of the discussion. Efforts should be made not just to define cyber warfare but also to reach an internationally supported consensus on what the implications of this definition are for the norms and rules of cyber warfare. It was also stressed that the search for consensus needs to account for multi-national as well as multi-stakeholder views, including academia, citizens, netizens, corporations, and other internet stakeholders. It is clear that cyber warfare is not an isolated phenomenon: it connects many different political, economic, social and military spheres . Hence, any definition needs to be determined using a multi-stakeholder approach.

During the discussion it was mentioned that a possible way to deal with contention in cyberspace – awaiting a solid definition of cyber warfare – would be to develop consensus regarding the definition of the term 'cyber weapon'. This definition is particularly important because it is a necessary prerequisite for nation states to adequately apply international humanitarian law and develop future ethical norms for cyber warfare. It was argued that defining cyber weapon as a concept would be a solid initial step towards reaching a larger consensus.

A more political realist view of the situation would involve a partial acceptance that, at least under the status quo, it is not feasible to prevent cyber warfare. We should therefore focus our energy on developing means of guaranteeing a sense of cyber stability. This stability refers to the ebbs and flows of what is going on in the world, mapping existing conflicts. If we are unable to prevent cyber warfare then perhaps we should focus our efforts on ensuring that inter- and intra-state conflicts in cyberspace cannot trigger kinetic warfare in the physical world.

Participants provided examples of areas where common ground on cyber warfare norms could be found. Two of the areas repeatedly proposed were existing International Humanitarian Law, and forums for multinational discussion such as the United Nations. It became clear that there is a tension in cyberspace resulting from the fact that the lack of definition both helps and hinders the ability of different political actors to leverage this newfound space of international relations.

13.2 Law Section – Summary

The second part of the conference investigated the relevance of existing legal mechanisms of warfare to cyberspace. Special attention was paid to deterrence, proportionality, perfidy and casus belli. Increased transparency and improving attribution were identified as vital to developing ethical norms for cyber warfare.

The discussion started with recognition of the importance of the United Nations Group of Governmental Experts' (UNGGE) consensus that International Law applies to cyberspace. According to some experts, this should be seen as a confirmation of the intent of countries to remain bound to the existing legal, political and ethical frameworks for warfare, whether it refers to kinetic or cyber action. The perspectives on these issues were clearly informed by the professional and academic backgrounds of the various speakers.

Building on this, it was proposed that an alternative approach to developing an ethical framework for cyber warfare could be based on existing legal structures. The experts critically discussed the validity of a system that was not created with 'cyber' in mind, because it complicates application of the existing system to a digitised world. One expert asserted that laws of targeting apply in their current form, which involves the principles of distinction and proportionality and the duty to take precautions when conducting an attack. Another expert mentioned that we need to be careful when employing analogies between kinetic and cyber warfare. Instead of fighting the new implications of this type of warfare, we need to start focusing on devising solutions to deal with them. A consensus was reached that, for the sake of practicality, pressing cyber warfare issues should be addressed using existing legal framework however imperfect their fit to the new situation. The preference for a faulty system over no system is based on the assumption that if the alternative is a cyber 'Wild West' in which actors perceive they are not bound by any of the rules associated with traditional warfare, a flawed system is preferred to none at all. As one expert pointed out using the example of Stuxnet, unregulated cyber actions often not only harm the intended military, but also end up heavily damaging civilian infrastructures.

As the discussion continued, it became clear that using existing legal frameworks as the basis for filling regulatory gaps in cyber warfare raises more questions than it answers. Some of the main questions were related to the way in which the impact of actions of cyber warfare should be measured. Several options for measurement were offered. Certain experts argued for an effects-based approach, where an attempt is made to measure the impact of an attack using indicators associated with kinetic warfare, such as civilian casualties and proportionality. Others argued for an approach that examines the methods used by the cyber-attackers, because as cyber attacks are non-physical, many of the standard theories of casus belli are less applicable and states are too easily enticed to engage in unilateral defensive responses to perceived cyber attacks. An alternative possibility would be to combine the two approaches. A particular issue surrounding the UNGGE consensus is that the report does not contain a specific definition of how international law should apply to cyber warfare. This gap became particularly clear in the discussion about the applicability of the concepts of deterrence, proportionality and perfidy in cyber warfare . Nevertheless the principles of proportionality and perfidy were subject to Tallinn Manual research and are addressed in the book.

13.3 Deterrence and Proportionality

Participants provided examples of areas where common ground on cyber norms might be found. One of those areas repeatedly proposed was deterrence. Although it does not provide a solid guarantee of cyber peace, it is a principle that has the potential to unify different actors and build further consensus.

According to the experts, the main problem we face is the way in which we can make cyberspace subject to the ethical, political and legal rules of warfare as they pertain to kinetic action. One of the possible solutions to this conundrum is adapting the concept of deterrence to cyberspace by assessing an adversary's most valuable assets and threatening those with attack, as is done with nuclear deterrence. This requires a recalibration of what constitutes prized assets in cyberspace. One expert argued this could be done when cyber assets such as anonymity, deniability, uncertainty, and culpability are turned into liabilities. The idea of deterrence by association was introduced: if attackers can be associated with a cyber attack they will be less inclined to execute it. The fear for reputational damage and sanctions which might follow from being associated with an attack could work as a cyber deterrent, stressing the need for the development of international norms that would make highly disadvantageous for any state to take part in a cyber attack. Another expert argued that conflict in cyberspace, like the possibility of nuclear warfare in the sixties, is transforming military strategy. The Cold War era led to a shift from the science of military strategy to game theory. Cyber war could change this paradigm again by moving game theory to a situation of 'blue war': the threat and gradual occurrence of public and proportionate cyber attacks to resolve disputes between nations without an all-out escalation of violence. In this scenario, a digital arms race is inevitable; but it might also have many positive effects in terms of new technological spin-offs. This particular scenario, for the time being, remains to be paired with conventional kinetic and nuclear deterrence, as societies are not (yet) fully dependent for their functioning on ICT.

Proportionality was also highlighted as a useful concept to regulate cyber warfare. Again, the problem identified by the experts concerned its applicability to the case of cyber warfare. In order to apply this particular principle two parameters need to be weighed: the expected incidental damage to civilians and civilian objects on one side, and the anticipated concrete and direct military advantage on the other. It was in the weighing of these two that the experts and participants diverged on the practical usefulness of this concept. The main problems in this discussion centred on defining incidental damage and military advantage in a cyber scenario. In addition, the question was raised of what to do about non-violent military attacks, because non-violent attacks are not sufficient to trigger the application of the law of targeting under the Geneva Additional Protocol I. Some experts argued that this does not raise a substantial issue, as the civilian damage in cyber attacks is potentially less harmful to civilians because it involves less loss of life. Others argued that cyber attacks are potentially more dangerous because, by default, they involve civilian infrastructure. Taking into account society's dependency on information technology, we need to recognise that it has become possible to do significant harm through non-destructive means. A third group argued that incidental damage and military advantage are good examples of the 'unknown unknowns'. There are too many undetermined variables involved in cyber attacks to make a good prediction of these two parameters.

The discussion on deterrence also prompted a debate on retaliation for and escalation of a cyber attack. The question in this case was whether a cyber attack could trigger a kinetic response, and if so what guidelines should regulate such a scenario. During the discussion it was argued that the answer to these questions can be found in the U.N. Charter and hinges on the extent to which an attack is proportionate.

The discussion on proportionality and retaliation also prompted a debate on the need to reconsider the definition of the concepts of violence, harm, and attack so as to incorporate the ways in which new technology has enabled us to cause damage by, for example, limiting or removing functionality. The Tallinn Manual, which was mentioned often as a good foundation on which to build our understanding of cyber warfare, does not include functionality loss as an attack in the sense of the law of armed conflict. One of the participants gave the example of a cyber attack leading to functionality loss of the energy grid.

13.4 Perfidy and Dual Use

Discussing perfidy – the illegal act of military forces impersonating civilians in time of war – brought out some of the difficulties in measuring the impact of cyber attacks. The experts agreed that perfidy was seen as a measure through which to regulate cyber warfare. The discussion focused on the strategic role of perfidy in war, and its potential role in cyber warfare. Perfidy in kinetic warfare is considered a breach of the laws of war. Yet, in cyber warfare, those engaging in the attack often see perfidy as a necessary evil. One expert detailed that, common belief notwithstanding, computers are relatively resistant to manipulation. This means that the best hope for a successful attack lies in preventing computers from recognising the attacker – which, it can be argued, is perfidy. Dual use in cyber warfare also emerged as a key problematic point. All speakers agreed that dual-use targets should not be considered legitimate military targets. There is a clear need to define norms that would ensure proper identification of targets, the acknowledgement of the attack, the creation of cyber havens, and rules and norms to ensure attacks are proportionate. This last issue was the source of disagreement amongst the speakers and led to a broader discussion on attribution and transparency. Some speakers argued in favour of enhanced transparency, on the basis that it might deter the attackers for fear of retaliation, others that increased transparency also increases the information about attacks that might be used for improving the impact of existing cyber attacks.

13.5 Attribution and Transparency

Attribution and transparency in attacks are important in kinetic warfare, but their place and role in cyber warfare is less well established. Attribution is particularly complicated due to the nature of cyber attacks. The experts argued that attribution is important in order to apply ethical rules to cyber warfare because, without it, it is unclear to whom rules, and sometimes penalties, should be applied. During this discussion, several specific issues were mentioned that countered the assumption that attribution is important. The first focused on the fact that attribution in the physical world is often not as straightforward as assumed; examples of the current conflict in Ukraine were mentioned.

The debate on attribution turned out to be closely linked to the question of transparency. Some experts argued that there needs to be more transparency surrounding cyber attacks, others that increased transparency has several negative side-effects. Those arguing in favour of increased transparency held that it was a necessary requirement in order to make cyberspace political, in the sense that it is seen as a legitimate space where political, legal, and ethical norms and rules apply. This point was the source of much discussion, as some of the experts and participants argued against transparency over attacks – mainly to ensure that cyber criminals did not abuse the code of attack vehicles, as it was the case with the leaked Stuxnet code.

In the closing round-table session, the experts and participants maintained that common ground to shape ethical principles for cyber warfare can be found in existing International Humanitarian Law. There was consensus amongst the experts that even if cyber warfare could be construed as a new phenomenon, this does not imply that current laws do not apply. There is a clear need to work with the current legislative framework and build from there. In this process it will become clear which laws can be applied to cyberspace, which need to be adapted, and which must be created anew.

13.6 Academia Section – Summary

The third session of the workshop addressed the role of academia in developing solutions to the regulatory gaps and lack of ethical standards. The experts described how, where, and by whom solutions should be developed, what the main obstacles were, and what were the possible options for going forward.

Panellists and participants agreed that the traditional governance entities, such as the nation-state and international bodies, such as the United Nations, dominate the development of regulations and ethical norms guiding cyber warfare. However, the increased influence of the private sector and the need for cooperation between different types of stakeholders was also recognised. The experts identified several issues facing the development of an ethical framework for cyber warfare.

A specific issue was the problem of the lack of a shared understanding of what cyberspace is. Several of the experts proposed the development of a model that would allow decision-makers to use the same reference point and compare different considerations so as to guide stake-holders to make a fair decision based on a shared understanding of the problem. This model would bring together different stakeholders, and approach cyber attack from a holistic perspective. Although the panellists agreed that the model brought together different views highlighting the circumstances in which miscommunication about cyber issues arises, there was some concern over the practical applicability of this model. In particular, one participant argued that the problem with the model was that it was created without taking into consideration the needs of the individuals it is intended to support. It assumes that key actors in the debate on cyber attacks have a clear stake in developing strong ethical norms, which is not always the case. Cyber attackers also benefit from the lack of a consolidated set of norms and rules, and political actors benefit from pushing their particular definition. It is this ideological divide that has been limiting the development of a global consensus. Considering alternative models for guiding cyberspace, although not always politically feasible, could produce interesting results and novel approaches that might benefit the discussion.

Another issue that was considered was the need to develop a better understanding of how Just War Theory can be applied to cyber attacks. One of the participants argued that we should be careful when trying to force new phenomena to fit old paradigms. The analogy of pushing square pegs into round holes was repeatedly mentioned, and it was argued that if we attempt to force cyber warfare into the shapes we readily recognise, we might not be doing justice to the complicated nature of this issue. Some practical examples of this were discussed. For instance, the questions of when a cyber attack can be considered a use of force, and when a cyber operation amounts to an armed attack arose, especially as many aspects of such an attack will be intangible and hence difficult to quantify.

There was general agreement that Just War Theory is still a good framework for regulating war, including cyber war. But as Just War Theory presupposes knowledge of identities, intentions and trust, there is a clear need to update it to the cyber era in which we live, as identities, intentions and trust are not necessarily transparent in cyber warfare. One of the possible solutions presented was to approach Just War Theory as a necessary but not sufficient instrument in developing norms for cyber warfare. It was argued that to develop a suitable ethical framework for analysis of cyber warfare, Just War Theory needs to be merged with Information Ethics so as to expand the scope of the analysis and regulation to include also the cyber realm, artificial agents and intangible objects.

The existence of a hiatus between the ontology of the entities involved in traditional warfare and those involved in cyber warfare was stressed. As a consequence of the hiatus, it was argued that Just War Theory does not provide the sufficient conceptual tools to regulate cyber warfare, hence posing the need for the development of an ethical set of guidelines that can simultaneously weigh cyber risks, rights associated with information technology, and issues of determining responsibility for actions of cyber warfare.

It was also stressed that, with the so-called information revolution, some of the pivotal categories used in regulating society, such as private vs public, military vs civilian, are being reshaped and the very distinction between such categories is blurring. Warfare provides a particularly clear example of how lines become blurred as information technologies are increasingly used to deliver military strategy. The theatre of war has moved online, where attacks are aimed at physical and non-physical targets and increasingly focused on disrupting the critical ICT structures of enemies. This type of warfare merges the non-physical domain, often associated with attacks on digital agents and targets and low levels of violence, with the physical domain which focuses on attacking physical targets and agents and has high levels of violence. In this context, the discussion focused once more on the definition of cyber attack and on the requirements that it needs to meet to qualify as a use of force or an armed attack. Disagreement amongst the participants and experts on how to define this remained unresolved. However, by focusing on the disconnect between the current tests, laws and ethical frameworks that focus on the definition of war as physical and violent, and the current technological capabilities of cyber warfare, steps were made towards bridging the regulatory and ethical gaps.

13.7 Conclusions

The session ended with a call to action for the experts and panellists to ensure that the conversations about these important topics continue beyond the workshop. In particular, emphasis was put on the importance of ensuring that the development of a solution to the current regulatory gap in cyber warfare, and the ethical issues underpinning it, is developed in an interdisciplinary way and involves all stakeholders;

a networked problem can only be addressed in a networked fashion. Participants and experts agreed that there was an increased need for dialogue between the stakeholders involved in cyberspace. At the moment, there are too few opportunities for discussion where governments, legal scholars, and public and private sectors are all represented. This event offered one such occasion and should be seen as a step in the right direction; however, many more such steps need to be taken in order to address some of our time's most pressing issues concerning cyber warfare.

Corinne N.J. Cath worked at the intersection of Internet governance, tech-policy and human rights for various years. She joined the Oxford Internet Institute in 2015 as an MSc student and is currently working for Article 19, a human rights organization focused on the defence and promotion of freedom of expression and freedom of information worldwide. Previously, she worked as a policy advisor for a Democratic congressman in Washington, DC. She has a BA in anthropology, an MA in international relations and an MSc in internet studies. Her past research at the Oxford Internet Institute focused on the (im)possibility of instantiating human rights principles in Internet standards and protocols, as well as values by design, 'responsibility by design', ethics in technology and human rights online.

Ludovica Glorioso is Captain (Cpt ITA A) of the Italian Army and a legal advisor to the Italian Armed Forces, currently working at the Defence General Staff. Prior to her current position, she served as Legal Researcher at the NATO Cooperative Cyber Defence Centre of Excellence (NATO CCD CCOE, 2012–2016), as legal advisor in NATO Peacekeeping Operations in the Balkans and Afghanistan. She holds an MA in Law from University of Palermo (Italy), an LL.M in European and Transitional Law from The University of Trento (Italy) and she is admitted to the Italian bar. Her research and activities focus on International Humanitarian Law, Law of Armed Conflict, and International Law applied to cyber warfare.

Mariarosaria Taddeo works at the Oxford Internet Institute, University of Oxford and Faculty Fellow at the Alan Turing Institute. Her recent work focuses mainly on the ethical analysis of cyber security practices and information conflicts. Her area of expertise is Information and Computer Ethics, although she has worked on issues concerning Philosophy of Information, Epistemology, and Philosophy of AI. She published several papers focusing on online trust, cyber security and cyber warfare and guest-edited a number of special issues of peer-reviewed international journals: Ethics and Information Technology, Knowledge, Technology and Policy, Philosophy & Technology. She also edited (with L. Floridi) a volume on 'The Ethics of Information Warfare' (Springer, 2014) and is currently writing a book on 'The Ethics of Cyber Conflicts' under contract for Routledge. Dr. Taddeo is the 2010 recipient of the Simon Award for Outstanding Research in Computing and Philosophy and of the 2013 World Technology Award for Ethics. She serves editor-in-chief of Minds & Machines, in the executive editorial board of Philosophy & Technology. Since 2016, Dr Taddeo is Global Future Council Fellow for the Council on the Future of Cybersecurity of the World Economic Forum.

Index

A
Academia section, 239–240
Active Cyber Defence, 157
Advanced analyst (LEA-AA), 147
Advanced Persistent Threats (APTs), 130, 172
A History of Warfare, 10
Albanian government, 119, 120
Alberts, D.S., 176
Alliance for Cybersecurity, 127
Anarchical Society, 3
Antolin-Jenkins, V., 89
ARCO rights (access, rectification, cancellation and objection), 146
Armed conflict, 13
Arquilla, J., 160
Article 49(1) of Additional Protocol I, 102
Article 52 of Additional Protocol, 107
Article 52(2) of Additional Protocol I, 111
Artificial socio-cognitive systems, 144
Attacks, 100–102
Attributes of attack, 177–179
Attribution and transparency, 238
Australian government, 133
Automated exploit tools, 221
AVOIDIT methodology, 173

B
Bank of England, 200
Barkham, J., 89
Barrett, E.T., 7, 77
Baylon, C., 213–229
Begby, E., 11
Bellamy, A.J., 11
Bishop, M., 173
Boundary problem
 CNE/CNA, 27
 computer network's structure, 26, 27
 consequential harm, 27
 hacking back, 26
 hostile system, 26
 legitimate target, 25
 malicious software, 26
 military commanders, 27
 preparatory exploitation, 26
 self-propagating software, 26
Bradbury, S.G., 89
Brenner, S., 89
Brown, D., 95
Bundesamt für Sicherheit in der Informationstechnik (BSI) standards, 127, 130, 133
Budapest Convention, 117, 118
Bull, H., 3
Burdon, M., 153

C
Cabinet Office, 197
Cameron, K., 148
CAPER, 146–152, 154, 156
CAPER crawling system, 148
CAPER regulatory model (CRM), 142, 149, 152–162
Card, S., 176
Casanovas, P., 139–162
Cath, C.N.J., 231–241
Caveats, 123–124
Cavoukian, A., 148
Centre for Protection of National Infrastructure (CPNI), 174, 197, 199, 202

China, 128–130
China General Nuclear Power Corporation (CGN), 226
China Internet Network Information Centre (CNNIC), 109
Chinese deal, 226
CIA, 142
Civilians
  attacks, 37–38
  cybercriminals, 40
  cyberspace, 35
  desirable targets, 36
  dual-use resources, 36
  dual-use targets, 37
  easy targets, 35–36
  ethical justification, 37
  hardware, 35
  intermediate steps, 38–39
  military activities, 35
  military operations, 35
  side effects, 37, 39–40
  spoofing, 41–42
  unreliability of cyberwarfare, 39
Clarke, R., 218
Classical antiquity, 11
Classical legal theory, 154
Coates, A.J., 8
Cohen, F., 173
Collateral damage, 43
Collier, J., 187–211
Committee on the Protection of Critical Infrastructure, 194
Common Law, 141
Companies selling vulnerabilities
  cyber attacks, 219
  industrial control systems, 219
  penetration testing tool, 219
  SCADA zero-days, 219
  security, 220
  TechTarget, 220
  vendors, 220
  zero days, 219
Computer network attacks (CNA)
  bricking, 23
  CNE, 23
  cyber Pearl Harbors, 23
  international humanitarian law, 23
  national jurisdictions, 24
  physical damage, 24
  violent attacks, 23
Computer network exploitation (CNE)
  and CNA, 23, 28
  electronic attacks, 23
  illegitimate access, 25

interpretation/distinction, 25
legitimate target, 26
non-human agency, 20
software tools, 20
system malfunctioning, 27
targeted network, 20
Cooperative Cyber Defence Centre of Excellence (CCDCOE), 91
Corfu Channel, 119–121, 123
Corfu Channel case, 116
Cornish, P., 1–16
Corporate Insider Threat Detection (CITD), 175
Council of European Convention on Cybercrime, 117
Critical infrastructure (CI), 126–130
Customary International Cybersecurity Law, 117–118, 121, 124, 133
Customary International Humanitarian Law, 106
Cyber attacks, 122, 132, 133, 142, 156, 157, 161, 170–172, 177–183, 189, 191, 192, 196–198, 200–203, 205, 210
  attack methods, 34
  Australian sewage, 38
  automated exploit tools, 221
  automatic propagation, 40–41
  catastrophic destruction, 29
  computer networks, 22
  computer system, 19
  core model
    attributes of attack, 177–179
    impact, 180
    mitigation, 180–181
    values, 179
  cybercrime, 35
  cyber-operations, 63
  DDoS's attempt, 64
  definition, 20, 170
  ethics, 30
  human interests, 64
  intentional disruption, 19
  international armed conflicts, 21
  international humanitarian law, 20
  LOGIC BOMB, 64
  mail system, 37
  Metasploit Framework, 221
  military target, 35
  physical harm and damage, 19
  physical operation, 22
  political conflict, 30
  public policy requirements, 64
  reformations, 23
  related work, 172–175

Index

245

search engines, 221, 222
sequential manner, 24
service attacks, 19, 64
side effects, 37
Stuxnet, 34
Tallinn Manual, 19
technical restrictions, 222
violence, 28
vulnerabilities, 36
Cyber coercion, 46
Cyber commands, 227, 228
Cyber conflict
   aggression, 6
   attribution problem, 7
   conflict-focused ethical constraints, 5
   and coercion, 6
   double effect principle, 6
   ethical debate, 2
   ethical governance, 8
   ethically-guided judgement, 3
   international humanitarian law, 2
   political leaders, 7
   pre-emption debate, 8
   public debate, 2
   relationship, 3
   sensational attacks, 2
   technological challenges, 6
Cybercrimes, 145, 160, 161
Cybercriminals, 39, 216
Cyber crisis management, 188–192, 194, 196–201
   comparison between Estonia and United Kingdom, 201–203
   explanatory variables
      digital dependence, 206–207
      history, 205
      political philosophy, 206
      size, 203–204
      threat landscape, 205–206
   lessons, strategy and policy, 207–209
Cyber defence, 173, 174, 181, 182
Cyber Defence League (CDL), 193–195, 202
Cyberharm
   ability and opportunity, 56
   bilateral and dynamic, 56
   computer-information systems, 51
   cyberattacks, 55, 57
   cyber-ethics, 49
   cyberwarfare, 50
   economic damage, 50
   economic injury, 55
   economic loss/physical injury, 54
   environmental and informational systems, 54
   hacker's experimentations, 54
   human agents, 53
   human interests, 50, 55–57
   human values, 57
   information systems, 52
   initial plausibility, 52
   instrumental view, 50, 51, 54
   intentional disruption, 51
   interests and entitlements, 54
   intrinsic view, 53, 54
   legal entitlements, 55
   legitimate moral entitlements, 50
   manumission, 51
   mediators, 56
   metaethical and metaphysical theses, 54
   moralized conception, 50
   non-living systems, 52
   non-sentient entities, 53
   ontological commonalities, 52
   psychological consequences, 56
   psychological discomfort, 55
   public elements, 55
   public honor, 53
   sentient animals, 52
   social equilibrium, 52
   theoretical judgments, 56
Cyber operations
   active defense, 92
   amendment, 96
   armed attack, 92
   biases and blindspots, 97
   categories, 89
   civilian and military targets, 94
   cyberspace, 94
   cyber-specific arrangement, 95
   cyber weapons, 94
   human experience, 97
   human rights, 88
   international law, 88, 89, 96
   international legal rules, 94
   international legal system, 95
   international peace and security, 90
   international regulation, 97
   international relations, 90
   international rules and structures, 90
   jus in bello, 93, 94
   law-making role, 89
   military objective, 93
   proportionality, 93
   revision and amendment, 98
   revolution, 88
   self-defense, 92
   software corporations, 95
   stakeholders and resistance, 89

Cyber operations (cont.)
   Tallinn Manual, 91
   technological advancement, 88
   technology corporations, 89
   vulnerabilities, 92
   warfare, 93
   weapons and capabilities, 97
Cyber-security, 116–133, 159, 172, 174–176, 181–183, 188, 190–203, 205–210, 225–228
Cyber-security due diligence
   nation regulations
      China, 128–130
      Germany, 127–128
      United States, 125–126
   polycentric approach, 132
Cyber Security Information Sharing Partnership (CISP), 198
Cyber-security policy, 181
Cyber security strategy, 196
Cyber situational awareness and understanding models, 174, 175
Cyberspace, 152, 157, 158, 160, 170, 172, 174–177, 180, 187–189, 210
   communication and information transfer, 5
   cyber debate, 15
   human invention, 4
   innovation and progressive development, 88
   moral framework, 4
   nuclear proliferation, 14
   nuclear threat, 15
   partitioning, 45
   political organisation, 5
   reverse-engineering, 15
   technologically skilled, 4
   vulnerabilities, 4
Cybersphere, 233
Cyber warfare, 142, 156–162
   Article 52, 34
   collateral damage, 34
   confidential information, 226
   cyberattack, 60
   cyberspace, 34
   definition, 232
   deflationary feature, 61
   human interest, 61
   information systems, 60
   international community, 232
   international legal norms, 232
   military and civilian targets, 34
   participants, 227
   principles, 44
   Stuxnet, 34
   technical issues, 231
   universal principles, 233
Cyber weapon, 7, 214
   during peacetime, 218
   ethical problems with Stuxnet, 215–217
   Stuxnet, 215

D
Danks, D., 2
Danks, J.H., 2
Data, 151
   collection and storage, 151
   reuse and transfer, 151
Decision-making process, 171, 172, 177, 183
Definite military advantage, 107, 108
de Maizière, T., 127
Democratic Control of Armed Forces (DCAF), 158
Department of Health, 196
Department of Transport, 196
Deterrence, 233, 236, 237
Digital by Default, 195
Digital dependence, 206, 207
Digital systems, 188–190, 200, 206, 207, 210
Dinstein, Y., 89
Dipert, R., 2, 49, 73
Dipert, R.R., 2
Distinction
   cyberspace, 35
   ethical problems, 46
   military and civilian, 34
Distributed denial of service (DDoS), 102, 191, 192
Domain Name System (DNS), 108
Domestic institutional organisation, 191–192
Donohue, L., 159
Dual-use
   civilian entities, 37
   civilians and militaries, 36
   command-and-control, 37
   government's actions, 37
   military and civilian users, 45
   military parts, 37
Duffield, M., 7

E
Eberle, C.J., 5
Electricite de France (EDF), 226
Electronic attacks
   CNE, 20
   computer networks, 20, 30
   computer system's function, 19

Index

Jensen, E.T., 106
Jinping, Xi., 129
Joint Cyber Unit (JCU), 200
Judaeo-Christian just war, 5
jus ad bellum
    armed conflict, 89
    armed hostilities, 95
    cyber operations, 90
    epochal shift, 96
    and *jus in bello*, 90
    power dynamics, 94
jus in bello
    armed conflict, 89, 93
    hostilities, 93
Just war theory (JWT), 233
    attacked computer system, 74
    categorization, 21
    CNE, 23
    combatants and non-combatants, 9
    communications networks, 10
    computer systems, 22
    cruising computer virus, 73
    cyber attacks, 22
    cyber conflict, 9
    domain, 77
    ethical analysis, 73
    ethical discussion, 21
    ethical no-go zone, 10
    existing apparatus of laws, 77
    hostile action, 22
    human behaviour, 9
    human life and liberty, 9
    intelligence collection, 22
    international contexts, 73
    jurisprudence and practice, 9
    jus in bello, 72
    knowledge formation, 22
    knowledge generation, 23
    legitimate targets, 21
    military activity, 22
    moral analysis, 73
    moral discourse, 74
    ontological hiatus, 74
    ontology, 73
    pacifism/realism, 72
    personal and political responsibility, 21
    sanguinary war, 75
    storage devices, 22
    surveillance, 22
    targeted killings, 22
    tense relations, 75
    universal nature, 77
    war as last resort, 74

K
Kang, M.J., 144
Keegan, J., 10
Kennedy, D., 90
Kerry, J., 195
Killchain, C., 173
Kilovaty, I., 157
Kirgis, F.L., 118
Knopf, J.W., 14
Koops, B.-J., 148, 149
Kornmaier, A., 158
Kuehn, A., 115–133

L
Lauristin, M., 141
Law of armed conflict, 11, 99, 100
Law section, 235, 236
Leahy, S.P., 88
LEA's External User (LEU), 147
Leenes, P., 148
Legal Enforcement Agents (LEAs), 140, 144–149, 151, 152, 156, 160
Legg, P., 175
Lessig, L., 153
lex feranda, 123
Lin, H., 89
Lin, P., 159
Lough, D.L., 173

M
Mahabharata, 11
Margaret Thatcher, 206
McDonald, J., 17–30
Measurement and signatures intelligence (MASINT), 143
Mental models, 170, 171, 175, 176, 178, 182
Meta-rule of law, 142, 144–146, 152–154, 156, 158, 160–162
Metasploit Framework, 221
Microsoft, 133
Military objectives
    Article 52(2) of Additional Protocol I, 104
    carbon filaments, 102
    circumstances, 104
    cloud computing, 106
    cyber attacks, 104
    cyber context, 104
    cyber operations, 102, 103
    dual-use cyber infrastructure, 106
    Geneva Conventions, 103
    internet, 109
    military manuals, 103

Military objectives (cont.)
  partial destruction, 103
  pro-Russian propaganda, 102
  Protocol I, 101, 106
  software components, 109
  transmission of orders, 105
  transportation-related networks, 107
  uranium enrichment, 104
  US definition, 106
  US government communications, 105
Ministry of Defence (MoD), 197
Mirror model, 173
Mirroring effect, 181
Mission integration framework, 174
Mitigation of Attack, 180–181
Mobile technology, 144
Moral reflection, 11, 13
Moral vocabulary, 12
Morality and war, 29
Multi-level protection scheme (MLPS), 129
Multi-lingual Crime Ontology (MCO), 148
Murkens, J.E.M., 152

N
National Cyber Security and Communications Center, 195
National Defence Autorization Act, 143
National grid, 199
National Institute of Standards and Technology (NIST) Framework, 116, 118, 125–128, 130–133
National Security Agency (NSA), 220
NATO Cooperative Cyber Defence Centre of Excellence (NATO CCDCOE), 190, 191, 201
  cyber attacks, 233
  cyber warfare, 232
  cyber weapon, 232
  cyberspace, 232, 233
  ethical principles, 232
  legal mechanisms, 232
  Oxford, 231
  perfidy, 233
  policy-makers, 232
  stakeholders, 233
NATO cyber security framework manual, 174
Networks
  command-and-control, 37
  cyberattack, 35
  military operations, 35
  protocols, 41
  web browsers, 35
Nicaragua, 117, 121–122

Nicaragua v. United States, 117
Non-tangible terms, 180, 183
Nordic-Baltic, 195, 205, 207
Norman, D.A., 170
Normative-SWRM, 155, 156
North American Treaty Organization (NATO), 11, 91, 143, 188, 190, 195, 201, 203–205, 208
Nuclear facilities
  commercial off-the-shelf systems, 223
  cyber security, 224
  foreign components, 224
  IAEA, 223
  public internet, 223
Nuclear security, 214
Nussbaum, M., 56

O
Obama, B., 125, 126, 128
Oceanian flagship, 64
O'Connor, T.R., 145
One-size-fits-all policy, 209
Ontological design patterns (ODP), 155
Open social intelligence (OSI), 145
Open source (OS), 140
Open Source Intelligence (OSINT), 139, 140, 142–146, 148, 149, 152–157, 161
Open Social Intelligence (OSI), 140, 143–145
Opinio juris, 118
Organization of American States, 118
OSINT cybercrime crawling, 161
Ostrom, E., 132, 152, 156
Ostrom, V., 132

P
Paet, U., 195
Palombella, G., 152
Perfidy, 41, 237
Pirolli, P., 176
Political philosophy, 206
Political section, 233–235
Politics
  armed conflict, 12
  cyberspace, 3
  ethical project, 13
  moral consciousness, 12
  moral language, 13
  strategy and ethics, 3
  technological challenges, 12
  and war, 13

Index

251

Prikk, K., 192
Primary privacy law, 159
Principles of Fair Information Practices (FIPs), 148
Privacy by Default (PbD), 140
Privacy by Design (PbD), 140, 142, 147–152, 159
Privacy Impact Assessment (PIA), 147, 149, 151, 152
Product tampering, 42
Propagation
　cyberattacks, 40–41
　worm-based, 34
Psychological damage, 42
Pudong New Area of Shanghai, 105
Pythian, M., 142

R
Radio Television of Serbia (RTS), 109
RAND paper, 160
Regulations on Classified Protection of Information Security, 129
Reichberg, G.M., 11
Reparations, 45
ReVuln customers, 219
Rid, T., 23
Right of data access, 151
Rights Expression Languages (REL), 155
Ritsch, M., 152
Roig, A., 149
Ronfeldt, D., 160
Roscini, M., 99–111
Rosenzweig, P., 89
Rowe, N.C., 33–46
Russell, S., 115–133
Russia, 207, 208

S
Sanders, J.W., 80
SCADA systems, 24
Schmitt, M., 89
Schmitt, M.N., 157
Schneier, B., 23
Schreier, F., 158
Secondary privacy law, 159, 160
Self-defense, 91
　Amy Pascal's privacy, 58
　annexation, 59
　beneficial features, 58
　cyberattacks, 57, 58, 60
　cyberharm, 57
　damaging effects, 59
　eastasian occoupation, 59
　instrumentalist view, 58
　interpersonal relationships, 58
　invasion, 58
　rare earth minerals, 60
　state action, 59
　Stuxnet, 58
　target nation, 59
　temporal proximity, 60
Self-determination
　computer programs, 61
　cyberharm, 60
　cyberwarfare, 62
　epistemic factor, 62
　goals, 62
　highwayman, 62
　human interests, 61, 63
　individual's entitlement, 61
　interests and entitlements, 61
　magnitude, 62
　military action, 60–61, 63
　political action, 62
　resistance, 63
　self-defense, 62
Semantic Web Regulatory Models (SWRM), 155
Sensemaking model, 175, 176, 181
Shackelford, S.J., 115–133
Shanghai Cooperation Organization, 128
Shaw, M.N., 8
Signals intelligence (SIGINT), 143
Smith, P.T., 49–64
Social and people-centric models, 174
Social engineering, 174
Social intelligence, 144
Social media, 146
Steinhauer, J., 88
Strachan, H., 13
Strategic intelligence, 145
Stuxnet, 28, 101, 122
　collateral damage, 216
　cyber arms race, 216
　cyber domain, 216
　cyber weapon, 213
　high-profile cyber weapon, 217
　Iranian nuclear facilities, 215
　nuclear weapons, 215
　sovereign state, 215
Supervisory Control and Data Acquisition (SCADA) systems, 105, 157, 179, 218
Swarming, 161
Syse, H., 11

T

Taddeo, M., 10, 67–84, 231–241
Tallinn Manual, 116, 118
Tallinn Manual on International Law applied to Cyber Warfare, 157
Tallinn Manual on the International Law Applicable to Cyber Warfare, 101
Tangible terms, 179, 180, 183
Technology-centric models, 173
Tene, O., 159
Terrorists, 217
Total Defence model, 202, 204
Trachtman, J., 88
Trail Smelter, 121, 123
Transnational rule of law, 152
Tranter, K., 88

U

UN General Assembly, 6
UN General Assembly Resolution, 118
UN Security Council, 6
United Kingdom (UK), 188–190, 195–211
   COBRA, 196
   Cyber Security Strategy, 196
   domestic institutional organisation, 196–198
   DSTL, 174
   vs.Estonia, 201–203
   GCHQ, 196, 197, 199
   international engagement, 200–201
   JCU, 208
   stakeholder mobilisation, 198–200
United Kingdom Computer Emergency Response Team (CERT-UK), 197, 198, 201, 202, 208
United Nations (UN), 195
United Nations Group of Governmental Experts (UNGGE), 232, 235
United States (U.S.), 117, 121, 125–126, 128, 130, 195, 204
USAF Intelligence Targeting Guide, 111

US Air Force Pamphlet, 107
U.S. Computer Emergency Readiness Team (US-CERT), 195
US CYBERCOM former, 107
US Department for Homeland Security, 217
US National Security Agency (NSA), 228
Usmani, Z-ul-H., 145

V

Validation exposure randomness de-allocation improper conditions taxonomy (VERDICT), 173
Values of Attack, 179
Vendors, 224
Visual analytics module (VA), 148
von Clausewitz, C., 13
Vulnerabilities, 222

W

Walzer, M., 10
Waxman, M., 89
Web 2.0, 142, 153
Weick, K.E., 181
Weinstein, J., 131
Westin, A., 148
Whole of nation approach, 192
Windsor, P., 13
Wireless network standard (WAPI), 129
World Trade Organization (WTO), 124

X

Xiao, L., 176, 177
X-road, 191

Z

Zeadally, S., 157
Zittrain, J.L., 153